Mechanical Properties
of
Solid Polymers

Mechanical Properties
of
Solid Polymers

I. M. Ward

Department of Physics,
The University of Leeds

WILEY-INTERSCIENCE
a division of John Wiley & Sons Ltd
London · New York · Sydney · Toronto

Library of Congress catalog card number 79-149575

ISBN 0 471 91995 0

Set on Monophoto filmsetter and
printed by J. W. Arrowsmith Ltd., Bristol, England

Preface

This text-book is a somewhat extended version of a lecture course given to M.Sc. students studying the Physics of Materials in the Physics Department of the University of Bristol during the period 1965–1970. It is therefore written for those whose background is a first degree in physics, chemistry, engineering, metallurgy or some related discipline. Although it is intended to be self-contained, there is much contingent reading particularly in mechanics and in polymer science if the reader is to derive full benefit.

The book follows the approach of developing the mechanics of the behaviour first, and then discussing molecular and structural interpretations. The relative detail with which this is done reflects the current state of the art. In this respect, then, the treatment of the different aspects of mechanical properties is often very different in style. For example, the expositions of rubber-like behaviour and linear viscoelastic behaviour are mostly didactic, as befits subjects of long standing. On the other hand, the discussions of yield and fracture take the form of extensive reviews of progress.

Each chapter has been written as a separate unit and I hope that this will appeal to those who would not wish to read the book as a whole in the order of presentation. In this respect it is hoped that the book may prove helpful to research workers in Polymer Science who are seeking an introduction to a part of the subject which is other than their own speciality.

Many of my colleagues have given considerable assistance by reading and commenting on the draft manuscript. The whole text was considered by Professors G. Allen, D. W. Saunders and R. S. Stein who made many helpful suggestions. In addition I wish to thank the following for reading various chapters: G. R. Davies, R. A. Duckett, J. S. Foot, J. S. Harris, N. H. Ladizesky, S. Rabinowitz, E. B. Ranby, T. G. Rogers and J. Scanlan.

I owe particular debt of gratitude to Mr. Ralph Davies, who undertook the production of the figures.

Finally, I am especially grateful to my wife, without whose constant encouragement this book would not have been written.

I. M. WARD

Acknowledgements

Grateful acknowledgement is made to the following publishers and Societies for permission to reproduce illustrations.

Book or Journal	Publisher	Figure
British Plastics	Iliffe Industrial Publications	6.3, 6.4, 12.19a, 12.19b
Fibres from Synthetic Polymers ed. by R. Hill	Elsevier Publishing Company	1.11
Memoirs of the Faculty of Engineering	Faculty of Engineering, Kyushu University, Fukuoka, Kyushu, Japan.	1.13, 8.11
Progress Reports in Physics	Institute of Physics and Physical Society	1.12
Proc. Phys. Soc.	Institute of Physics and Physical Society	6.18, 6.19, 6.20, 10.9, 10.10, 12.10
Ind. Eng. Chem.	American Chemical Society	8.4
Journal of Colloid Science	Academic Press Inc.	7.3
Journal of Applied Physics	American Institute of Physics	8.8, 8.15, 8.16, 10.13, 10.14, 12.23, 12.24a, 12.24b, 12.32
Journal of Materials Science	Chapman & Hall	11.19, 11.20, 11.22
Kolloid Zeitschrift	Steinkopff Darmstadt	7.5
Research Report R63-49	Materials Research Laboratory, Massachusetts Institute of Technology	11.31, 11.32, 11.33, 11.34 11.35
Materie Plastiche ed Elastomeri	Editrice I'Industria	6.1
Stress Waves in Solids by H. Kolsky	Dover Publications Inc.	2.4
The Physics of Rubber Elasticity by L. R. G. Treloar	Oxford University Press	3.6, 4.1, 4.7, 4.8
Philosophical Magazine	Taylor & Francis Ltd.	6.21
Physics of Plastics ed. Ritchie	The Plastics Institute	8.2, 8.3, 12.17, 12.18 Table 8.1, 8.2
Plastics	Temple Press Ltd.	12.4a, 12.8

Book or Journal	Publisher	Figure
Polymer	Iliffe Science and Technology Publications Ltd.	11.7, 12.5
Polymer Engineering & Science	Society of Plastics Engineers Inc.	9.1, 9.2, 9.3, 9.4, 9.5
S. Glasstone, Textbook of Physical Chemistry, 2nd Edition.	MacMillan & Co. Ltd.	7.9
L. E. Nielson, Mechanical Properties of Polymers 1962	Van Nostrand–Reinhold Co. Ltd.	8.5, 8.6
The Review of Scientific Instruments	American Institute of Physics	6.7
Phil. Trans.	The Royal Society	3.4, 3.7
Proc. Roy. Soc.	The Royal Society	1.10, 6.6, 8.17, 11.13
"Fracture" Eds. B. L. Averbach et al. 1959	The Massachusetts Institute of Technology	12.21, 12.22, 12.25, 12.27, 12.28
Transactions of the Faraday Society	The Faraday Society	7.6
Journal of Research of National Bureau of Standards	U.S. Government Printing Office	7.1
Physics of Non-Crystalline Solids 1965	North Holland Publishing Co.	8.1
Progress in Solid Mechanics Ed. Sneddon & Hill 1961	North Holland Publishing Co.	3.8
Proc. 2nd Int. Cong. of Rheology, London 1954	Butterworths	7.2
Kunststoffe 1951	Carl Hanser	8.7
Publication Scientifiques et Techniques de Ministère de l'Air	Centre de Documentation de L'Armament, Paris	12.20

We also wish to give the origins of the following diagrams.
Fig. 7.3 McLoughlin and Tobolsky, J. Colloid Sci., 7, 555 (1952)
Fig. 12.4a Vincent, Plastics, 26, 141 (1961)
Fig. 12.13 Sternstein and Cessna, J. Polymer Sci., 33, 825 (1965)
Fig. 12.17 Adams and Jackson, Soc. Plastics Engrs J., 12, 3, 13 (1956)

Much of Chapter 12 originally appeared in 1968 as an Internal Report for Imperial Chemical Industries, Ltd., and I am grateful to the Directorate of the Petrochemical and Polymer Laboratory, Runcorn, for permission to publish it here.

Contents

1 **The Structure of Polymers** 1
   1.1 Chemical Composition 1
      1.1.1 Polymerization 1
      1.1.2 Cross-linking and Chain Branching . . . 3
      1.1.3 Molecular Weight and Molecular-Weight Distri-
           bution 3
      1.1.4 Chemical and Steric Isomerism and Stereoregu-
           larity 5
      1.1.5 Blends, Grafts and Copolymers . . . 6
   1.2 Physical Structure 7
      1.2.1 Rotational Isomerism 7
      1.2.2 Orientation and Crystallinity 7

2 **The Mechanical Properties of Polymers:**
  General Considerations 15
   2.1 Objectives 15
   2.2 The Different Types of Mechanical Behaviour . . 15
   2.3 The Elastic Solid and the Behaviour of Polymers . 17
   2.4 Stress and Strain 19
      2.4.1 The State of Stress 19
      2.4.2 The State of Strain 21
   2.5 The Generalized Hooke's Law 24

3 **The Behaviour of Polymers in the Rubber-Like State: Finite**
  Strain Elasticity 29
   3.1 The Generalized Definition of Strain . . . 29
   3.2 The Definition of Components of Stress . . . 35
   3.3 The Stress–Strain Relationships 35
   3.4 The Use of a Strain-Energy Function . . . 41
      3.4.1 Thermodynamic Considerations . . . 41
      3.4.2 The Form of the Strain-Energy Function . . 44

3.4.3 The Strain Invariants 45
3.4.4 The Stress–Strain Relations 48
3.5 Experimental Studies of Finite-Elastic Behaviour in
Rubbers 51

4 The Statistical Molecular Theories of the Rubber-Like State . 61
4.1 Thermodynamic Considerations 61
4.1.1 The Thermoelastic-inversion effect . . . 62
4.1.2 The Statistical Theory 63
4.2 The Internal-Energy Contribution to Rubber Elasticity 72
4.3 The Molecular Interpretation of Higher Order Terms in
the Strain-Energy Function 75

5 Linear Viscoelastic Behaviour 77
5.1 Viscoelastic Behaviour 77
5.1.1 Linear Viscoelastic Behaviour 77
5.1.2 Creep 79
5.1.3 Stress Relaxation 82
5.2 Mathematical Treatment of Linear Viscoelastic
Behaviour 83
5.2.1 The Boltzmann Superposition Principle and the
Definition of Creep Compliance . . . 84
5.2.2 The Stress-Relaxation Modulus . . . 87
5.2.3 The Formal Relationship Between Creep and
Stress Relaxation 87
5.2.4 Mechanical Models, Relaxation and Retardation-
Time Spectra 88
5.2.5 The Maxwell Model 89
5.2.6 The Kelvin or Voigt Model 90
5.2.7 The Standard Linear Solid 92
5.2.8 Relaxation-Time Spectra and Retardation-Time
Spectra 94
5.3 Dynamical Mechanical Measurements: The Complex
Modulus and Complex Compliance 95
5.3.1 Experimental Patterns for G_1, G_2, etc. as a Function of Frequency 97
5.4 The Relationships between the Complex Moduli and
the Stress-Relaxation Modulus 101
5.4.1 Formal Representations of the Stress–Relaxation
Modulus and the Complex Modulus . . . 102

5.4.2 Formal Representations of the Creep Compliance
and the Complex Compliance 105
5.4.3 The Formal Structure of Linear Viscoelasticity . 106
5.5 The Relaxation Strength 107

6 **The Measurement of Viscoelastic Behaviour** . . . 110
6.1 Creep and Stress Relaxation 111
6.1.1 Creep: Conditioning 111
6.1.2 Extensional Creep 111
6.1.3 Extensometers 112
6.1.4 High-Temperature Creep. 114
6.1.5 Torsional Creep 115
6.1.6 Stress Relaxation 115
6.2 Dynamic Mechanical Measurements; The Torsion
Pendulum 118
6.3 Resonance Methods 121
6.3.1 The Vibrating-Reed Technique . . . 121
6.4 Forced-Vibration Non-Resonance Methods . . 124
6.4.1 Measurement of Dynamic Extensional Modulus . 125
6.5 Wave-Propagation Methods 127

7 **Experimental Studies of the Linear Viscoelastic Behaviour of**
Polymers 133
7.1 General Introduction 133
7.1.1 Amorphous Polymers 133
7.1.2 Temperature Dependence of Viscoelastic
Behaviour 136
7.1.3 Crystalline Polymers 138
7.2 Time–Temperature Equivalence and Superposition . 140
7.3 Transition-State Theories 142
7.3.1 The Site-Model Theory 145
7.4 The Time–Temperature Equivalence of the Glass-Transition Viscoelastic Behaviour in Amorphous Polymers and
the Williams, Landel and Ferry (WLF) Equation . 148
7.4.1 The Williams, Landel and Ferry Equation, the
Free-Volume Theory and other Related Theories . 155
7.4.2 The Free-Volume Theory of Cohen and Turnbull 156
7.4.3 The Statistical Thermodynamic Theory of Adam
and Gibbs 156

7.4.4 An Objection to Free-Volume Theories . . 158
7.5 Normal-Mode Theories Based on Motion of Isolated
Flexible Chains 158

**8 Relaxation Transitions and Their Relationship to Molecular
Structure** 166
8.1 Relaxation Transitions in Amorphous Polymers: Some
General Features 166
8.2 Glass Transitions in Amorphous Polymers: Detailed
Discussion 168
8.2.1 Effect of Chemical Structure 168
8.2.2 Effect of Molecular Weight and Cross-Linking . 171
8.2.3 Blends, Grafts and Copolymers . . . 173
8.2.4 Effect of Plasticizers 175
8.3 The Crankshaft Mechanism for Secondary Relaxations . 176
8.4 Relaxation Transitions in Crystalline Polymers . . 177
8.4.1 General Discussion. 177
8.4.2 Relaxation Processes in Polyethylene . . 181
8.4.3 Application of the Site Model 188
8.4.4 Use of Mechanical Anisotropy. . . . 192
8.4.5 Conclusion 194

9 Non-Linear Viscoelastic Behaviour 196
9.1 General Introduction 196
9.2 The Design Engineer's Approach to Non-Linear Viscoelasticity 197
9.2.1 Use of the Isochronous Stress–Strain Curves . 197
9.2.2 Power Laws for Non-Linear Viscoelasticity . 204
9.3 The Molecular Approach 205
9.4 The Rheologist's Approach to Non-Linear Viscoelasticity 207
9.4.1 Large-Strain Behaviour of Elastomers . . 207
9.4.2 Creep and Recovery of Plasticized Polyvinyl
Chloride 210
9.4.3 Empirical Extension of Boltzmann Superposition
Principle 211
9.5 Formal Extension of the Boltzmann Superposition
Principle 214
9.5.1 Mathematical Rigour 218
9.5.2 The Practical Application of the Multiple-Integral
Representation 219

Contents

10 Anisotropic Mechanical Behaviour 225
10.1 The Description of Anisotropic Mechanical Behaviour 225
10.2 Mechanical Anisotropy in Polymers . . . 226
    10.2.1 The Elastic Constants for Specimens Possessing Fibre Symmetry 226
    10.2.2 The Elastic Constants for Specimens Possessing Orthorhombic Symmetry 229
10.3 Measurement of Elastic Constants 230
    10.3.1 Measurements on Films or Sheets . . . 230
    10.3.2 Measurements on Fibres and Monofilaments . 234
10.4 Experimental Studies of Mechanical Anisotropy in Polymers 239
    10.4.1 Low-Density Polyethylene Sheets . . . 240
    10.4.2 Nylon and Polyethylene Terephthalate Monofilaments 243
    10.4.3 Polyethylene Terephthalate, Nylon, Polyethylene and Polypropylene Monofilaments . 243
    10.4.4 Rolled and Annealed Polyethylene Sheets of Orthorhombic Symmetry 249
10.5 Correlation of the Elastic Constants of an Oriented Polymer with those of an Isotropic Polymer. The Aggregate Model 253
10.6 Mechanical Anisotropy of Fibres and Films of Intermediate Molecular Orientation 257
10.7 The Sonic Velocity 262
10.8 Other Theories of Mechanical Anisotropy . . 264
10.9 Anisotropic Viscoelastic Behaviour 264

11 The Yield Behaviour of Polymers 270
11.1 Discussion of Load–Elongation curve . . . 271
    11.1.1 Necking and the Ultimate Stress . . . 272
    11.1.2 Necking and Cold-Drawing: A Phenomenological Discussion 274
    11.1.3 Use of the Considère Construction . . 276
    11.1.4 Definition of Yield Stress 278
11.2 Ideal Plastic Behaviour 278
    11.2.1 The Yield Criterion: General Considerations . 279
    11.2.2 The Tresca Yield Criterion 281
    11.2.3 The Von Mises Yield Criterion . . . 281
    11.2.4 The Coulomb Yield Criterion . . . 283

11.2.5 Geometrical Representations of the Tresca, Von Mises and Coulomb Yield Criteria . . 283

11.2.6 Combined Stress States 286

11.2.7 Yield Criteria for Anisotropic Materials . . 287

11.2.8 The Stress–Strain Relations for Isotropic Materials 288

11.2.9 The Plastic Potential 289

11.2.10 The Stress–Strain Relations for an Anisotropic Material of Orthorhombic Symmetry . . 289

11.3 The Yield Process 290

11.3.1 The Adiabatic Heating Explanation . . 290

11.3.2 The Isothermal Yield Process; the Nature of the Load Drop. 295

11.4 Experimental Evidence for Yield Criteria in Polymers . 297

11.4.1 Isotropic Glassy Polymers 297

11.4.2 Anisotropic Polymers 301

11.4.3 The Band Angle in the Tensile Test . . 307

11.5 Simple Shear-Stress Yield Tests 309

11.5.1 The Bauschinger Effect 312

11.6 Influence of Hydrostatic Pressure on Yield . . 313

11.7 The Temperature and Strain-Rate Dependence of Yield and Drawing Processes 314

11.8 The Molecular Interpretations of Yield and Cold-Drawing 318

11.8.1 The Yield Process 318

11.8.2 Cold-Drawing 322

11.9 Deformation Bands in Oriented Polymers . . 326

12 Breaking Phenomena 330

12.1 Definition of Tough and Brittle Behaviour in Polymers 330

12.1.1 Factors Influencing Brittle–Ductile Behaviour . 332

12.2 Brittle Fracture of Polymers 340

12.2.1 The Continuum Approach: Application of the Griffith Theory 340

12.2.2 The Molecular Approach 348

12.3 Craze Formation 349

12.3.1 Stress Criterion for Craze Formation . . 350

12.4 Impact Strength of Polymers 352

12.4.1 High-Impact Polyblends 355

12.4.2 Crazing and Stress-Whitening . . . 357

12.5 The Nature of the Fracture Surface in Polymers . 358

12.6 Crack Propagation 360
12.7 Brittle Fracture by Stress Pulses 361
12.8 The Tensile Strength and Tearing of Polymers in the Rubbery State 361
    12.8.1 The Tearing of Rubbers: Application of Griffith Theory 361
    12.8.2 The Tensile Strength of Rubbers . . . 363
    12.8.3 Molecular Theories of the Tensile Strength of Rubbers 365
12.9 Effect of Strain Rate and Temperature . . . 368

Index 373

1

The Structure of Polymers

Although this book is primarily concerned with the mechanical properties of polymers, it seems desirable at the outset to introduce a few elementary ideas concerning their structure.

1.1 CHEMICAL COMPOSITION

1.1.1 Polymerization

The first essential point is that polymers consist of long molecular chains of covalently bonded atoms. One of the simplest polymers is polyethylene, which consists of long chains of the $-CH_2$ repeat unit (Figures 1.1a and b). This is an *addition* polymer and is made by polymerizing the monomer ethylene $CH_2=CH_2$ to form the polymer $[-CH_2-CH_2-]_n$.

A well-known class of polymers is made from the compounds $CH_2=CH$ (with an attached group X) where X represents a chemical group. These are the *vinyl* polymers and familiar examples are polypropylene

$$[-CH_2-\underset{\underset{CH_3}{|}}{CH}-]_n$$

polystyrene

$$[-CH_2-\underset{\underset{C_6H_5}{|}}{CH}-]_n$$

and polyvinylchloride

$$[-CH_2-\underset{\underset{Cl}{|}}{CH}-]_n$$

Condensation polymers are made by reacting difunctional molecules, with the elimination of water. One example is the formation of

1

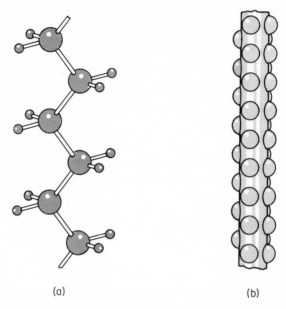

(a)   (b)

Figure 1.1. (a) The polyethylene chain $(CH_2)_n$ in schematic form (Larger spheres, carbon; smaller spheres, hydrogen); (b) Sketch of a molecular model of a polyethylene chain.

polyethylene terephthalate (Terylene or Dacron) from ethylene glycol and terephthalic acid.

$$nHO{-}CH_2{-}CH_2{-}OH + nHOOC \left\langle\!\!\bigcirc\!\!\right\rangle COOH$$

$$= H[{-}O{-}CH_2{-}CH_2{-}O{-}\overset{\displaystyle O}{\underset{\displaystyle O}{C}}\left\langle\!\!\bigcirc\!\!\right\rangle\overset{\displaystyle O}{C}{-}]_n OH + nH_2O$$

Nylon 66, $-NH{-}(CH_2)_6{-}NH{-}\overset{\displaystyle O}{\underset{\displaystyle O}{C}}{-}(CH_2)_4{-}\overset{\displaystyle O}{C}{-}$

is another condensation polymer.

1.1.2 Cross-Linking and Chain-Branching

We have so far considered linear chains where the monomer units are joined into a single continuous ribbon, and each molecular chain is a separate unit. In some cases, however, the chains are joined by other chains at points along their length to make a cross-linked structure (Figure 1.2) which is the situation for rubbers and for a thermoset such as bakelite.

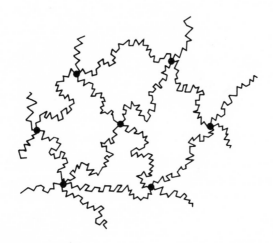

Figure 1.2. Schematic diagram of a cross-linked polymer.

A similar but less extreme complication is chain-branching, where a secondary chain initiates from a point on the main chain. This is illustrated in Figure 1.3 for polyethylene. Low-density polyethylene, as distinct from the high-density or linear polyethylene shown in Figure 1.1, possesses on average one long branch per molecule and a larger number of small branches, mainly ethyl $[-CH_2-CH_3]$ or butyl $[-(CH_2)_3-CH_3]$ side-groups. We will see that the presence of these branch points leads to considerable differences between the mechanical behaviour of low- and high-density polyethylene.

1.1.3 Molecular Weight and Molecular-Weight Distribution

Each sample of a polymer will consist of molecular chains of varying lengths, i.e. of varying molecular weight (Figure 1.4). The molecular-weight distribution is of importance in determining polymer properties,

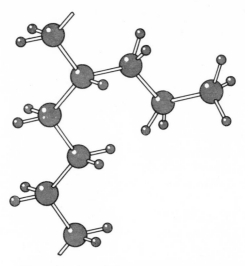

Figure 1.3. A chain branch in polyethylene.

but until the recent advent of gel permeation chromatography[1,2], this could only be determined by extremely tedious fractionation procedures. Most investigations were therefore concerned with different types of

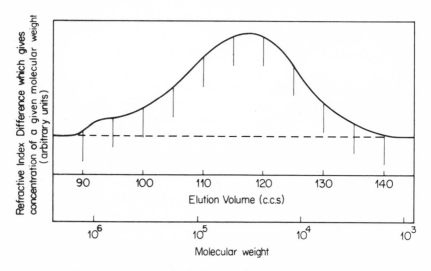

Figure 1.4. The gel permeation chromatograph trace gives a direct indication of the molecular distribution (result obtained in Marlex 6009 by Dr. T. Williams).

average molecular weight, the number average $\overline{M}_n$ and the weight average $\overline{M}_w$, which are determined by osmotic pressure and light-scattering measurements respectively, on dilute solutions of the polymers. These are:

$$\text{number average } \overline{M}_n = \frac{\Sigma N_i M_i}{\Sigma N_i}; \qquad \text{weight average } \overline{M}_w = \frac{\Sigma (N_i M_i) M_i}{\Sigma N_i M_i}$$

where N_i is the number of molecules of molecular weight M_i and Σ denotes summation over all i molecular weights.

The molecular-weight distribution is important in determining flow properties. It may therefore affect the mechanical properties of a solid polymer indirectly by influencing the final physical state. Direct correlations of molecular weight to viscoelastic behaviour and brittle strength have also been obtained, and this is a developing area of polymer science.

1.1.4 Chemical and Steric Isomerism and Stereoregularity

A further complication in the chemical structure of polymers is the possibility of different chemical isomeric forms for the repeat unit, or for a series of repeat units. A particularly simple example is presented by the possibility of either head-to-head or head-to-tail addition of vinyl mono-

mer units $-CH_2-\overset{\displaystyle X}{\underset{\displaystyle |}{C}H}-$. We can have:

$$-CH_2-\overset{X}{\underset{|}{C}H}-CH_2-\overset{X}{\underset{|}{C}H}- \quad \text{(head-to-tail)}$$

or

$$-CH_2-\overset{X}{\underset{|}{C}H}-\overset{X}{\underset{|}{C}H}-CH_2 \quad \text{(head-to-head)}$$

Most vinyl polymers show predominantly head-to-tail substitution, but the presence of the alternative situation leads to a loss of regularity which can reflect itself in a reduced degree of crystallization and hence affect mechanical properties.

A rather more complex case is that of steric isomerism and the question of stereoregularity. Consider the simplest type of vinyl polymer in which a substituent group X is attached to every alternative carbon atom. For illustrative purposes let us suppose that the polymer chain is a planar zig-zag. Then as shown in Figures 1.5a and b there are two very simple regular polymers which can be constructed. In the first of these the substituent

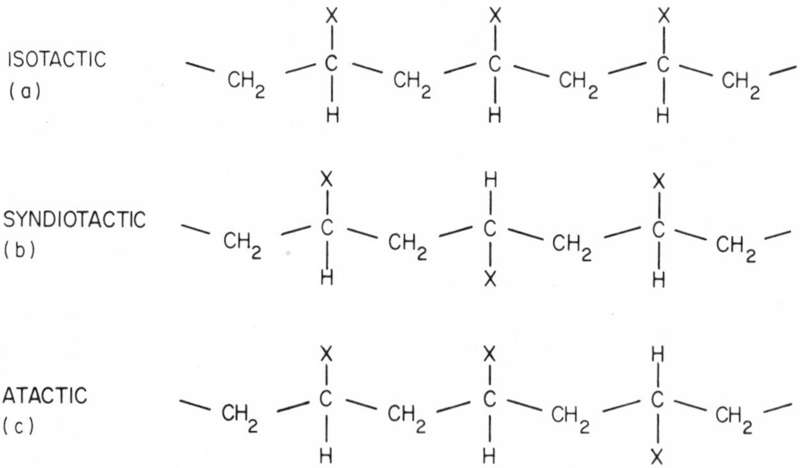

ISOTACTIC (a)

$$-\ CH_2\ -\overset{\overset{\displaystyle X}{|}}{\underset{\underset{\displaystyle H}{|}}{C}}-\ CH_2\ -\overset{\overset{\displaystyle X}{|}}{\underset{\underset{\displaystyle H}{|}}{C}}-\ CH_2\ -\overset{\overset{\displaystyle X}{|}}{\underset{\underset{\displaystyle H}{|}}{C}}-\ CH_2\ -$$

SYNDIOTACTIC (b)

$$-\ CH_2\ -\overset{\overset{\displaystyle X}{|}}{\underset{\underset{\displaystyle H}{|}}{C}}-\ CH_2\ -\overset{\overset{\displaystyle H}{|}}{\underset{\underset{\displaystyle X}{|}}{C}}-\ CH_2\ -\overset{\overset{\displaystyle X}{|}}{\underset{\underset{\displaystyle H}{|}}{C}}-\ CH_2\ -$$

ATACTIC (c)

$$-\ CH_2\ -\overset{\overset{\displaystyle X}{|}}{\underset{\underset{\displaystyle H}{|}}{C}}-\ CH_2\ -\overset{\overset{\displaystyle X}{|}}{\underset{\underset{\displaystyle H}{|}}{C}}-\ CH_2\ -\overset{\overset{\displaystyle H}{|}}{\underset{\underset{\displaystyle X}{|}}{C}}-\ CH_2\ -$$

Figure 1.5. A substituted α olefine can take three stereosubstituted terms.

groups are all added in an identical manner as we proceed along the chain (Figure 1.5a). This is the isotactic polymer. In the second regular polymer, there is an inversion of the manner of substitution for each monomer unit. In this planar model, the substituent groups alternate regularly on opposite sides of the chain. This polymer is called *syndiotactic* (Figure 1.5b). The importance of this regular addition is that it makes it possible for crystallization to occur. This is stereoregularity, and stereoregular polymers are crystalline and can possess high melting points. This can extend their working range appreciably, as compared with amorphous polymers, whose range is limited by the lower softening point.

The two structures which we have considered are those for ideal isotactic or syndiotactic polymers. In practice, it is found that although stereospecific catalysts can produce polymer chains which are predominantly isotactic or syndiotactic, there are always a considerable number of faulty substitutions. Thus a number of chains are produced with an irregular substitution pattern. These can be separated from the rest of the polymer by solvent extraction and this fraction is called *atactic* (see Figure 1.5c).

1.1.5 Blends, Grafts and Copolymers

Blending, grafting and copolymerization are commonly used to increase the ductility and toughness of brittle polymers or to increase the stiffness of rubbery polymers.

A BLEND is a mixture of two or more polymers.

A GRAFT is where long side chains of a second polymer are chemically attached to the base polymer.

A COPOLYMER is where chemical combination exists in the main chain between two polymers $[A]_n$ and $[B]_n$. A copolymer can be a BLOCK copolymer [AAA ...] [BB ...] or a RANDOM copolymer ABAABAB, the latter having no long sequences of A or B units.

1.2 PHYSICAL STRUCTURE

When the chemical composition of a polymer has been determined there remains the important question of the arrangement of the molecular chains in space. This has two distinct aspects.

(1) The arrangement of a single chain without regard to its neighbours: rotational isomerism.

(2) The arrangement of chains with respect to each other: orientation and crystallinity.

1.2.1 Rotational Isomerism

The arrangement of a single chain relates to the fact that there are alternative conformations for the molecule because of the possibility of hindered rotation about the many single bonds in the structure. Rotational isomerism has been carefully studied in small molecules using spectroscopic techniques[3] and the *trans* and *gauche* isomerism of disubstituted ethanes Figure 1.6 (see Plate I, facing p. 176) is a familiar example. Similar considerations apply to polymers, and the situation for polyethylene terephthalate[4] where there are two possible conformations of the glycol residue is illustrated in Figure 1.7.

To pass from one rotational isomeric form to another requires that an energy barrier be surmounted (Figure 1.8). The possibility of the chain molecules changing their conformations therefore depends on the relative magnitude of the energy barrier compared with thermal energies and the perturbing effects of applied stresses. We can therefore see the possibilities of linking molecular flexibility to deformation mechanisms, a theme which will be developed in detail subsequently.

1.2.2 Orientation and Crystallinity

When we consider the arrangement of molecular chains with respect to each other there are again two rather separate aspects, those of molecular

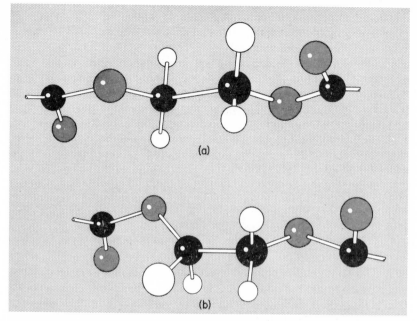

Figure 1.7. Polyethylene terephthalate the crystalline *trans* conformation (a) and the *gauche* conformation (b) which is present in 'amorphous' regions (after Grime and Ward).

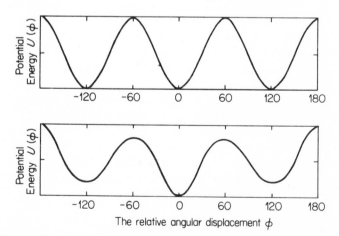

Figure 1.8. Potential energy for rotation (a) around the C–C bond in ethane (b) around the central C–C bond in *n*-butane (after McCrum, Read and Williams, 1967).

orientation and *crystallinity*. In semicrystalline polymers this distinction may at times be an artificial one.

Many polymers when cooled from the molten state form a disordered structure, which is termed the amorphous state. At room temperature these polymers may be of high modulus such as polymethylmethacrylate, polystyrene and melt-quenched polyethylene terephthalate, or of low modulus such as rubber or atactic polypropylene. Amorphous polymers are usually considered to be a random tangle of molecules (Figure 1.9a). It is, however, apparent that completely random packing cannot occur. This follows even from simple arguments based on the comparatively high density[5] but there is no distinct structure as, for example, revealed by X-ray diffraction techniques.

Polymethylmethacrylate, polystyrene and melt-quenched polyethylene terephthalate are examples of amorphous polymers. If such a polymer is stretched the molecules may be preferentially aligned along the stretch direction. In polymethylmethacrylate and polystyrene such molecular

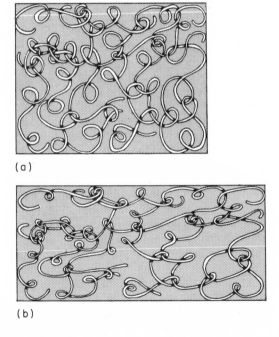

(a)

(b)

Figure 1.9. Schematic diagrams of (a) unoriented amorphous polymer (b) oriented amorphous polymer.

orientation may be detected by optical measurements; but X-ray diffraction measurements still reveal no sign of three-dimensional order. The structure is therefore regarded as a somewhat elongated tangled skein (Figure 1.9b). We would say that such a structure is oriented amorphous, but not crystalline.

In polyethylene terephthalate, however, stretching produces both molecular orientation and small regions of three-dimensional order,

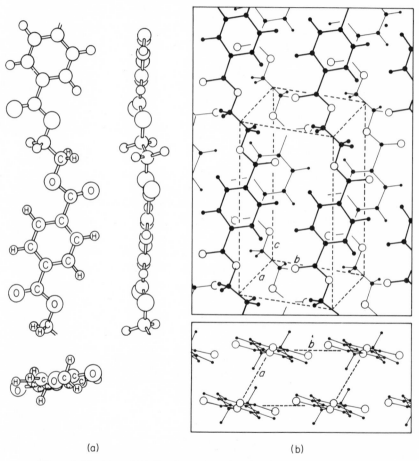

(a)   (b)

Figure 1.10. (a) and (b) Configuration of molecules and arrangement of molecules in crystalline polyethylene terephthalate. Above, projection normal to 010 plane; below, projection along c axis (larger dots carbon; smaller dots, hydrogen; open circles, oxygen) (after Daubeny, Bunn and Brown).

namely crystallites. The simplest explanation of such behaviour is that the orientation processes have brought the molecules into adequate juxtaposition for them to take up positions of three-dimensional order, and hence crystallize.

Many polymers, including polyethylene terephthalate, also crystallize if they are cooled slowly from the melt. In this case we may say that they are crystalline but unoriented. Although such specimens are unoriented in the macroscopic sense, i.e. they possess isotropic bulk mechanical properties, they are not homogeneous in the microscopic sense and often show a spherulitic structure under a polarizing microscope.

The crystal structures of the crystalline regions can be determined from the wide angle X-ray diffraction patterns of polymers in the stretched-crystalline form. Such structures have been obtained for all the well-known crystalline polymers, e.g. polyethylene, nylon, polyethylene terephthalate, polypropylene (Figure 1.10).

Such information is extremely valuable in gaining an understanding of the structure of a polymer. It was early recognized, however, that in addition to the discrete reflections from the crystallites the diffraction pattern of a semicrystalline polymer also shows much diffuse scattering and this was attributed to the 'amorphous' regions.

This lead to the so-called fringed micelle model (Figure 1.11) for the structure of a semicrystalline polymer, which is a natural development of the imagined situation in an amorphous polymer. The molecular chains alternate between regions of order (the crystallites) and disorder (the amorphous regions).

This model has undergone considerable revision in recent years as a result of the discovery of the chain-folding of molecules in polymer

Figure 1.11. The fringed micelle representation of crystalline polymers (from the article by Bunn in *Fibres From Synthetic Polymers*).

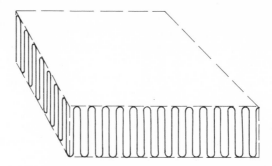

Figure 1.12. Diagrammatic representation of chain folding in polymer crystals with the folds drawn sharp and regular (after Keller, 1969).

crystals grown from dilute solution[6,7,8]. Linear polyethylene, when crystallized from dilute solution, forms single-crystal lamellae, with lateral dimensions of the order of 10–20 microns and of the order of 100 Å thick. Electron diffraction shows that the molecular chains are approximately normal to the lamellar surface and since the molecules are usually of the order of 10,000 Å in length, it can be deduced that they must be folded back and forth within the crystals (Figure 1.12). Single crystals have been isolated for most crystalline polymers, including nylon and polypropylene.

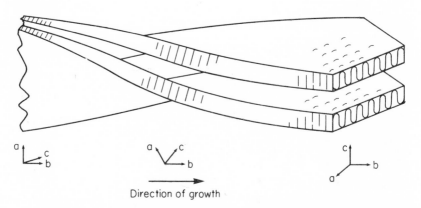

Direction of growth

Figure 1.13. A model of the lamellar arrangement in a polyethylene spherulite (after Takayanagi, 1963). The small diagrams of the *a, b, c* axes show the orientation of the unit cell at various points.

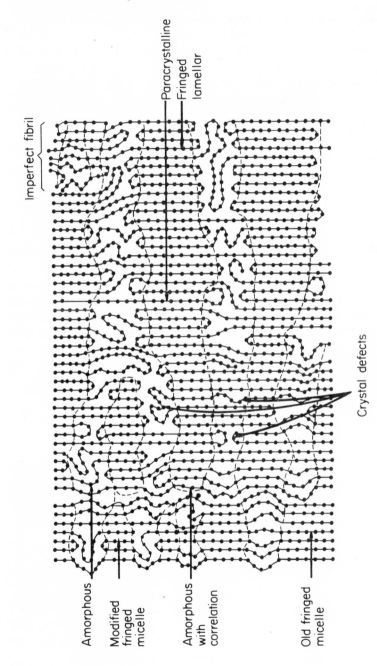

Figure 1.14. Schematic composite diagram of different types of order and disorder in oriented polymers (after Hearle).

The relevance of chain-folding to the structure of bulk crystalline polymers has not so far been established unequivocally, although the single crystal experiments have led to the belief that the basic structural block is a lamella with folded chains[9]. When a polymer melt is cooled, crystallization is initiated at nuclei in different points of the specimen. Crystallization is believed to proceed from a central nucleus, in all directions, by a twisting of these lamellae along fibrils, as shown schematically in Figure 1.13. The largest unit growing out of a nucleus is the spherulite.

It therefore seems most likely that in crystalline polymers chain-folding occurs in addition to the more conventional threading of molecules through the crystalline regions so that a compromise model must be proposed. This is shown schematically in Figure 1.14, which also shows other types of irregularity.

REFERENCES

1. M. F. Vaughan, *Nature*, **188**, 55 (1960).
2. J. C. Moore, *J. Polymer Sci.*, **A2**, 835 (1964).
3. S.-I. Mizushima, *Structure of Molecules and Internal Rotation*, Academic Press, New York, 1954.
4. D. Grime and I. M. Ward, *Trans. Faraday Soc.*, **54**, 959 (1958).
5. R. E. Robertson, *J. Phys. Chem.*, **69**, 1575 (1965).
6. E. W. Fischer, *Naturforsch*, **12a**, 753 (1957).
7. A. Keller, *Phil. Mag.*, **2**, 1171 (1957).
8. P. H. Till, *J. Polymer Sci.*, **24**, 301 (1957).
9. A. Keller and S. Sawada, *Makromol. Chem.*, **74**, 190 (1964).

Further Reading

F. W. Billmeyer, *Textbook of Polymer Science*, Wiley, New York, 1963.
P. H. Geil, *Polymer Single Crystals*, Interscience Publishers, New York, 1963.
P. Meares, *Polymers: Structure and Bulk Properties*, Van Nostrand, London, 1965.

2

The Mechanical Properties of Polymers: General Considerations

2.1 OBJECTIVES

Discussions of the mechanical properties of solid polymers often contain two interrelated objectives. The first of these is to obtain an adequate macroscopic description of the particular facet of polymer behaviour under consideration. The second objective is to seek an explanation of this behaviour in molecular terms, which may include details of the chemical composition and physical structure. In this book we will endeavour, where possible, to separate these two objectives and, in particular, to establish a satisfactory macroscopic or phenomenological description before discussing molecular interpretations.

This should make it clear that many of the established relationships are purely descriptive, and do not necessarily have any implications with regard to an interpretation in structural terms. For engineering applications of polymers this is sufficient, because a description of the mechanical behaviour under conditions which simulate their end use is often all that is required, together with empirical information concerning their method of manufacture.

2.2 THE DIFFERENT TYPES OF MECHANICAL BEHAVIOUR

It is difficult to classify polymers as particular types of materials such as a glassy solid or a viscous liquid, since their mechanical properties are so dependent on the conditions of testing, e.g. the rate of application of load, temperature, amount of strain.

A polymer can show all the features of a glassy, brittle solid or an elastic rubber or a viscous liquid depending on the temperature and time scale of measurement. Polymers are usually described as *viscoelastic* materials, a generic term which emphasizes their intermediate position between viscous liquids and elastic solids. At low temperatures, or high frequencies of

15

measurement, a polymer may be glass-like with a Young's modulus of 10^{10}–10^{11} (dyn cm^{-2}) and will break or flow at strains greater than 5%. At high temperatures or low frequencies, the same polymer may be rubber-like with a modulus of 10^{7}–10^{8} (dyn cm^{-2}) withstanding large extensions ($\sim 100\%$) without permanent deformation. At still higher temperatures, permanent deformation occurs under load, and the polymer behaves like a highly viscous liquid.

In an intermediate temperature or frequency range, commonly called the glass-transition range, the polymer is neither glassy nor rubber-like. It shows an intermediate modulus, is viscoelastic and may dissipate a considerable amount of energy on being strained. The glass transition manifests itself in several ways, for example by a change in the volume coefficient of expansion, which can be used to define a glass-transition temperature T_g. The glass transition is central to a great deal of the mechanical behaviour of polymers for two reasons. First, there are the attempts to link the time–temperature equivalence of viscoelastic behaviour with the glass-transition temperature T_g. Secondly, glass transitions can be studied at a molecular level by such techniques as nuclear magnetic resonance and dielectric relaxation. In this way it is possible to gain an understanding of the molecular origins of the viscoelasticity.

The different features of polymer behaviour such as creep and recovery, brittle fracture, necking and cold-drawing are usually considered separately, by comparative studies of different polymers. It is customary, for example, to compare the brittle fracture of polymethylmethacrylate, polystyrene, and other polymers which show similar behaviour at room temperature. Similarly comparative studies have been made of the creep and recovery of polyethylene, polypropylene and other polyolefines. Such comparisons often obscure the very important point that the whole range of phenomena can be displayed by a single polymer as the temperature is changed. Figure 2.1 shows load–elongation curves for the polymer at four different temperatures. At temperatures well below the glass transition (curve A), where brittle fracture occurs, the load rises to the breaking point linearly with increasing elongation, and rupture occurs at low strains ($\sim 10\%$). At high temperatures (curve D), the polymer is rubber-like and the load rises to the breaking point with a sigmoidal relationship to the elongation, and rupture occurs at very high strains (~ 30–1000%).

In an intermediate temperature range below the glass transition (curve B), the load–deformation relationship resembles that of a ductile metal, showing a load maximum, i.e. a yield point before rupture occurs. At slightly higher temperatures (curve C), still below the glass transition, the

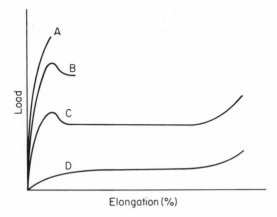

Figure 2.1. Load–elongation curves for a polymer at different temperatures. Curve A: Brittle fracture. Curve B: Ductile failure. Curve C: Cold-drawing. Curve D: Rubber-like behaviour.

remarkable phenomenon of necking and cold-drawing is observed. Here the conventional load–elongation curve again shows a yield point and a subsequent decrease in conventional stress. However, with a further increase in the applied strain, the load falls to a constant level at which deformations of the order of 300–1000% are accomplished. At this stage a neck has formed and the strain in the specimen is not uniform. (This is discussed in detail in Chapter 11.) Eventually the load begins to rise again and finally fracture occurs.

It is usual to discuss the mechanical properties in the different temperature ranges separately, because different approaches and mathematical formalisms are adopted for the different features of mechanical behaviour. This conventional treatment will be followed here, although recognizing that it somewhat arbitrarily isolates particular facets of the mechanical properties of polymers.

2.3   THE ELASTIC SOLID AND THE BEHAVIOUR OF POLYMERS

Mechanical behaviour is in most general terms concerned with the deformations which occur under loading. In any specific case the deformations depend on details such as the geometrical shape of the specimen or the way in which the load is applied. Such considerations are the province of the plastics engineer, who is concerned with predicting the performance of a polymer in a specified end use. In our discussion of the mechanical properties of polymers we will ignore such questions as these, which relate to solving particular problems of behaviour in practice. We will concern ourselves only with the generalized equations termed *constitutive relations*, which relate stress and strain for a particular type of material. First it will be necessary to find constitutive relations which give an adequate description of the mechanical behaviour. Secondly, where possible, we will obtain a molecular understanding of this behaviour by a molecular model which predicts the constitutive relations.

One of the simplest constitutive relations is Hooke's Law, which relates the stress σ to the strain e for the uniaxial deformation of an ideal elastic isotropic solid. Thus

$$\sigma = Ee, \qquad \text{where } E \text{ is the Young's modulus.}$$

There are five important ways in which the mechanical behaviour of a polymer may deviate from that of an ideal elastic solid obeying Hooke's Law. First, in an elastic solid the deformations induced by loading are independent of the history or rate of application of the loads, whereas in a polymer the deformations can be drastically affected by such considerations. This means that the simplest constitutive relation for a polymer should in general contain time or frequency as a variable in addition to stress and strain. Secondly, in an elastic solid all the situations pertaining to stress and strain can be reversed. Thus, if a stress is applied, a certain deformation will occur. On removal of the stress, this deformation will disappear exactly. This is not always true for polymers. Thirdly, in an elastic solid obeying Hooke's Law, which in its more general implications is the basis of small-strain elasticity theory, the effects observed are *linearly* related to the influences applied. This is the essence of Hooke's Law; stress is exactly proportional to strain. This is not generally true for polymers, but applies in many cases only as a good approximation for very small strains; in general the constitutive relations are non-linear. It is important to note that *non-linearity* is not related to *recoverability*. In contrast to metals, polymers may recover from strains beyond the proportional limit without any permanent deformation.

Fourthly, the definitions of stress and strain in Hooke's Law are only valid for small deformations. When we wish to consider larger deformations a new theory must be developed in which both stress and strain are defined more generally.

Finally, in many practical applications (such as films and synthetic fibres) polymers are used in an oriented or anisotropic form, which requires a considerable generalization of Hooke's Law.

It will be convenient to discuss these various aspects separately as follows:

(1) Behaviour at large strains in Chapters 3 and 4 (finite elasticity and rubber-like behaviour, respectively).

(2) Time-dependent behaviour (viscoelastic behaviour) in Chapters 5, 6, 7 and 8.

(3) Non-linearity in Chapter 9 (non-linear viscoelastic behaviour).

(4) The behaviour of oriented polymers in Chapter 10 (mechanical anisotropy).

(5) The non-recoverable behaviour in Chapter 11 (plasticity and yield).

It should be recognized, however, that we cannot hold to an exact separation and that there are many places where these aspects overlap and can be brought together by the physical mechanisms which underlie the phenomenological description.

2.4 STRESS AND STRAIN

It is desirable at this juncture to outline very briefly the concepts of stress and strain. For a more comprehensive discussion the reader is referred to standard text books on the theory of elasticity.[1,2]

2.4.1 The State of Stress

The components of stress in a body can be defined by considering the forces acting on an infinitesimal cubical volume element (Figure 2.2) whose edges are parallel to the coordinate axes x, y, z. In equilibrium, the forces per unit area acting on the cube faces are

$$P_1 \text{ on the } yz \text{ plane}$$

$$P_2 \text{ on the } zx \text{ plane}$$

$$P_3 \text{ on the } xy \text{ plane}$$

These three forces are then resolved into their nine components in the x, y and z directions as follows:

$$P_1: \quad \sigma_{xx}, \quad \sigma_{xy}, \quad \sigma_{xz}$$

$$P_2: \quad \sigma_{yx}, \quad \sigma_{yy}, \quad \sigma_{yz}$$

$$P_3: \quad \sigma_{zx}, \quad \sigma_{zy}, \quad \sigma_{zz}$$

The first subscript refers to the direction of the normal to the plane on which the stress acts, and the second subscript to the direction of the

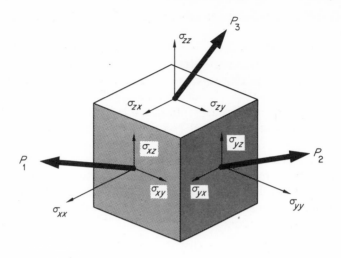

Figure 2.2. The stress components.

stress. In the absence of body torques the total torque acting on the cube must also be zero, and this implies three further equalities:

$$\sigma_{xy} = \sigma_{yx}, \qquad \sigma_{xz} = \sigma_{zx}, \qquad \sigma_{yz} = \sigma_{zy}$$

The components of stress are therefore defined by six independent quantities, σ_{xx}, σ_{yy} and σ_{zz}, the normal stresses, and σ_{xy}, σ_{yz} and σ_{zx} the shear stresses. These form the six independent components of the stress tensor σ_{ij}.

$$\sigma_{ij} = \begin{bmatrix} \sigma_{xx} & \sigma_{xy} & \sigma_{xz} \\ \sigma_{yx} & \sigma_{yy} & \sigma_{yz} \\ \sigma_{zx} & \sigma_{zy} & \sigma_{zz} \end{bmatrix}$$

The state of stress at a point in a body is determined when we can specify the normal components and the shear components of stress acting on a plane drawn in any direction through the point. If we know these six components of stress at a given point the stresses acting on any plane through this point can be calculated. (See References 1, Section 67, and 2, Section 47.)

2.4.2   The State of Strain

The Engineering Components of Strain

The displacement of any point P_1 (see Figure 2.3) in the body may be resolved into its components u, v and w parallel to x, y and z (Cartesian-coordinate axes chosen in the undeformed state) so that if the coordinates

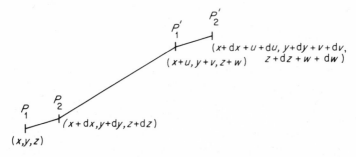

Figure 2.3. The displacements produced by deformation.

of the point in the undisplaced position were (x, y, z) they become $(x+u, y+v, z+w)$ on deformation. In defining the strains we are not interested in the *displacement* or rotation but in the *deformation*. The latter is the displacement of a point relative to adjacent points. Consider a point P_2, very close to P_1, which in the undisplaced position had coordinates $(x+dx, y+dy, z+dz)$ and let the displacement which it has undergone have components $(u+du, v+dv, w+dw)$. The quantities required are then du, dv and dw, the *relative* displacements.

If dx, dy and dz are sufficiently small, i.e., infinitesimal

$$du = \frac{\partial u}{\partial x}dx + \frac{\partial u}{\partial y}dy + \frac{\partial u}{\partial z}dz$$

$$dv = \frac{\partial v}{\partial x}dx + \frac{\partial v}{\partial y}dy + \frac{\partial v}{dz}dz$$

$$dw = \frac{\partial w}{\partial x}dx + \frac{\partial w}{\partial y}dy + \frac{\partial w}{\partial z}dz$$

Thus we require to define the nine quantities

$$\frac{\partial u}{\partial x}, \frac{\partial u}{\partial y}, \dots \text{etc.}$$

For convenience these nine quantities are regrouped and denoted as follows

$$e_{xx} = \frac{\partial u}{\partial x}, \qquad e_{yy} = \frac{\partial v}{\partial y}, \qquad e_{zz} = \frac{\partial w}{\partial z}$$

$$e_{yz} = \frac{\partial w}{\partial y} + \frac{\partial y}{\partial z}, \qquad e_{zx} = \frac{\partial u}{\partial z} + \frac{\partial w}{\partial x}, \qquad e_{xy} = \frac{\partial v}{\partial x} + \frac{\partial u}{\partial y}$$

$$2\bar{\omega}_x = \frac{\partial w}{\partial y} - \frac{\partial v}{\partial z}, \qquad 2\bar{\omega}_y = \frac{\partial u}{\partial z} - \frac{\partial w}{\partial x}, \qquad 2\bar{\omega}_z = \frac{\partial v}{\partial x} - \frac{\partial u}{\partial y}$$

The first three quantities e_{xx}, e_{yy} and e_{zz} correspond to the fractional expansions or contractions along the x, y and z axes of an infinitesimal element at P_1. The second three quantities e_{yz}, e_{zx}, and e_{xy} correspond to the components of shear strain in the yz, zx and xy planes respectively. The last three quantities $\bar{\omega}_x$, $\bar{\omega}_y$, and $\bar{\omega}_z$ do not correspond to a deformation of the element at P_1, but are the components of its rotation as a rigid body.

The concept of shear strain can be conveniently illustrated by a diagram showing the two-dimensional situation of shear in the yz plane (Figure 2.4).

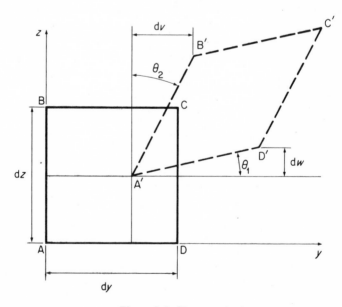

Figure 2.4. Shear strains.

ABCD is an infinitesimal square which has been displaced and deformed into the rhombus A'B'C'D', θ_1 and θ_2 being the angles A'D' and A'B' make with the y and z axes respectively.

Now

$$\tan \theta_1 \doteqdot \theta_1 = \frac{dw}{dy} \rightarrow \frac{\partial w}{\partial y}$$

$$\tan \theta_2 \doteqdot \theta_2 = \frac{dv}{dz} \rightarrow \frac{\partial v}{\partial z}$$

The shear strain

$$e_{yz} = \frac{\partial w}{\partial y} + \frac{\partial v}{\partial z} = \theta_1 + \theta_2$$

$2\bar{\omega}_x = \theta_1 - \theta_2$ does not correspond to a deformation of ABCD but to twice the angle through which AC has been rotated.

The deformation is therefore defined by the first six quantities e_{xx}, e_{yy}, e_{zz}, e_{yz}, e_{zx}, e_{xy}, which are called the components of strain. It is important to note that *engineering strains* have been defined (see below).

The Tensor Components of Strain

In the tensor notation the components of strain are defined as the components of the strain tensor

$$\varepsilon_{ij} = \frac{1}{2}\left(\frac{\partial u_i}{\partial x_j} + \frac{\partial u_j}{\partial x_i}\right)$$

where i, j can take values 1, 2, 3 and we sum over all possible values.

In terms of the notation of the last section

$$x_1 = x, \qquad x_2 = y, \qquad x_3 = z$$

and

$$u_1 = u, \qquad u_2 = v, \qquad u_3 = w$$

Then

$$\varepsilon_{ij} = \begin{bmatrix} \varepsilon_{xx} & \varepsilon_{xy} & \varepsilon_{xz} \\ \varepsilon_{xy} & \varepsilon_{yy} & \varepsilon_{yz} \\ \varepsilon_{xz} & \varepsilon_{yz} & \varepsilon_{zz} \end{bmatrix}$$

$$
= \begin{bmatrix}
\dfrac{\partial u}{\partial x} & \dfrac{1}{2}\left(\dfrac{\partial v}{\partial x}+\dfrac{\partial u}{\partial y}\right) & \dfrac{1}{2}\left(\dfrac{\partial w}{\partial x}+\dfrac{\partial u}{\partial z}\right) \\[2ex]
\dfrac{1}{2}\left(\dfrac{\partial v}{\partial x}+\dfrac{\partial u}{\partial y}\right) & \dfrac{\partial v}{\partial y} & \dfrac{1}{2}\left(\dfrac{\partial v}{\partial z}+\dfrac{\partial w}{\partial y}\right) \\[2ex]
\dfrac{1}{2}\left(\dfrac{\partial w}{\partial x}+\dfrac{\partial u}{\partial z}\right) & \dfrac{1}{2}\left(\dfrac{\partial v}{\partial z}+\dfrac{\partial w}{\partial y}\right) & \dfrac{\partial w}{\partial z}
\end{bmatrix}
$$

in our previous notation and

$$
\varepsilon_{ij} = \begin{bmatrix}
e_{xx} & \tfrac{1}{2}e_{xy} & \tfrac{1}{2}e_{xz} \\[1.5ex]
\tfrac{1}{2}e_{xy} & e_{yy} & \tfrac{1}{2}e_{yz} \\[1.5ex]
\tfrac{1}{2}e_{xz} & \tfrac{1}{2}e_{yz} & e_{zz}
\end{bmatrix}
$$

in terms of the engineering components of strain.

2.5 THE GENERALIZED HOOKE'S LAW

The most general *linear* relationship between stress and strain is obtained by assuming that each of the tensor components of stress is linearly related to all the tensor components of strain and vice-versa. Thus

$$
\sigma_{xx} = a\varepsilon_{xx}+b\varepsilon_{yy}+c\varepsilon_{zz}+d\varepsilon_{xz}+ \dots \text{etc.}
$$

and

$$
\varepsilon_{xx} = a'\sigma_{xx}+b'\sigma_{yy}+c'\sigma_{zz}+d'\sigma_{xz} + \dots \text{etc.}
$$

where $a, b, \dots, a', b', \dots$ are constants. This is the generalized Hooke's Law.

In tensor notation we relate the second-rank tensor σ_{ij} to the second-rank strain tensor ε_{ij} by fourth-rank tensors c_{ijkl} and s_{ijkl}. Thus

$$
\sigma_{ij} = c_{ijkl}\varepsilon_{kl}
$$

or equivalently

$$
\varepsilon_{ij} = s_{ijkl}\sigma_{kl}
$$

where

$$
\sigma_{ij} = \sigma_{xx}, \sigma_{yy}, \dots \text{etc.}
$$

and

$$
\varepsilon_{ij} = \varepsilon_{xx}, \varepsilon_{yy} \dots \text{etc.}
$$

The fourth-rank tensors s_{ijkl} and c_{ijkl} contain the compliance and stiffness constants respectively, with i, j, k, l taking values $1, 2, 3$. These equations use $1, 2, 3$ as being synonymous with x, y, z respectively. The compliances and stiffnesses are written in terms of 1, 2, 3 and the stresses and strains in terms of x, y, z.

It is customary to adopt an abbreviated nomenclature in which the generalized Hooke's Law relates the six independent components of stress to the six independent components of the engineering strains. We have

$$\sigma_p = c_{pq}e_q$$

and

$$e_p = s_{pq}\sigma_q$$

where σ_p represents σ_{xx}, σ_{yy}, σ_{zz}, σ_{xz}, σ_{yz} or σ_{xy}, and e_q represents e_{xx}, e_{yy}, e_{zz}, e_{xz}, e_{yz} or e_{xy}. We form matrices for c_{pq} and s_{pq} in which p and q take the values 1, 2 ... 6. In the case of the stiffness constants the p's and q's are obtained in terms of i, j, k, l by substituting 1 for 11, 2 for 22, 3 for 33, 4 for 23, 5 for 13 and 6 for 12. For the compliance constants rather more complicated rules apply owing to the occurrence of the factor-2 difference between the definition of the tensor shear strain components and the definition of engineering-shear strains. Thus

$$s_{ijkl} = s_{pq} \text{ when } p \text{ and } q \text{ are 1, 2 or 3}$$

$$2s_{ijkl} = s_{pq} \text{ when either } p \text{ or } q \text{ are 4, 5 or 6}$$

$$4s_{ikjl} = s_{pq} \text{ when both } p \text{ and } q \text{ are 4, 5 or 6}$$

A typical relationship between stress and strain is now written as

$$\sigma_{xx} = c_{11}e_{xx}+c_{12}e_{yy}+c_{13}e_{zz}+c_{14}e_{xz}+c_{15}e_{yz}+c_{16}e_{xy}$$

The existence of a strain-energy function (see Reference 2, p. 149) provides the relationships

$$c_{pq} = c_{qp}$$

$$s_{pq} = s_{qp}$$

and reduces the number of independent constants from 36 to 21. We then have

$$c_{pq} = \begin{vmatrix} c_{11} & c_{12} & c_{13} & c_{14} & c_{15} & c_{16} \\ c_{12} & c_{22} & c_{23} & c_{24} & c_{25} & c_{26} \\ c_{13} & c_{23} & c_{33} & c_{34} & c_{35} & c_{36} \\ c_{14} & c_{24} & c_{34} & c_{44} & c_{45} & c_{46} \\ c_{15} & c_{25} & c_{35} & c_{45} & c_{55} & c_{56} \\ c_{16} & c_{26} & c_{36} & c_{46} & c_{56} & c_{66} \end{vmatrix}$$

and similarly

$$
S_{pq} = \begin{pmatrix}
s_{11} & s_{12} & s_{13} & s_{14} & s_{15} & s_{16} \\
s_{12} & s_{22} & s_{23} & s_{24} & s_{25} & s_{26} \\
s_{13} & s_{23} & s_{33} & s_{34} & s_{35} & s_{36} \\
s_{14} & s_{24} & s_{34} & s_{44} & s_{45} & s_{46} \\
s_{15} & s_{25} & s_{35} & s_{45} & s_{55} & s_{56} \\
s_{16} & s_{26} & s_{36} & s_{46} & s_{56} & s_{66}
\end{pmatrix}
$$

These matrices define the relationships between stress and strain in a general elastic solid, whose properties vary with direction, i.e. an anisotropic elastic solid. In most of this book we will be concerned with isotropic polymers; all discussion of anisotropic mechanical properties will be reserved for Chapter 10.

It is then most straightforward to use the compliance-constants matrix, and note that measured quantities such as the Young's modulus E, Poisson's ratio v and the torsional or shear modulus G, relate directly to the compliance constants.

For an isotropic solid, the matrix s_{pq} reduces to

$$
\begin{pmatrix}
s_{11} & s_{12} & s_{12} & 0 & 0 & 0 \\
s_{12} & s_{11} & s_{12} & 0 & 0 & 0 \\
s_{12} & s_{12} & s_{11} & 0 & 0 & 0 \\
0 & 0 & 0 & 2(s_{11}-s_{12}) & 0 & 0 \\
0 & 0 & 0 & 0 & 2(s_{11}-s_{12}) & 0 \\
0 & 0 & 0 & 0 & 0 & 2(s_{11}-s_{12})
\end{pmatrix}
$$

It can be shown that the Young's modulus $E = 1/S_{11}$, the Poisson's ratio

$$
v = \frac{-s_{12}}{s_{11}}
$$

and the torsional modulus

$$
G = \frac{1}{2(s_{11}-s_{12})}
$$

Thus we obtain the stress–strain relationships which are the starting point in many elementary textbooks of elasticity (Reference 1, pp. 7–9).

$$e_{xx} = \frac{1}{E}\sigma_{xx} - \frac{v}{E}(\sigma_{yy} + \sigma_{zz})$$

$$e_{yy} = \frac{1}{E}\sigma_{yy} - \frac{v}{E}(\sigma_{xx} + \sigma_{zz})$$

$$e_{zz} = \frac{1}{E}\sigma_{zz} - \frac{v}{E}(\sigma_{xx} + \sigma_{yy})$$

$$e_{xz} = \frac{1}{G}\sigma_{xz}$$

$$e_{yz} = \frac{1}{G}\sigma_{yz}$$

$$e_{xy} = \frac{1}{G}\sigma_{xy}$$

where

$$G = \frac{E}{2(1+v)}$$

Another basic quantity is the bulk modulus K, which determines the dilatation $\Delta = e_{xx} + e_{yy} + e_{zz}$ produced by a uniform hydrostatic pressure. Using the stress–strain relationships above, it can be shown that the strains produced by a uniform hydrostatic pressure p are given by

$$e_{xx} = (s_{11} + 2s_{12})p$$
$$e_{yy} = (s_{11} + 2s_{12})p$$
$$e_{zz} = (s_{11} + 2s_{12})p$$

Then

$$K = \frac{p}{\Delta} = \frac{1}{3(s_{11} + 2s_{12})}$$

$$= \frac{E}{3(1 - 2v)}$$

This completes our introduction to linear elastic behaviour at small strains. The extension to large strains will be considered in the next chapter on finite elasticity.

REFERENCES

1. S. Timoshenko and J. N. Goodier, *Theory of Elasticity*, McGraw–Hill, New York, 1951.
2. A. E. H. Love, *A Treatise on the Mathematical Theory of Elasticity*, 4th ed., Macmillan, New York, 1944.

3

The Behaviour of Polymers in the Rubber-like State: Finite Strain Elasticity

In the rubber-like state, a polymer may be subjected to large deformations and still show complete recovery. The behaviour of a rubber band stretching to two or three times its original length and, when released, recovering essentially instantaneously to its original shape, is a matter of common experience. To a good approximation this is elastic behaviour at large strains. The first stage in developing an understanding of this behaviour is to seek a generalized definition of strain which will not suffer the restriction of that derived in Section 2, that the strains are small. This is followed by a rigorous definition of stress for the situation where the deformations are not small. These considerations are the basis of finite elasticity theory. This subject has been considered in several notable texts[1,2] and it is not intended to duplicate the elegant treatments presented elsewhere. For the most part the development of finite-strain elasticity theory has been made using tensor calculus. In this book the treatment will be at a more elementary level, and it is hoped that in this way the exposition will be clear to those who only require an outline of the subject in order to understand the relevant mechanical properties of polymers.

3.1 THE GENERALIZED DEFINITION OF STRAIN

The generalized theory of strain considers the ratio of the length of the line joining two points in the undeformed solid to the length of the line joining the same two points in the deformed solid. It will be shown that this comparison, suitably defined, can lead to a strain tensor with six independent components which reduce to the expressions of Section 2.4.2 for small strains when the deformation is small.

Consider a system of rectangular coordinate axes x, y, and z (Figure 3.1). The point P has coordinates (x, y, z) and the neighbouring point Q

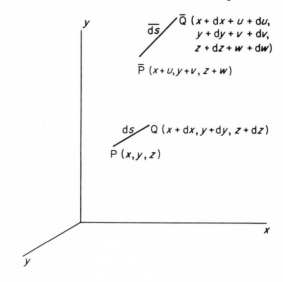

Figure 3.1. Finite deformation transforms the line PQ to the line $\overline{PQ}$.

coordinates $(x+\mathrm{d}x, y+\mathrm{d}y, z+\mathrm{d}z)$. When the body is deformed, P and Q become $\overline{P}$ and $\overline{Q}$ with coordinates $(x+u, y+v, z+w)$ and $(x+\mathrm{d}x+u+\mathrm{d}u, y+\mathrm{d}y+v+\mathrm{d}v, z+\mathrm{d}z+w+\mathrm{d}w)$ respectively, where

$$\mathrm{d}u = \frac{\partial u}{\partial x}\mathrm{d}x + \frac{\partial u}{\partial y}\mathrm{d}y + \frac{\partial u}{\partial z}\mathrm{d}z$$

$$\mathrm{d}v = \frac{\partial v}{\partial x}\mathrm{d}x + \frac{\partial v}{\partial y}\mathrm{d}y + \frac{\partial v}{\partial z}\mathrm{d}z \tag{3.1}$$

and

$$\mathrm{d}w = \frac{\partial w}{\partial x}\mathrm{d}x + \frac{\partial w}{\partial y}\mathrm{d}y + \frac{\partial w}{\partial z}\mathrm{d}z$$

The line PQ has length $\mathrm{d}s$ where $\mathrm{d}s^2 = \mathrm{d}x^2 + \mathrm{d}y^2 + \mathrm{d}z^2$. The line $\overline{PQ}$ has length $\overline{\mathrm{d}s}$ where $\overline{\mathrm{d}s}^2 = (\mathrm{d}x+\mathrm{d}u)^2 + (\mathrm{d}y+\mathrm{d}v)^2 + (\mathrm{d}z+\mathrm{d}w)^2$.

We wish to relate $\overline{\mathrm{d}s}$, the length of the line $\overline{PQ}$ in the deformed state to $\mathrm{d}s$, the length of the line in the undeformed state PQ. Then

$$\frac{\overline{\mathrm{d}s}^2}{\mathrm{d}s^2} = \frac{(\mathrm{d}x+\mathrm{d}u)^2 + (\mathrm{d}y+\mathrm{d}v)^2 + (\mathrm{d}z+\mathrm{d}w)^2}{\mathrm{d}s^2}$$

$$= \left(\frac{\mathrm{d}x}{\mathrm{d}s} + \frac{\mathrm{d}u}{\mathrm{d}s}\right)^2 + \left(\frac{\mathrm{d}y}{\mathrm{d}s} + \frac{\mathrm{d}v}{\mathrm{d}s}\right)^2 + \left(\frac{\mathrm{d}z}{\mathrm{d}s} + \frac{\mathrm{d}w}{\mathrm{d}s}\right)^2$$

Substituting for du/ds, dv/ds and dw/ds from Equations (3.1)

$$\left(\frac{\overline{ds}}{ds}\right)^2 = \left(\frac{dx}{ds} + \frac{\partial u}{\partial x}\frac{dx}{ds} + \frac{\partial u}{\partial y}\frac{dy}{ds} + \frac{\partial u}{\partial z}\frac{dz}{ds}\right)^2$$

$$+ \left(\frac{dy}{ds} + \frac{\partial v}{\partial x}\frac{dx}{ds} + \frac{\partial v}{\partial y}\frac{dy}{ds} + \frac{\partial v}{\partial z}\frac{dz}{ds}\right)^2$$

$$+ \left(\frac{dz}{ds} + \frac{\partial w}{\partial x}\frac{dx}{ds} + \frac{\partial w}{\partial y}\frac{dy}{ds} + \frac{\partial w}{\partial z}\frac{dz}{ds}\right)^2 \tag{3.2}$$

The deformation is being described with respect to a system of coordinates in the *undeformed* state. The line ds has direction cosines

$$l = \frac{dx}{ds}, \qquad m = \frac{dy}{ds}, \qquad n = \frac{dz}{ds}$$

Equation (3.2) can therefore be written as

$$\left(\frac{\overline{ds}}{ds}\right)^2 = \left\{ l\left(1 + \frac{\partial u}{\partial x}\right) + m\frac{\partial u}{\partial y} + n\frac{\partial u}{\partial z} \right\}^2$$

$$+ \left\{ l\frac{\partial v}{\partial x} + m\left(1 + \frac{\partial v}{\partial y}\right) + n\frac{\partial v}{\partial z} \right\}^2$$

$$+ \left\{ l\frac{\partial w}{\partial x} + m\frac{\partial w}{\partial y} + n\left(1 + \frac{\partial w}{\partial z}\right) \right\}^2$$

Finally we write

$$\left(\frac{\overline{ds}}{ds}\right)^2 = (1 + 2e_{xx})l^2 + (1 + 2e_{yy})m^2 + (1 + 2e_{zz})n^2$$

$$+ 2e_{yz}mn + 2e_{zx}nl + 2e_{xy}lm \tag{3.3}$$

where e_{xx}, e_{yy}, etc., are the six components of finite strain and are given by:

$$e_{xx} = \frac{\partial u}{\partial x} + \frac{1}{2}\left\{ \left(\frac{\partial u}{\partial x}\right)^2 + \left(\frac{\partial v}{\partial x}\right)^2 + \left(\frac{\partial w}{\partial x}\right)^2 \right\}$$

$$e_{yy} = \frac{\partial v}{\partial y} + \frac{1}{2}\left\{ \left(\frac{\partial u}{\partial y}\right)^2 + \left(\frac{\partial v}{\partial y}\right)^2 + \left(\frac{\partial w}{\partial y}\right)^2 \right\}$$

$$e_{zz} = \frac{\partial w}{\partial z} + \frac{1}{2}\left\{ \left(\frac{\partial u}{\partial z}\right)^2 + \left(\frac{\partial v}{\partial z}\right)^2 + \left(\frac{\partial w}{\partial z}\right)^2 \right\}$$

$$e_{yz} = \frac{\partial w}{\partial y} + \frac{\partial v}{\partial z} + \frac{\partial u}{\partial y}\frac{\partial u}{\partial z} + \frac{\partial v}{\partial y}\frac{\partial v}{\partial z} + \frac{\partial w}{\partial y}\frac{\partial w}{\partial z}$$

$$\mathbf{e}_{zx} = \frac{\partial u}{\partial z} + \frac{\partial w}{\partial x} + \frac{\partial u}{\partial z}\frac{\partial u}{\partial x} + \frac{\partial v}{\partial z}\frac{\partial v}{\partial x} + \frac{\partial w}{\partial z}\frac{\partial w}{\partial x}$$

$$\mathbf{e}_{xy} = \frac{\partial v}{\partial x} + \frac{\partial u}{\partial y} + \frac{\partial u}{\partial x}\frac{\partial u}{\partial y} + \frac{\partial v}{\partial x}\frac{\partial v}{\partial y} + \frac{\partial w}{\partial x}\frac{\partial w}{\partial y}$$

Equation (3.3) for $(\overline{ds}/ds)^2$, defines the ratio of the deformed to the un-deformed length of an elemental line originally at a point x, y, z, having any specified direction defined by direction cosines l, m, n in the undeformed state.

This is known as the Lagrangian measure of strain, as distinct from the Eulerian measure where the coordinates are convected with the deformation. $(1 + 2\mathbf{e}_{xx})$, $(1 + 2\mathbf{e}_{yy})$ and $(1 + 2\mathbf{e}_{zz})$ give the values of $(\overline{ds}/ds)^2$ for elements which have directions parallel to the x, y and z axes respectively in the undeformed state. This observation forms the link between the generalized definition of finite strain and the simple ideas of extension ratios which are used in the molecular theories of rubber elasticity.

A deformation is called pure strain if the three orthogonal lines which are chosen as the system of coordinates in the undeformed state are not rotated by the deformation (Reference 5, p. 68). This means that if we choose the system of Cartesian-coordinate axes to coincide with these three orthogonal lines the shear strain components are zero.

Thus for an homogeneous pure strain where λ_1, λ_2 and λ_3 are the lengths in the deformed state of linear elements parallel to the x, y and z axes respectively which have unit length in the undeformed state, we have

$$\lambda_1^2 = 1 + 2\mathbf{e}_{xx}, \quad \lambda_2^2 = 1 + 2\mathbf{e}_{yy}, \quad \text{and} \quad \lambda_3^2 = 1 + 2\mathbf{e}_{zz};$$

$$\text{and} \quad \mathbf{e}_{yz} = \mathbf{e}_{zx} = \mathbf{e}_{xy} = 0$$

In the general case the shear components of strain $\mathbf{e}_{yz}$, $\mathbf{e}_{zx}$, $\mathbf{e}_{xy}$, are not zero in the chosen system of coordinates. A series of suitable rotations is then required to find a system of coordinates in which the shear components of strain are zero, thus defining the principal axes of strain. λ_1, λ_2 and λ_3 are the principal extension ratios. This is the well-known eigenvalue/eigenvector problem in which λ_1, λ_2 and λ_3 are found by taking the roots of the determinant

$$\begin{vmatrix} 1 + 2\mathbf{e}_{xx} - \lambda^2 & \mathbf{e}_{yx} & \mathbf{e}_{zx} \\ \mathbf{e}_{xy} & 1 + 2\mathbf{e}_{yy} - \lambda^2 & \mathbf{e}_{zy} \\ \mathbf{e}_{xz} & \mathbf{e}_{yz} & 1 + 2\mathbf{e}_{zz} - \lambda^2 \end{vmatrix} = 0$$

(See Reference 3.)

A rubber has a very high bulk modulus compared with the other moduli. To a good approximation the changes in volume on deformation may therefore be neglected. This gives the important relationship between the principal extension ratios

$$\lambda_1 \lambda_2 \lambda_3 = 1$$

The finite-strain components include terms such as $(\partial u/\partial y)^2$, $(\partial v/\partial y)^2$, $\partial u/\partial y \, \partial u/\partial z$, $\partial v/\partial y \, \partial v/\partial z$, etc., which are second-order and can be neglected when the strains are small. The strain components then reduce to the expressions derived previously for small strains (Section 2.4.2 above), e.g.

$$\mathbf{e}_{xx} \rightarrow e_{xx} = \frac{\partial u}{\partial x} \qquad \mathbf{e}_{yz} \rightarrow e_{yz} = \frac{\partial w}{\partial y} + \frac{\partial v}{\partial z}$$

This can also be seen directly from Equation (3.3). Consider the case of simple elongation in the x direction.
Then $\overline{ds}^2 = ds^2(1 + 2\mathbf{e}_{xx})$, i.e. $(\overline{ds}/ds)^2 - 1 = 2\mathbf{e}_{xx}$.
For small strains

$$\frac{\overline{ds}}{ds} = 1 + e_{xx}$$

$$\text{and } (1 + e_{xx})^2 - 1 = 2\mathbf{e}_{xx}$$

i.e. $e_{xx} \simeq \mathbf{e}_{xx}$ when $e_{xx} \ll 1$ as required.

It is important to note that we have defined *engineering strains* as for the small-strain case in Chapter 2. The components of finite strain, defined as the components of the strain tensor, are given by $\mathbf{e}_{xx}$, $\mathbf{e}_{yy}$, $\mathbf{e}_{zz}$, $\frac{1}{2}\mathbf{e}_{xz}$, $\frac{1}{2}\mathbf{e}_{yz}$, $\frac{1}{2}\mathbf{e}_{xy}$, i.e. as for small strains, the tensor shear strains are one-half the engineering shear strains.

In much of the literature the index notation is used. This notation is extremely useful for the further development of the theory. We will therefore summarize briefly the definition of finite strain in this notation.

The point P (x, y, z) has coordinates x_1, x_2, x_3 which are written as x_i.

The neighbouring point Q $(x + dx, y + dy, z + dz)$ is similarly written as $x_i + dx_i$.

When the body is deformed these points become

$$x_i + u_i \text{ and } x_i + u_i + dx_i + du_i$$

which we have written for $x_1 + u_1$, $x_2 + u_2$, $x_3 + u_3$, etc. The line PQ has length ds where

$$ds^2 = (dx_1)^2 + (dx_2)^2 + (dx_3)^2$$
$$= dx_i \, dx_i = \delta_{ij} \, dx_i \, dx_j$$

where δ_{ij} is the Kronecker δ defined as

$$\delta_{ij} = 1 \quad \text{for} \quad i = j \quad \text{and} \quad \delta_{ij} = 0 \quad \text{for} \quad i \neq j$$

The summation convention is that a product involving several variables such as $\delta_{ij}\,dx_i\,dx_j$ represents the sum of each of the products involving all possible combinations of the repeated indices. In this case both i and j can take the values 1, 2 and 3. The line PQ has length $\overline{ds}$ where

$$\overline{ds}^2 = (dx_1 + du_1)^2 + (dx_2 + du_2)^2 + (dx_3 + du_3)^2$$

and

$$du_1 = \frac{\partial u_1}{\partial x_1}\,dx_1 + \frac{\partial u_1}{\partial x_2}\,dx_2 + \frac{\partial u_1}{\partial x_3}\,dx_3$$

and similarly for du_2 and du_3. Because u_i is quite generally a function of x_j i.e. each displacement u_1 etc., depends on x_1, x_2, x_3; we must write:

$$\overline{ds}^2 = \delta_{ij}(dx_i + du_i)(dx_j + du_j)$$

$$= \delta_{ij}\left(dx_i + \frac{\partial u_i}{\partial x_h}\,dx_h\right)\left(dx_j + \frac{\partial u_j}{\partial x_k}\,dx_k\right)$$

Note that h and k can take the values 1, 2 and 3 as in the use of i and j.

In the index notation

$$\frac{\partial u_i}{\partial x_h} = u_{i,h}$$

Then

$$\overline{ds}^2 = \delta_{ij}(dx_i + u_{i,h}\,dx_h)(dx_j + u_{j,k}\,dx_k)$$

$$= \delta_{ij}\,dx_i\,dx_j + \delta_{ij}\,dx_i u_{j,k}\,dx_k$$

$$+ \delta_{ij}u_{i,h}\,dx_h\,dx_j + \delta_{ij}u_{i,h}\,dx_h u_{j,k}\,dx_k$$

Changing some of the dummy indices this gives

$$\overline{ds}^2 = \delta_{hk}\,dx_h\,dx_k + \delta_{hj}\,dx_h u_{j,k}\,dx_k$$

$$+ \delta_{ik}u_{i,h}\,dx_h\,dx_k + \delta_{ij}u_{i,h}u_{j,k}\,dx_h\,dx_k$$

$$= [\delta_{hk} + u_{h,k} + u_{k,h} + u_{j,h}u_{j,k}]\,dx_h\,dx_k$$

remembering that $\delta_{hj} = 0$ unless $h = j$, etc. This gives

$$\overline{ds}^2 = [\delta_{hk} + 2e_{hk}]\,dx_h\,dx_k \tag{3.4}$$

where

$$2e_{hk} = u_{h,k} + u_{k,h} + u_{j,h}u_{j,k}$$

If $e_{hk} = 0$ then $\overline{ds}^2 = ds^2$, i.e. there is no deformation. Otherwise e_{hk} describes the deformation and the quantities e_{hk} are the strain components. There are nine strain components $e_{11}, e_{22}, e_{33}, e_{12}, e_{13}, e_{23}, e_{31}, e_{21}, e_{32}$. It is immediately apparent from the definition of e_{hk} above that $e_{12} = e_{21}$, etc., leaving only six independent components.

$$e_{11} = e_{xx}, \qquad e_{22} = e_{yy}, \qquad e_{33} = e_{zz},$$

$$e_{13} = \tfrac{1}{2}e_{xz}, \qquad e_{23} = \tfrac{1}{2}e_{yz}, \qquad e_{12} = \tfrac{1}{2}e_{xy}.$$

3.2 THE DEFINITION OF COMPONENTS OF STRESS

In small-strain elasticity theory, the components of stress in the deformed body are defined by considering the equilibrium of an elemental cube within the body. When the strains are small the dimensions of the body are to a first approximation unaffected by the strains. It is thus of no consequence whether the components of stress are referred to an elemental cube in the deformed body or to an elemental cube in the undeformed body. For finite strains, however, this is no longer true. We will *choose* to define the components of stress with reference to the equilibrium of a cube in the *deformed* body. We will also refer the stress components to a point of the material which is at (x, y, z) in the undeformed state, i.e. to a point which is at $x' = x+u, y' = y+v, z' = z+w$, in the deformed state. To distinguish these stress components from those in the small strain case we will write σ_{xx}, σ_{yy}, etc., instead of σ_{xx}, σ_{yy}, etc.

Thus the stress component σ_{xx} denotes the force parallel to the x-axis, per unit area of the deformed material, this area being normal to the x-axis in the deformed state. The terms σ_{yy} and σ_{zz} are defined similarly.

The stress component σ_{yz} denotes the force parallel to the z-axis, per unit area of the deformed material, which in the deformed state is normal to the y-axis. The stress components $\sigma_{zy}, \sigma_{zx}, \sigma_{xz}, \sigma_{xy}$ and σ_{yx} are similarly defined.

This defines nine stress components. Considering the equilibrium of the cube gives the condition that these forces have zero moment. Thus by taking moments about the edges of the elemental cube, it can be shown that $\sigma_{xy} = \sigma_{yx}, \sigma_{xz} = \sigma_{zx}, \sigma_{yz} = \sigma_{zy}$ as for the small-strain case in Section 2.4.1 above. There are therefore six independent stress components; three normal components of stress and three shear components of stress.

3.3 THE STRESS–STRAIN RELATIONSHIPS

Using these definitions of finite strain and of stress it is clearly possible to construct constitutive relations for finite-strain elastic behavior which are analogous to the generalized Hooke's Law for small-strain elastic behaviour. In principle, each component of stress can be a general function of each component of strain and vice-versa. The restriction similar to Hooke's Law would be that each component of stress is a *linear* function of each component of strain and vice-versa, e.g.

$$\sigma_{xx} = a\mathbf{e}_{xx} + b\mathbf{e}_{yy} + c\mathbf{e}_{zz} + d\mathbf{e}_{xz} + e\mathbf{e}_{yz} + f\mathbf{e}_{xy}$$

We could use this as a starting point for a theory of finite elasticity. It would be desirable to reduce the number of independent coefficients *a*, *b*, *c*, etc., by considerations such as that of material symmetry. In principle it is possible to develop a theory from this basis to solve problems in finite elasticity in a similar manner to those solved in small-strain elasticity. It would be necessary, for example, to satisfy conditions of stress equilibrium and compatibility of strain. The latter are more complex for finite strain than for small strain, as can be imagined from the inclusion of second-order terms in displacement derivatives, and involve the Riemann–Christoffel tensors[1].

In this text we will not be concerned with the development of a general theory of finite elasticity. Instead, we will seek to describe the phenomenological behaviour of rubbers from the most elementary considerations. There are then two bases for simplification.

(1) A rubber is isotropic in the undeformed state.

(2) The changes in volume on deformation are very small and may be neglected, i.e. a rubber is incompressible.

First consider the simplifications which these assumptions make to small-strain elasticity theory.

For an isotropic solid Hooke's Law can be written in terms of Young's modulus E and Poisson's ratio v.

$$e_{xx} = \frac{1}{E}\sigma_{xx} - \frac{v}{E}\sigma_{yy} - \frac{v}{E}\sigma_{zz}$$

$$e_{yy} = -\frac{v}{E}\sigma_{xx} + \frac{1}{E}\sigma_{yy} - \frac{v}{E}\sigma_{zz}$$

$$e_{zz} = -\frac{v}{E}\sigma_{xx} - \frac{v}{E}\sigma_{yy} + \frac{1}{E}\sigma_{zz}$$

$$e_{yz} = \frac{2}{E}(1+v)\sigma_{yz}$$

$$e_{xz} = \frac{2}{E}(1+v)\sigma_{xz}$$

$$e_{xy} = \frac{2}{E}(1+v)\sigma_{xy}$$

Rewrite the first three expressions in a more symmetrical form as

$$e_{xx} = \frac{1+v}{E}\left[\sigma_{xx} - \frac{v}{1+v}(\sigma_{xx} + \sigma_{yy} + \sigma_{zz})\right]$$

$$e_{yy} = \frac{1+v}{E}\left[\sigma_{yy} - \frac{v}{1+v}(\sigma_{xx}+\sigma_{yy}+\sigma_{zz})\right]$$

$$e_{zz} = \frac{1+v}{E}\left[\sigma_{zz} - \frac{v}{1+v}(\sigma_{xx}+\sigma_{yy}+\sigma_{zz})\right]$$

As a further simplification put

$$\frac{v}{1+v}(\sigma_{xx}+\sigma_{yy}+\sigma_{zz}) = p.$$

Because p is the same for each direction, it is equivalent to a hydrostatic pressure.

This gives

$$e_{xx} = \frac{1+v}{E}(\sigma_{xx}-p)$$

$$e_{yy} = \frac{1+v}{E}(\sigma_{yy}-p)$$

$$e_{zz} = \frac{1+v}{E}(\sigma_{zz}-p)$$

At this stage the second assumption of incompressibility is introduced.

$$e_{xx}+e_{yy}+e_{zz} = 0$$

gives

$$\frac{1+v}{E}\{\sigma_{xx}+\sigma_{yy}+\sigma_{zz}-3p\} = 0$$

Substituting the definition of p gives $v = \frac{1}{2}$. For an isotropic, incompressible elastic solid, Hooke's Law therefore reduces to

$$e_{xx} = \frac{3}{2E}(\sigma_{xx}-p)$$

$$e_{yy} = \frac{3}{2E}(\sigma_{yy}-p)$$

$$e_{zz} = \frac{3}{2E}(\sigma_{zz}-p)$$

$$e_{yz} = \frac{3}{E}\sigma_{yz}$$

$$e_{xz} = \frac{3}{E}\sigma_{xz}$$

and

$$e_{xy} = \frac{3}{E}\sigma_{xy}$$

It is to be noted that if the stresses are specified, p is determined by the relationship $p = \frac{1}{3}(\sigma_{xx} + \sigma_{yy} + \sigma_{zz})$. If on the other hand, the strains are specified, we can only find the quantities $\sigma_{xx} - p$, $\sigma_{yy} - p$, $\sigma_{zz} - p$, i.e. the normal components of stress are indeterminate to the extent of an arbitrary hydrostatic pressure p. On reflection this is to be expected; incompressibility just means that an arbitrary hydrostatic pressure produces no change in volume.

We will now propose by analogy the constitutive relations for the case where the deformations are finite. The form of these relations must be such that they reduce to those derived above when the strains are small.

To simplify matters we will consider the case where the shear-strain components are zero. As will be apparent from the previous discussion there is no loss of generality in so doing because any general strain can be reduced to three principal strains by suitable choice of the coordinate axes.

For small strains the constitutive relations are then

$$2e_{xx} = \frac{3}{E}(\sigma_{xx} - p)$$

$$2e_{yy} = \frac{3}{E}(\sigma_{yy} - p)$$

$$2e_{zz} = \frac{3}{E}(\sigma_{zz} - p)$$

and

$$e_{yz} = e_{zx} = e_{xy} = 0 \tag{3.5}$$

Following Rivlin[4] we propose that the analogous relations for the finite-strain case are

$$1 + 2\mathbf{e}_{xx} = \frac{3}{E}(\boldsymbol{\sigma}_{xx} - \mathbf{p})$$

$$1 + 2\mathbf{e}_{yy} = \frac{3}{E}(\boldsymbol{\sigma}_{yy} - \mathbf{p})$$

$$1 + 2e_{zz} = \frac{3}{E}(\sigma_{zz} - \mathbf{p})$$

and

$$e_{yz} = e_{zx} = e_{xy} = 0 \qquad (3.6)$$

The difference between the quantities such as the stress components σ_{xx} and σ_{xx}, and the hydrostatic pressures p and $\mathbf{p}$ are indicated by the use of bold type for the finite-deformation case. Again we have the situation that if the strains are specified we can only find the quantities $(\sigma_{xx} - \mathbf{p})$, etc., i.e. the normal components of stress are indeterminate to the extent of a hydrostatic pressure $\mathbf{p}$. It should be pointed out that although in the small-strain case, if the stresses are specified, $p = \frac{1}{3}(\sigma_{xx} + \sigma_{yy} + \sigma_{zz})$, the corresponding relationship does not hold for finite deformations, i.e. if σ_{xx}, σ_{yy} and σ_{zz} are given, $\mathbf{p} \neq \frac{1}{3}(\sigma_{xx} + \sigma_{yy} + \sigma_{zz})$. This is because the stress–strain relations are not the same for the two cases.

Consider the relationships (3.6) for the simple case of extension parallel to the x-axis under an applied stress σ^*. This gives a pure homogeneous deformation with $\lambda_1, \lambda_2, \lambda_3$ as the extension ratios in the x, y and z directions respectively. Then

$$\lambda_1^2 = 1 + 2e_{xx} \qquad \lambda_2^2 = 1 + 2e_{yy} \qquad \lambda_3^2 = 1 + 2e_{zz}$$

$$e_{xz} = e_{yz} = e_{xy} = 0$$

$$\sigma_{xx} = \sigma^*, \sigma_{yy} = \sigma_{zz} = 0$$

$$\sigma_{xz} = \sigma_{yz} = \sigma_{xy} = 0 \qquad (3.7)$$

Substituting in Equations (3.6) gives

$$\lambda_1^2 = \frac{3}{E}(\sigma^* - \mathbf{p})$$

$$\lambda_2^2 = -\frac{3}{E}\mathbf{p}$$

$$\lambda_3^2 = -\frac{3}{E}\mathbf{p}$$

The condition that the solid is incompressible gives $\lambda_1 \lambda_2 \lambda_3 = 1$

$$\therefore \quad \lambda_2^2 = \lambda_3^2 = -\frac{3\mathbf{p}}{E} = \frac{1}{\lambda_1}$$

and

$$\mathbf{p} = -\frac{E}{3}\frac{1}{\lambda_1}$$

$$\therefore \quad \sigma^* = \lambda_1{}^2\frac{E}{3} + \mathbf{p}$$

$$= \frac{E}{3}\left(\lambda_1{}^2 - \frac{1}{\lambda_1}\right)$$

It is sometimes convenient to consider the *nominal stress* f, the force per unit area of unstrained cross-section.

This gives

$$f = \frac{E}{3}\left(\lambda_1 - \frac{1}{\lambda_1{}^2}\right) \tag{3.8}$$

For small strains $e_{xx} = \lambda_1 - 1$ and the stress–strain relationship is

$$\sigma^* = \frac{E}{3}\left\{(1 + e_{xx})^2 - \frac{1}{1 + e_{xx}}\right\}$$

$$= Ee_{xx} \text{ (ignoring third-order terms)}$$

which is Hooke's Law. The constitutive relationships which we have proposed therefore satisfy the three major requirements of:

(1) Material isotropy

(2) Incompressibility

(3) Reduction to Hooke's Law for small strains.

For simplicity only the case where the shear strains are zero has been considered. The relationship may however readily be generalized.

We have already reached an important conclusion. This is that the familiar relationship:

$$f = \text{const}\left(\lambda_1 - \frac{1}{\lambda_1{}^2}\right)$$

which is more usually presented as a consequence of the molecular theories of rubber networks, follows from purely phenomenological considerations as a simple constitutive equation for the finite deformation of an isotropic, incompressible solid. Materials which obey this relationship are sometimes called neo-Hookeian.

This discussion also hints at the possibility that the most general constitutive relationships for a rubber-like material might in fact be more complex than Equations (3.6). Just as some materials deviate from Hooke's Law at small strains and show non-linear behaviour, it must now be asked what is the equivalent situation in finite elasticity. A better starting point for the development of a more complex theory is the introduction of the stored energy or strain-energy function and this will now be considered.

3.4 THE USE OF A STRAIN-ENERGY FUNCTION

3.4.1 Thermodynamic Considerations

An alternative approach to the theory of elasticity can be made in terms of a strain-energy function. As in our previous treatment originating in the generalized Hooke's Law, we will consider first the case of small-strain elasticity. Because we will wish to follow the phenomenological treatment by one based on statistical mechanics, it is important at the outset to examine the different types of strain-energy function which can be defined, depending on the experimental conditions. This introduces thermodynamic considerations.

In the first place we are required to relate the work done in the deformation of an elastic solid to the components of stress and strain in the solid. For the one-dimensional situation this can be done in the following manner. Consider the uniaxial extension of an elastic solid of length l and cross-sectional area A, fixed at one end. The origin of a system of Cartesian coordinates is situated at this fixed end and the x-axis is chosen to coincide with the length direction of the solid.

In equilibrium under the action of a tensile force f the other end of the solid is displaced from x_0 to x_1. The solid is subjected to a homogeneous extensional strain. In terms of the components of strain we have

$$e_{xx} = \frac{x_1 - x_0}{l}.$$

The stress in the solid is also homogeneous and is defined as in Section 2.4.1 in terms of the equilibrium of a cubical element at a given point. There is only one non-vanishing component of stress, the normal component $\sigma_{xx} = f/A$.

Now imagine a change in strain which involves only an infinitesimal change in the length of the solid. This is achieved by movement of the tensile force f, through a distance $x_2 - x_1$ parallel to the x-axis. The total work done is $f(x_2 - x_1)$, and the work done per unit volume is

$$\left(\frac{f}{A}\right)\frac{(x_2 - x_1)}{l}.$$

(We assume that $x_2 \sim x_1 \ll 1$.)

The new extensional strain is

$$e'_{xx} = \frac{x_2 - x_0}{l}$$

The infinitesimal change in strain is therefore

$$de_{xx} = e'_{xx} - e_{xx} = \frac{(x_2 - x_0)}{l} - \frac{(x_1 - x_0)}{l} = \frac{x_2 - x_1}{l}$$

But

$$\sigma_{xx} = f/A$$

The work done by external forces *per unit volume* may therefore be written as $\sigma_{xx}\, de_{xx}$. For a general deformation of an elastic solid this may be generalized to give the work done by external forces per unit volume as

$$\sigma_{xx}\, de_{xx} + \sigma_{yy}\, de_{yy} + \sigma_{zz}\, de_{zz} + \sigma_{yz}\, de_{yz} + \sigma_{xz}\, de_{xz} + \sigma_{xy}\, de_{xy}$$

Now consider a small-strain deformation of unit volume of an elastic solid occurring under adiabatic conditions at constant volume. The First Law of Thermodynamics gives

$$dW = dU - dQ$$

relating the work done on the solid dW to the increase in internal energy dU and the mechanical value of the heat supplied dQ. For an adiabatic change of state $dQ = 0$ and $dW = dU$. We can imagine that the deformation is produced by independent changes in each of the components of strain, i.e.

$$dW = dU = \frac{\partial U}{\partial e_{xx}} de_{xx} + \frac{\partial U}{\partial e_{yy}} de_{yy} + \frac{\partial U}{\partial e_{zz}} de_{zz} + \frac{\partial U}{\partial e_{yz}} de_{yz} + \frac{\partial U}{\partial e_{xz}} de_{xz} + \frac{\partial U}{\partial e_{xy}} de_{xy}$$

But dW is the work done by the external forces, i.e.

$$dW = dU = \sigma_{xx}\, de_{xx} + \sigma_{yy}\, de_{yy} + \sigma_{zz}\, de_{zz} + \sigma_{yz}\, de_{yz} + \sigma_{xz}\, de_{xz} + \sigma_{xy}\, de_{xy}$$

This means that

$$\sigma_{xx} = \frac{\partial U}{\partial e_{xx}} \qquad \sigma_{yy} = \frac{\partial U}{\partial e_{yy}} \qquad \sigma_{zz} = \frac{\partial U}{\partial e_{zz}}$$

$$\sigma_{yz} = \frac{\partial U}{\partial e_{yz}} \qquad \sigma_{xz} = \frac{\partial U}{\partial e_{xz}} \qquad \sigma_{xy} = \frac{\partial U}{\partial e_{xy}}$$

We can therefore define a strain-energy function or stored-energy function which defines the energy stored in the body as a result of the strain. For an adiabatic change of state at constant volume we will call this U_1 and note that it is identical with the thermodynamic internal-energy function U.

Then for changes at constant volume under adiabatic conditions

$$\sigma_{xx} = \frac{\partial U_1}{\partial e_{xx}} \qquad \sigma_{yy} = \frac{\partial U_1}{\partial e_{yy}} \qquad \sigma_{zz} = \frac{\partial U_1}{\partial e_{zz}}$$

$$\sigma_{yz} = \frac{\partial U_1}{\partial e_{yz}} \qquad \sigma_{xz} = \frac{\partial U_1}{\partial e_{xz}} \qquad \sigma_{xy} = \frac{\partial U_1}{\partial e_{xy}}$$

It is more usual experimentally to observe changes of state at constant pressure rather than at constant volume. It is then customary to introduce the thermodynamic enthalpy function $H = U + PV$.

For an adiabatic change of state at constant pressure

$$dW = dU + P\,dV = d(U + PV)_P = dH$$

and we can define a second strain-energy function U_2 which is identical with the enthalpy H.

Thus for deformations occurring at constant pressure under adiabatic conditions we have

$$\sigma_{xx} = \frac{\partial U_2}{\partial e_{xx}} \qquad \sigma_{yy} = \frac{\partial U_2}{\partial e_{yy}} \qquad \sigma_{zz} = \frac{\partial U_2}{\partial e_{zz}}$$

$$\sigma_{yz} = \frac{\partial U_2}{\partial e_{yz}} \qquad \sigma_{xz} = \frac{\partial U_2}{\partial e_{xz}} \qquad \sigma_{xy} = \frac{\partial U_2}{\partial e_{xy}}$$

Next, consider *isothermal* changes of state, first at constant volume.

Here $dW = dU - dQ$ but $dQ \neq 0$.

To deal with these changes of state the Helmholtz Free Energy $A = U - TS$ is introduced. For an isothermal change of state at constant volume

$$dW = dU - dQ = (dU - T\,dS) = d(U - TS)_{T,V} = dA$$

where $dQ = T\,dS$, S being the entropy.

A further strain-energy function U_3 can then be defined which is identical with the Helmholtz Free Energy A and the components of stress are the derivatives of U_3 with respect to the corresponding components of strain. It will be shown that this strain-energy function can be calculated for a polymer in the rubbery state from statistical mechanical considerations in terms of the strains existing in the material.

Finally we note that the most usual experimental procedure will be to measure the stress–strain relations at constant temperature and pressure. To deal with this situation the Gibbs Free Energy Function $G = U + PV - TS$ is introduced.

For isothermal changes of state at constant pressure

$$dW = dU + P\,dV - T\,dS = d(U + PV - TS)_{T,P} = dG$$

This leads to the definition of a strain energy function U_4 which is identical with the Gibbs Free Energy.

We will see that it is important to distinguish clearly between these various strain-energy functions when we come to relate the mechanics to the statistical molecular theories of the rubber-like state. Experimentally the stress–strain relations usually give the components of stress as derivatives of the strain-energy function U_4 with respect to the corresponding strains whereas it is most straightforward to calculate the value of U_3. In our subsequent discussion of the mechanics we will refer to the strain-energy function as U with no subscript, bearing in mind that for a particular experimental procedure we obtain U_1, U_2, U_3 or U_4.

3.4.2   The Form of the Strain-Energy Function

The generalized Hooke's Law states that the stress components are linear functions of the strain components:

$$\sigma_{xx} = c_{11}\,e_{xx} + c_{12}\,e_{yy} + \dots \text{ etc.,}$$

where c_{11}, c_{12}, etc., are the stiffness constants. It therefore follows that the strain-energy function U must be a homogeneous quadratic function of the strain components.

For a general anisotropic solid there are twenty-one independent co-efficients (Section 2.5 above). For an isotropic solid these are reduced to two. This result is forced upon us by the symmetry of the material, and does not depend on the existence of a strain-energy function[5].

It may also be obtained from considerations of the form of the strain-energy function. We make two premises:

(1) The strain-energy function must be a homogeneous quadratic function of the strain components.

(2) For an isotropic solid the strain-energy function must take a form which does not depend on the choice of the direction of the coordinate axes. This means that the strain-energy function must be a function of the strain invariants.

3.4.3 The Strain Invariants

For a second-rank tensor such as the strain tensor e_{ij}

$$\begin{bmatrix} e_{xx} & \frac{1}{2}e_{xy} & \frac{1}{2}e_{xz} \\ \frac{1}{2}e_{yx} & e_{yy} & \frac{1}{2}e_{yz} \\ \frac{1}{2}e_{zx} & \frac{1}{2}e_{zy} & e_{zz} \end{bmatrix}$$

there are three quantities which are independent of our choice of the directions of the Cartesian coordinate axes. In the case of the strain tensor these quantities are called the strain invariants. They are given in Table 3.1 both in the Cartesian-coordinate notation and in the index notation, the latter being more convenient in some of the further development of the theory.

Table 3.1. The strain invariants

Index notation	Cartesian notation	Number of terms
e_{ii}	$e_{xx}+e_{yy}+e_{zz}$	3
$e_{ij}e_{ji}$	$e_{xx}^2+e_{yy}^2+e_{zz}^2$ $+\frac{1}{4}e_{xy}^2+\frac{1}{4}e_{yz}^2+\frac{1}{4}e_{zx}^2$ $+\frac{1}{4}e_{yx}^2+\frac{1}{4}e_{zy}^2+\frac{1}{4}e_{xz}^2$	9
$e_{ij}e_{jk}e_{ki}$	$e_{xx}^3+\frac{1}{4}e_{xx}e_{xy}^2+\frac{1}{4}e_{xx}e_{xz}^2$ $+\frac{1}{4}e_{xy}^2e_{xx}+\frac{1}{4}e_{xy}^2e_{yy}+\frac{1}{8}e_{xy}e_{yz}e_{zx}$ etc.	27

The strain invariants are sometimes given as:

$$e_{xx}+e_{yy}+e_{zz}, \qquad e_{yy}e_{zz}+e_{zz}e_{xx}+e_{xx}e_{yy}-\tfrac{1}{4}(e_{yz}^2+e_{zx}^2+e_{xy}^2),$$

and

$$e_{xx}e_{yy}e_{zz}+\tfrac{1}{4}(e_{yz}e_{zx}e_{xy}-e_{xx}e_{yz}^2-e_{yy}e_{zx}^2-e_{zz}e_{xy}^2)$$

which follows from adding or subtracting simple functions of the three invariants of Table 3.1.

For an isotropic elastic solid the strain-energy function U must be a homogeneous quadratic function of the strains and a function of the strain invariants. It follows that U is a function of the first two strain invariants of Table 3.1 only, i.e.

$$2U = A'(e_{xx}+e_{yy}+e_{zz})^2 + B'[(e_{xx}^2+e_{yy}^2+e_{zz}^2)+\tfrac{1}{2}(e_{xy}^2+e_{yz}^2+e_{zx}^2)]$$

where A' and B' are constants possessing the dimensions of modulus, i.e. dyn/cm^2.

It can be shown that this equation may be written as:

$$2U = A(e_{xx}+e_{yy}+e_{zz})^2 + B(e_{yz}^2+e_{zx}^2+e_{xy}^2-4e_{yy}e_{zz}-4e_{zz}e_{xx}-4e_{xx}e_{yy})$$

where A and B are new constants and we have constructed a new strain invariant

$$(e_{yz}^2+e_{zx}^2+e_{xy}^2-4e_{yy}e_{zz}-4e_{zz}e_{xx}-4e_{xx}e_{yy}) = 2[(e_{xx}^2+e_{yy}^2+e_{zz}^2)+\tfrac{1}{2}(e_{xy}^2+e_{yz}^2+e_{zx}^2)]$$
$$-2(e_{xx}+e_{yy}+e_{zz})^2$$

This is the form of the strain-energy function given by Love (Reference 5, p. 102).

So far we have only considered small-strain elasticity, but it can readily be appreciated that a formal treatment of finite-strain elasticity can be made along similar lines. For an isotropic solid, the strain-energy function U is again a function of the strain invariants, and the algebra will be formally identical, since this part of the argument is based on the invariants of a second-rank tensor.

Thus for finite strains we have that the quantities $\mathbf{e}_{ii}$, $\mathbf{e}_{ij}\mathbf{e}_{ji}$ and $\mathbf{e}_{ij}\mathbf{e}_{jk}\mathbf{e}_{ki}$ are still invariant.

For reasons which will shortly be made apparent, it is convenient to define three alternative strain invariants in terms of these three basic invariants.

These are:

$$I_1 = 3+2\mathbf{e}_{rr}$$

$$I_2 = 3+4\mathbf{e}_{rr}+2(\mathbf{e}_{rr}\,\mathbf{e}_{ss}-\mathbf{e}_{rs}\,\mathbf{e}_{sr}),$$

and

$$I_3 = |\delta_{rs}+2\mathbf{e}_{rs}|$$

Our argument is then summarized by the equation:

$$U = f(I_1,I_2,I_3)$$

We will now evaluate the invariants I_1,I_2,I_3 for the case of a homogeneous pure strain in which the extension ratios parallel to the three coordinate axes are $\lambda_1,\lambda_2,\lambda_3$ respectively.

From the initial discussion (Equations 3.7)

$$\lambda_1{}^2 = 1+2\mathbf{e}_{xx}, \qquad \lambda_2{}^2 = 1+2\mathbf{e}_{yy}, \qquad \lambda_3{}^2 = 1+2\mathbf{e}_{zz}$$

then

$$I_1 = 3+2\mathbf{e}_{rr} = 3+2(\mathbf{e}_{xx}+\mathbf{e}_{yy}+\mathbf{e}_{zz})$$
$$= 3+(\lambda_1{}^2-1)+(\lambda_2{}^2-1)+(\lambda_3{}^2-1)$$

giving

$$I_1 = \lambda_1{}^2+\lambda_2{}^2+\lambda_3{}^2 \qquad (3.9)$$

Secondly,

$$I_2 = 3 + 4e_{rr} + 2(e_{rr}e_{ss} - e_{rs}e_{sr})$$

For homogeneous pure strain $e_{rs} = e_{sr} = 0$. (The shear-strain components are zero.)

$$I_2 = 3 + 4e_{rr} + 2e_{rr}e_{ss}$$

$$= 3 + 4(e_{xx} + e_{yy} + e_{zz}) + 4(e_{xx}e_{yy} + e_{xx}e_{zz} + e_{yy}e_{zz})$$

$$= 3 + 4[\tfrac{1}{2}(\lambda_1{}^2 - 1) + \tfrac{1}{2}(\lambda_2{}^2 - 1) + \tfrac{1}{2}(\lambda_3{}^2 - 1)]$$

$$\quad + 4[\tfrac{1}{4}(\lambda_1{}^2 - 1)(\lambda_2{}^2 - 1) + \tfrac{1}{4}(\lambda_1{}^2 - 1)(\lambda_3{}^2 - 1) + \tfrac{1}{4}(\lambda_2{}^2 - 1)(\lambda_3{}^2 - 1)]$$

giving

$$I_2 = \lambda_1{}^2\lambda_2{}^2 + \lambda_1{}^2\lambda_3{}^2 + \lambda_2{}^2\lambda_3{}^2 \tag{3.10}$$

and

$$I_3 = |\delta_{rs} + 2e_{rs}| = \begin{vmatrix} 1+2e_{xx} & & \\ & 1+2e_{yy} & \\ & & 1+2e_{zz} \end{vmatrix}$$

$$= \begin{vmatrix} \lambda_1{}^2 & & \\ & \lambda_2{}^2 & \\ & & \lambda_3{}^2 \end{vmatrix} = \lambda_1{}^2\lambda_2{}^2\lambda_3{}^2 \tag{3.11}$$

If we make the further simplifying assumption that there is no change in volume on deformation

$$I_3 = \lambda_1{}^2\lambda_2{}^2\lambda_3{}^2 = 1$$

Thus the strain-energy function U for an isotropic incompressible solid undergoing a pure homogeneous deformation is given by

$$U = \text{Function}\left(I_1 = \lambda_1{}^2 + \lambda_2{}^2 + \lambda_3{}^2, \; I_2 = \frac{1}{\lambda_1{}^2} + \frac{1}{\lambda_2{}^2} + \frac{1}{\lambda_3{}^2}\right)$$

with

$$\lambda_1{}^2\lambda_2{}^2\lambda_3{}^2 = 1$$

Let us represent U as a polynomial function of the strain invariants. If U is to vanish at zero strain this implies that

$$U = \sum_{i=0,j=0}^{\infty} C_{ij}(I_1 - 3)^i (I_2 - 3)^j \quad \text{with} \quad C_{00} = 0, \tag{3.12}$$

i.e. the strain-energy function involves powers of $(I_1 - 3)$ and $(I_2 - 3)$.

A neo-Hookeian material, that corresponding to the statistical theory for a Gaussian network (see p. 69), corresponds to the first term in this series, i.e.

$$U = C_1(I_1 - 3) = C_1(\lambda_1^2 + \lambda_2^2 + \lambda_3^2 - 3) \tag{3.13}$$

which corresponds exactly to the network theory if we put:

$$C_1 = \tfrac{1}{2}NKT$$

Another form of particular interest is:

$$U = C_1(I_1 - 3) + C_2(I_2 - 3) \tag{3.14}$$

This is the most general first-order relationship for U in terms of I_1 and I_2. This form of the strain-energy function was first derived by Mooney[6] on the assumption of a linear stress–strain relationship in shear. (Shear deformation will be considered below.) In general the Mooney equation gives a closer approximation to the actual behaviour of rubbers than the simpler equation involving only I_1.

3.4.4 The Stress–Strain Relations

Assuming that a satisfactory strain-energy function has been defined, it is now necessary to consider how the stress–strain relations are obtained.

For the small-strain situation the stress components are the first derivatives of the appropriate strain-energy function with respect to the corresponding strain components, i.e.

$$\sigma_{xx} = \frac{\partial U}{\partial e_{xx}}, \qquad \sigma_{yy} = \frac{\partial U}{\partial e_{yy}}, \text{ etc.,}$$

and using the strain-energy function for an elastic solid with any anisotropy

$$U = c_{11}e_{xx}^2 + 2c_{12}e_{xx}e_{yy} + 2c_{13}e_{xx}e_{zz} + \dots$$

the generalized Hooke's Law is obtained. To solve a specific problem in small-strain elasticity we find stress components which satisfy the equilibrium conditions, and the stress components and strain components which satisfy the boundary conditions and the compatibility conditions. This approach can be called the *direct* method of solution.

Solutions to problems in finite elasticity, on the other hand, are most easily obtained by *inverse* methods. The strain components are specified, and the stress components obtained using the strain-energy function. If we assume that the material is incompressible the stress components are then indeterminate with respect to an arbitrary hydrostatic pressure.

If the problem is treated in its most general form, a considerable amount of algebra is involved in obtaining the stress components σ_{ij} in terms of derivatives of the strain-energy function with respect to the strain components e_{ij} defined with respect to the *undeformed* body. The reader is referred to standard texts for the development of these procedures[1,2].

For most purposes, however, it is only necessary to derive the result for homogeneous pure strain, with extension ratios $\lambda_1, \lambda_2, \lambda_3$, and this is obtained in the following elementary manner.

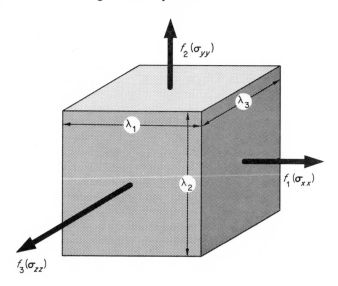

Figure 3.2. A cube of unit dimensions transforms to a rectangular parallelepiped of edges λ_1, λ_2 and λ_3 under the applied loads f_1, f_2 and f_3.

A cube of unit dimensions in the undeformed state deforms under the applied loads to the rectangular parallelepiped shown in Figure 3.2, which has edges $\lambda_1, \lambda_2, \lambda_3$, in the x, y, z directions respectively. In the deformed state the forces acting on the faces are f_1, f_2, f_3 with $f = $ force/unit of undeformed cross-section, i.e. the forces are calculated in terms of the applied loads per unit cross-section in the undeformed state.

The corresponding stress components as defined above in the deformed state are $\sigma_{xx}, \sigma_{yy}, \sigma_{zz}$, where

$$\sigma_{xx} = \frac{f_1}{\lambda_2\lambda_3} = \lambda_1 f_1 \qquad \sigma_{yy} = \frac{f_2}{\lambda_1\lambda_3} = \lambda_2 f_2 \qquad \sigma_{zz} = \frac{f_3}{\lambda_1\lambda_2} = \lambda_3 f_3 \quad (3.15)$$

The work done (per unit of initial undeformed volume) in an infinitesimal displacement from the deformed state where $\lambda_1, \lambda_2, \lambda_3$ change to $\lambda_1 + d\lambda_1$, $\lambda_2 + d\lambda_2, \lambda_3 + d\lambda_3$ is

$$dU = f_1\, d\lambda_1 + f_2\, d\lambda_2 + f_3\, d\lambda_3$$

$$= \frac{\sigma_{xx}}{\lambda_1} d\lambda_1 + \frac{\sigma_{yy}}{\lambda_2} d\lambda_2 + \frac{\sigma_{zz}}{\lambda_3} d\lambda_3 \qquad (3.16)$$

$$U = f(I_1, I_2, I_3)$$

where

$$I_1 = \lambda_1{}^2 + \lambda_2{}^2 + \lambda_3{}^2$$

$$I_2 = \frac{1}{\lambda_1{}^2} + \frac{1}{\lambda_2{}^2} + \frac{1}{\lambda_3{}^2}$$

$$I_3 = \lambda_1{}^2 \lambda_2{}^2 \lambda_3{}^2$$

$$\therefore \quad dU = \frac{\partial U}{\partial I_1}\frac{\partial I_1}{\partial \lambda_1} d\lambda_1 + \frac{\partial U}{\partial I_1}\frac{\partial I_1}{\partial \lambda_2} d\lambda_2 + \frac{\partial U}{\partial I_1}\frac{\partial I_1}{\partial \lambda_3} d\lambda_3$$

$$+ \frac{\partial U}{\partial I_2}\frac{\partial I_2}{\partial \lambda_1} d\lambda_1 + \frac{\partial U}{\partial I_2}\frac{\partial I_2}{\partial \lambda_2} d\lambda_2 + \frac{\partial U}{\partial I_2}\frac{\partial I_2}{\partial \lambda_3} d\lambda_3$$

$$+ \frac{\partial U}{\partial I_3}\frac{\partial I_3}{\partial \lambda_1} d\lambda_1 + \frac{\partial U}{\partial I_3}\frac{\partial I_3}{\partial \lambda_2} d\lambda_2 + \frac{\partial U}{\partial I_3}\frac{\partial I_3}{\partial \lambda_3} d\lambda_3$$

Substituting

$$\frac{\partial I_1}{\partial \lambda_1} = 2\lambda_1, \text{ etc.,}$$

$$\frac{\partial I_2}{\partial \lambda_1} = -\frac{2}{\lambda_1{}^3}, \text{ etc.,}$$

and

$$\frac{\partial I_3}{\partial \lambda_1} = 2\lambda_1 \lambda_2{}^2 \lambda_3{}^2, \text{ etc.,}$$

we have

$$
\begin{aligned}
dU = 2 &\left\{ \lambda_1 \frac{\partial U}{\partial I_1} d\lambda_1 + \lambda_2 \frac{\partial U}{\partial I_1} d\lambda_2 + \lambda_3 \frac{\partial U}{\partial I_1} d\lambda_3 \right\} \\
-2 &\left\{ \frac{1}{\lambda_1{}^3} \frac{\partial U}{\partial I_2} d\lambda_1 + \frac{1}{\lambda_2{}^3} \frac{\partial U}{\partial I_2} d\lambda_2 + \frac{1}{\lambda_3{}^3} \frac{\partial U}{\partial I_2} d\lambda_3 \right\} \\
+2I_3 &\left\{ \frac{1}{\lambda_1} \frac{\partial U}{\partial I_3} d\lambda_1 + \frac{1}{\lambda_2} \frac{\partial U}{\partial I_3} d\lambda_2 + \frac{1}{\lambda_3} \frac{\partial U}{\partial I_3} d\lambda_3 \right\}
\end{aligned} \tag{3.17}
$$

In Equations (3.16) and (3.17), λ_1, λ_2 and λ_3 are independent variables.

We can therefore equate the coefficients of $d\lambda_1$, $d\lambda_2$, and $d\lambda_3$ in these equations to find the stress components.

This gives

$$
\sigma_{xx} = 2 \left\{ \lambda_1{}^2 \frac{\partial U}{\partial I_1} - \frac{1}{\lambda_1{}^2} \frac{\partial U}{\partial I_2} + I_3 \frac{\partial U}{\partial I_3} \right\}, \text{ etc.}
$$

If the solid is incompressible $I_3 = 1$ and $U = f(I_1, I_2)$ only. In this case the stresses are now indeterminate with respect to an arbitrary hydrostatic pressure, $\mathbf{p}$, because this pressure does not produce any changes in the deformation variables λ_1, λ_2, λ_3. Then

$$
\sigma_{xx} = 2 \left\{ \lambda_1{}^2 \frac{\partial U}{\partial I_1} - \frac{1}{\lambda_1{}^2} \frac{\partial U}{\partial I_2} \right\} + \mathbf{p} \tag{3.18}
$$

In index notation the stresses are given as

$$
\sigma_{ii} = 2 \left\{ \lambda_i{}^2 \frac{\partial U}{\partial I_1} - \frac{1}{\lambda_i{}^2} \frac{\partial U}{\partial I_2} \right\} + \mathbf{p}
$$

$$
\sigma_{ij} = 0 \tag{3.19}
$$

Because any homogeneous strain can be produced by a homogeneous pure strain followed by a suitable rotation, we do not lose generality by restricting the discussion to pure homogeneous strain.

3.5 EXPERIMENTAL STUDIES OF FINITE–ELASTIC BEHAVIOUR IN RUBBERS

The experimental applications of finite elasticity are primarily confined to the behaviour of rubbers. One of the most definitive series of experiments was undertaken by Rivlin and Saunders[7] on sheets of vulcanized rubber. They examined homogeneous pure strain in a thin sheet with

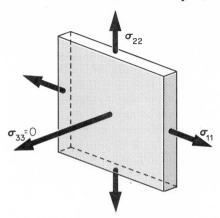

Figure 3.3. The rubber sheet is subjected to normal stresses σ_{11} and σ_{22} in the plane of the sheet.

stresses maintained along the edges only (Figure 3.3), and assumed that the rubber was incompressible. For this situation $\sigma_{33} = 0$, which enables the arbitrary hydrostatic pressure to be obtained in terms of known quantities. Thus

$$\sigma_{11} = 2\left\{\lambda_1^2\frac{\partial U}{\partial I_1} - \frac{1}{\lambda_1^2}\frac{\partial U}{\partial I_2}\right\} + \mathbf{p}$$

$$\sigma_{22} = 2\left\{\lambda_2^2\frac{\partial U}{\partial I_1} - \frac{1}{\lambda_2^2}\frac{\partial U}{\partial I_2}\right\} + \mathbf{p}$$

$$\sigma_{33} = 2\left\{\lambda_3^2\frac{\partial U}{\partial I_1} - \frac{1}{\lambda_3^2}\frac{\partial U}{\partial I_2}\right\} + \mathbf{p} = 0$$

Substituting for $\mathbf{p}$ we have

$$\sigma_{11} = 2\left\{\lambda_1^2 - \frac{1}{\lambda_1^2\lambda_2^2}\right\}\left\{\frac{\partial U}{\partial I_1} + \lambda_2^2\frac{\partial U}{\partial I_2}\right\}$$

$$\sigma_{22} = 2\left\{\lambda_2^2 - \frac{1}{\lambda_1^2\lambda_2^2}\right\}\left\{\frac{\partial U}{\partial I_1} + \lambda_1^2\frac{\partial U}{\partial I_2}\right\} \tag{3.20}$$

Solving these two equations for $\partial U/\partial I_1$ and $\partial U/\partial I_2$

$$\frac{\partial U}{\partial I_1} = \frac{\left\{\dfrac{\lambda_1^2\sigma_{11}}{\lambda_1^2 - 1/\lambda_1^2\lambda_2^2} - \dfrac{\lambda_2^2\sigma_{22}}{\lambda_2^2 - 1/\lambda_1^2\lambda_2^2}\right\}}{2(\lambda_1^2 - \lambda_2^2)}$$

and

$$\frac{\partial U}{\partial I_2} = \frac{\left\{ \dfrac{\sigma_{11}}{\lambda_1{}^2 - 1/\lambda_1{}^2\lambda_2{}^2} - \dfrac{\sigma_{22}}{\lambda_2{}^2 - 1/\lambda_1{}^2\lambda_2{}^2} \right\}}{2(\lambda_2{}^2 - \lambda_1{}^2)} \tag{3.21}$$

If λ_1, λ_2, are varied such that I_2 remains constant, we obtain values of $\partial U/\partial I_1$ and $\partial U/\partial I_2$ for various values of I_1 and constant I_2. If on the other hand, λ_1, λ_2 are varied such that I_1 remains constant, we obtain values of $\partial U/\partial I_1$ and $\partial U/\partial I_2$ for various I_2 at constant I_1. The results obtained by Rivlin and Saunders for vulcanized rubber are shown in Figure 3.4. It can be seen that $\partial U/\partial I_1$ is approximately constant and independent of both I_1 and I_2. $\partial U/\partial I_2$, however is independent of I_1 but decreases with increasing I_2. If we assume the polynomial form for the strain-energy function (p. 47) the terms which must be retained to satisfy

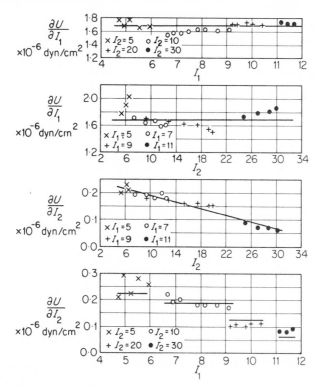

Figure 3.4. Dependence of $\dfrac{\partial U}{\partial I_1}$ and $\dfrac{\partial U}{\partial I_2}$ on I_1 and I_2 (Rivlin and Saunders).

these results give

$$U = C_1(I_1 - 3) + f(I_2 - 3)$$

i.e. the sum of the first-order neo-Hookeian term and a function of I_2. The second term is generally small compared with the first, and decreases as I_2 increases.

In this case

$$\frac{\partial U}{\partial I_1} \sim 1 \cdot 7 \times 10^6 \text{ dyn cm}^{-2},$$

whereas

$$\frac{\partial U}{\partial I_2} \sim 1 \cdot 5 \times 10^5 \text{ dyn cm}^{-2}$$

i.e. about one-tenth as large.

It is next important to consider whether this strain-energy function is consistent with results obtained for different types of deformation. Rivlin and Saunders also undertook measurements in pure shear and in simple elongation, and these will now be discussed in turn.

Pure Shear

Consider a sheet clamped between two edges AB, CD (Figure 3.5). If the width of the sheet (AB or CD) is sufficiently large compared with its length, the non-uniformity and strain arising because the outer edges AC, BD are unconstrained can be neglected. The strain parallel to the edges AB, CD can then be considered to remain constant in a deformation produced by moving AB and CD apart but keeping them parallel.

Shear is by definition a deformation in which the strain is zero in one direction and there is no volume change. If we make the simplifying assumption that the rubber is incompressible, the deformation which has been described (AB and CD displaced normal to their length) will be pure shear.

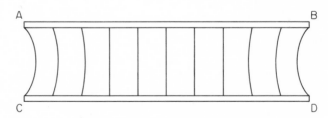

Figure 3.5. Pure shear deformation.

From the stress–strain relationship (Equations 3.20) we have:

$$\sigma_{11} = 2\left\{\lambda_1{}^2 - \frac{1}{\lambda_1{}^2\lambda_2{}^2}\right\}\left\{\frac{\partial U}{\partial I_1} + \lambda_2{}^2\frac{\partial U}{\partial I_2}\right\}$$

The pure-strain experiments suggested that

$$U = C_1(I_1 - 3) + f(I_2 - 3),$$

i.e. $\partial U/\partial I_1 = C_1$. For the pure shear experiment $\lambda_2 = \text{constant} = 1$ and

$$\sigma_{11} = 2\left(\lambda_1{}^2 - \frac{1}{\lambda_1{}^2}\right)\left(C_1 + \frac{\partial U}{\partial I_2}\right)$$

Thus if $\lambda_1{}^2$ is varied (and hence I_2) by measuring σ_{11} we obtain values of $\partial U/\partial I_2$ as a function of I_2.

Rivlin and Saunders then did a second experiment in which pure shear was superimposed on an initial extension of the sheet along the AB/CD direction with $\lambda_2 = 0\cdot776$.

They compared the calculated values of

$$\left(\frac{\partial U}{\partial I_1} + (0\cdot776)^2\frac{\partial U}{\partial I_2}\right)$$

using the results from the first pure-shear experiment, with these obtained from direct experiment (using Equations 3.21). Good agreement was obtained, confirming that the assumed form of the strain-energy function was a good mathematical model for the material.

Simple Elongation

A further key experiment is simple elongation. From Equation (3.20) above when

$$\sigma_{22} = \sigma_{33} = 0 \quad \text{and} \quad \lambda_1 = \frac{1}{\lambda_2{}^2} = \lambda \qquad (\text{since } \lambda_2 = \lambda_3 \text{ and } \lambda_1\lambda_2\lambda_3 = 1)$$

We have

$$\sigma_{11} = 2\left(\lambda^2 - \frac{1}{\lambda}\right)\left(\frac{\partial U}{\partial I_1} + \frac{1}{\lambda}\frac{\partial U}{\partial I_2}\right) \qquad (3.22)$$

The load f required to extend a specimen of initial cross-sectional area A is

$$f = \frac{A\sigma_{11}}{\lambda} = 2A\left(\lambda - \frac{1}{\lambda^2}\right)\left(\frac{\partial U}{\partial I_1} + \frac{1}{\lambda}\frac{\partial U}{\partial I_2}\right)$$

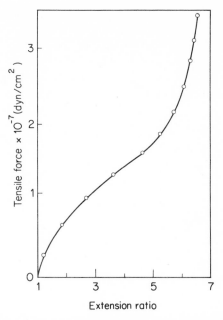

Figure 3.6. Typical force–extension curve for vulcanized rubber (after Treloar, 1958).

The well-known general shape of the load-extension curve is shown in Figure 3.6. A more revealing curve is obtained by plotting

$$\left(\frac{\partial U}{\partial I_1} + \frac{1}{\lambda}\frac{\partial U}{\partial I_2}\right) \quad \text{against} \quad \frac{1}{\lambda}$$

i.e.

$$\frac{f}{2A(\lambda - 1/\lambda^2)} \quad \text{against} \quad \frac{1}{\lambda}$$

This is shown in Figure 3.7. A nearly linear relationship is observed over the range of values $1/\lambda$ from 0·9 to 0·45. The obvious simple inference from this result taken by itself is that both $\partial U/\partial I_1$ and $\partial U/\partial I_2$ are constant in this range of extension, i.e. that the Mooney equation $U = C_1(I_1 - 3) + C_2(I_2 - 3)$ is adequate for $1/\lambda$ greater than 0·45. There is, however, a hidden snag in this simple interpretation. Values of C_1 and C_2 chosen in this way do not then agree with the values obtained from the two-dimensional extension and the pure-shear experiments. The explanation of this apparent contradiction is as follows: The Mooney equation is not

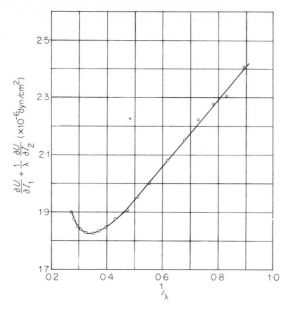

Figure 3.7. Plot of $[\partial U/\partial I_1 + (1/\lambda)\partial U/\partial I_2]$ against $1/\lambda$ from experiment in simple extension (after Rivlin and Saunders).

an adequate representation of the strain-energy function. Instead we should write:

$$U = C_1(I_1 - 3) + f(I_2 - 3)$$

as indicated by the two-dimensional extension experiment. Expanding in powers of $(I_2 - 3)$ and curtailing the expansion after the cubic term,

$$U = C_1(I_1 - 3) + C_2(I_2 - 3) + C_3(I_2 - 3)^2 + C_4(I_2 - 3)^3$$

For simple extension this gives:

$$f = 2A\left(\lambda - \frac{1}{\lambda^2}\right)\left\{\left[(C_1 + 4C_3 - 36C_4) + \frac{1}{\lambda}(C_2 - 6C_3 + 27C_4)\right]\right.$$
$$\left. + \left[\left(12\lambda + \frac{12}{\lambda^2} - \frac{18}{\lambda^3} + \frac{3}{\lambda^5}\right)C_4 + \frac{2}{\lambda^3}C_3\right]\right\}$$

Now C_1 and C_2 of the Mooney equation are replaced by $(C_1 + 4C_3 - 36C_4)$ and $(C_2 - 6C_3 + 27C_4)$ respectively. There are also additional groups of terms involving different powers of λ. If C_3 and C_4 are small, this latter group may well only give some scatter in the experimental data. But it could be that C_3 and C_4 are of different sign, e.g. C_3 could be negative as found from the two-dimensional extension experiment and C_4 positive. Close examination of the results of simple elongation and the two-dimensional extension confirms that this is the case, leading to a significant decrease in the apparent value of C_1 (if the Mooney equation is adopted) and a corresponding increase in the apparent value of C_2.

This example has been discussed in some detail because it emphasizes that any conclusions regarding the form of the strain-energy function in rubbers which are derived from a one-dimensional simple elongation experiment are necessarily suspect.

Recent work by Becker[8] has confirmed the validity of Rivlin and Saunders' work over the intermediate range of strains, but suggests that this formulation may not be adequate at low strains.

Simple Shear

The last situation to be discussed in rubber elasticity is that of simple shear. Consider an elemental cube, whose $x_3 = 0$ plane is shown in Figure 3.8, which is bounded initially by the planes $x_i = 0$, $x_i = a_i$ and subjected to simple shear in which planes parallel to the plane $x_2 = 0$ move parallel to the x_1 axis by amounts proportional to their x_2 coordinate.

For general finite deformation, points x_i suffer displacements u_i, the new coordinates being $x_i + u_i$. Put

$$x_i + u_i = x'_i$$

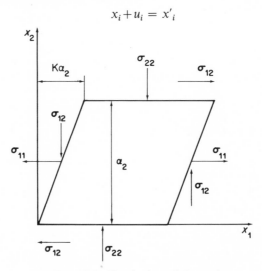

Figure 3.8. Simple shear deformation.

An element of length ds in the undeformed body changes to $\overline{ds}$ where

$$ds^2 = dx_i\, dx_i$$

and

$$\overline{ds}^2 = dx'_i\, dx'_i = \frac{\partial x'_i}{\partial x_r}\frac{\partial x'_i}{\partial x_s}\, dx_r\, dx_s$$

Summing over repeated suffices,

$$\overline{ds}^2 - ds^2 = 2e_{ij}\, dx_i\, dx_j$$

where

$$e_{ij} = \frac{1}{2}\left(\frac{\partial x'_r}{\partial x_i}\frac{\partial x'_r}{\partial x_j} - \delta_{ij}\right)$$

This definition of strain enables a direct evaluation of the strain components.

The coordinates x'_i for the deformation are given by:

$$x'_i = x_1 + Kx_2, \qquad x'_2 = x_2, \qquad x'_3 = x_3$$

where K is a constant defining the magnitude of the shear strain. This leads to strain invariants

$$I_1 = I_2 = 3 + K^2, \qquad (I_3 = 1, \text{ for incompressible solid})$$

and it can be shown that the stresses are given by:

$$\sigma_{11} = 2K^2\frac{\partial U}{\partial I_1}, \qquad \sigma_{22} = -2K^2\frac{\partial U}{\partial I_2}$$

and

$$\sigma_{12} = 2K\left(\frac{\partial U}{\partial I_1} + \frac{\partial U}{\partial I_2}\right) \qquad \sigma_{13} = \sigma_{23} = \sigma_{33} = 0$$

To produce a finite shear, therefore, it is necessary to apply *normal components of stress* σ_{22} to the surfaces $x_2 = 0, a_2$ and σ_{11} to the surfaces initially at $x_1 = 0, a_1$ in addition to the shear stress σ_{12}. This is equivalent to the situation in viscous flow called the Weissenberg effect[9]: It is to be noted that these normal stresses are second-order in K, which explains why they vanish for small strains.

The important point is that if the strain-energy function U is a linear function of the strain invariants I_1, I_2 the shear stress σ_{12} is proportional to the shear displacement K. This result was obtained by Mooney[6] and led to his proposal that the strain-energy function $U = C_1(I_1 - 3) + C_2(I_2 - 3)$. If we examine the problem of simple shear in detail, however, it turns out that for quite large amounts of shear the strain invariants I_1 and I_2 remain fairly small. Thus, if the higher order derivatives of U are fairly small compared with $\partial U/\partial I_1$ and $\partial U/\partial I_2$, which is actually the case for rubber, an approximately linear load–deformation relationship results.

The final conclusion of this discussion is that we must be very careful not to suppose that experiments on rubbers involving simple extension or simple shear only, will lead to a satisfactory understanding of the form of the strain-energy function.

REFERENCES

1. A. E. Green and W. Zerna, *Theoretical Elasticity*, Clarendon Press, Oxford, 1954.
2. A. E. Green and J. E. Adkins, *Large Elastic Deformations and Non-Linear Continuum Mechanics*, Clarendon Press, Oxford, 1960.
3. H. Goldstein, *Classical Mechanics*, Harvard University Press, Cambridge, 1959, p. 119.
4. R. S. Rivlin, *Phil. Trans. Roy. Soc.*, **A240**, 459 (1948); *Phil. Trans. Roy. Soc.*, **A240**, 491 (1948) and *Phil. Trans. Roy. Soc.*, **A241**, 379 (1948).
5. A. E. H. Love, *A Treatise on the Mathematical Theory of Elasticity*, Cambridge University Press, 1927.
6. M. Mooney, *J. Appl. Phys.*, **11**, 582 (1940).
7. R. S. Rivlin and D. W. Saunders, *Phil. Trans. Roy. Soc.*, **A243**, 251 (1951); *Trans. Faraday Soc.*, **48**, 200 (1952).
8. G. W. Becker, *J. Polymer Sci.*, **C16**, 2893 (1967).
9. K. Weissenberg, *Nature*, **159**, 310 (1947).

4

The Statistical Molecular Theories of the Rubber-like State

4.1 THERMODYNAMIC CONSIDERATIONS

The rubber-like state is the only part of polymer mechanical behaviour which can be understood in terms of a well-established molecular theory. In formal mathematical terms the aim is to predict the strain-energy function U in terms of the strain invariants, and the appropriate molecular parameters. The theory is based on statistical mechanical considerations and processes are considered to be reversible in the thermodynamic sense. We will therefore develop our arguments within the framework of Section 3.4.1 above.

For a reversible isothermal change of state at constant volume, the work done can be equated to the change in the Helmholtz Free Energy A.

For simplicity we will consider the uniaxial extension under a tensile force f of an elastic solid of initial length l.

The work done on the solid in an infinitesimal displacement dl is

$$dW = f\, dl = dA = dU - T\, dS \tag{4.1}$$

At constant volume we have,

$$f = \left(\frac{\partial A}{\partial l}\right)_T = \left(\frac{\partial U}{\partial l}\right)_T - T\left(\frac{\partial S}{\partial l}\right)_T \tag{4.2}$$

A further manipulation of the thermodynamic quantities is required to show that

$$\left(\frac{\partial U}{\partial l}\right)_T = f - T\left(\frac{\partial f}{\partial T}\right)_l \tag{4.3}$$

At constant volume

$$dU = f\, dl + T\, ds \tag{4.4}$$

Hence

$$dA = f\, dl - S\, dT \tag{4.5}$$

61

Then

$$\left(\frac{\partial A}{\partial l}\right)_T = f \quad \text{and} \quad \left(\frac{\partial A}{\partial T}\right)_l = -S \tag{4.6}$$

But

$$\frac{\partial}{\partial l}\left(\frac{\partial A}{\partial T}\right)_l = \frac{\partial}{\partial T}\left(\frac{\partial A}{\partial l}\right)_T$$

Substituting we have

$$\left(\frac{\partial S}{\partial l}\right)_T = -\left(\frac{\partial f}{\partial T}\right)_l \tag{4.7}$$

Using (4.2) we then find Equation (4.3).

The classic experiments of Meyer and Ferri[1] on the stress-temperature behaviour of rubber showed that the tensile force at constant length was very nearly proportional to the absolute temperature. The right-hand side of Equation (4.3) is therefore close to zero, showing that the internal-energy term is very small, and that the elasticity arises almost entirely from changes in entropy.

4.1.1   The Thermoelastic-inversion Effect

A typical set of results for the tensile force at constant length as a function of temperature are shown in Figure 4.1. The curves are linear at all

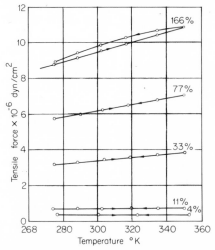

Figure 4.1. Force at constant length as a function of absolute temperature. Elongations as indicated (Meyer and Ferri, 1935).

elongations but it is to be noted that whereas above about 10% elongation the tensile force increases with increasing temperature, below this elongation it decreases slightly. This is called the thermoelastic-inversion effect. In physical terms it is caused by the thermal expansion of the rubber with increasing temperature. This increases the length in the unstrained state and hence reduces the effective elongation. It was therefore considered by Gee[2] and others that the more appropriate experiment is to measure the tensile force as a function of temperature at constant extension ratio where the expansion is corrected for. The experimental results obtained in this way by Gee[2] showed that the stress was directly proportional to absolute temperature and hence suggested that there was no internal-energy change (Equation 4.3).

This conclusion is based on certain approximations relating to the difference between constant-volume conditions and constant-pressure conditions which have subsequently been reconsidered (for a discussion of this see Section 4.2 below).

4.1.2 The Statistical Theory

The kinetic or statistical theory of rubber elasticity was originally proposed by Meyer, Susich and Valko[3] and subsequently developed by Guth and Mark[4], Kuhn[5] and others[6]. It is assumed that the individual molecules of the rubber exist in the form of very long chains, each of which is capable of assuming a variety of configurations in response to the thermal vibrations of 'micro-Brownian' motion of their constituent atoms.

Furthermore it is assumed that the molecular chains are interlinked so as to form a coherent network, but that the number of cross-links is relatively small and is not sufficient to interfere markedly with the motion of the chains.

The chain molecules will always tend to assume a set of crumpled configurations corresponding to a state of maximum entropy, unless constrained by external forces. Under such constraint the configurational arrangements of the chains will be changed to produce a state of strain.

Quantitative evaluation of the stress–strain characteristics of the rubber network then involves the calculation of the configurational entropy of the whole assembly of chains as a function of the state of strain. This calculation is considered in two stages, first the calculation of the entropy of a single chain, and, secondly, the change in entropy of a network of chains as a function of strain.

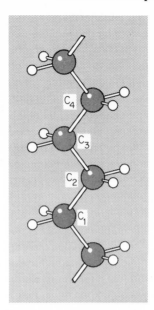

Figure 4.2. The polyethylene chain.

The Entropy of a Single Chain

The simplest consideration of the structure of a single chain can be made in terms of a polyethylene molecule $(CH_2)_n$. Fully extended, this takes the form of the planar zig-zag shown in Figure 4.2. If we allow free rotation from one conformation* to another, the local situation $C_1C_2C_3C_4$ can change from the planar zig-zag conformation to a variety of conformations, with the restriction that in each case the valence-bond angle between carbon atoms must remain at $109\frac{1}{2}°$. It is in principle possible to calculate the number of possible configurations which correspond to a chosen end-to-end distance and hence the entropy of such molecular chain. It is, however, easier to consider instead a mathematical abstraction—the 'freely jointed' chain. The freely jointed chain, as its name implies, consists of a chain of equal links jointed without the restriction that the valence angles should remain constant, i.e. random jointing is assumed.

This problem is more tractable mathematically, the first analysis of this type being undertaken by Kuhn[5] and by Guth and Mark[4].

---

* Conformation is used to denote differences in the immediate situation of a bond, e.g. *trans* and *gauche* conformations. Configuration is retained to refer to the arrangement of the whole molecular chain.

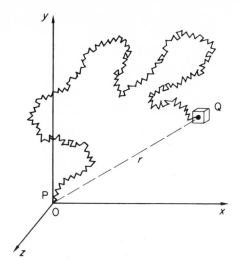

Figure 4.3. The freely jointed chain.

Consider a chain of n links each of length l, which has a configuration such that one end P is at the origin (Figure 4.3). The probability distribution for the position of the end Q is derived using approximations which are valid providing that the distance between the chain ends P and Q is much less than the extended chain length nl. The probability that Q lies within the elemental volume $dx\,dy\,dz$ at the point (x, y, z) can be shown to be

$$p(x, y, z)\,dx\,dy\,dz = \frac{b^3}{\pi^{3/2}} \exp\left(-b^2 r^2\right) dx\,dy\,dz$$

where

$$b^2 = \frac{3}{2nl^2}$$

and $r^2 = x^2 + y^2 + z^2$, i.e. the distribution of end-to-end vectors is defined by the Gaussian error function.

This shows immediately that the distribution is spherically symmetrical. It also shows that the most probable position of the end Q is at the origin. This does not mean that the most probable end-to-end distance is zero for the following reason. The probability that the chain end Q is in any given elemental volume between r and dr, irrespective of its direction, is the product of the probability distribution $p(r)$ and $4\pi r^2\,dr$, the volume of a concentric shell of radius r and thickness dr.

This probability is:

$$P(r)\,dr = p(r)4\pi r^2\,dr = \frac{b^3}{\pi^{3/2}}e^{-b^2r^2}4\pi r^2\,dr$$

$$= \left(\frac{4b^3}{\pi^{1/2}}\right)r^2\,e^{-b^2r^2}\,dr$$

and is shown in Figure 4.4.

It is seen that the most probable end-to-end distance, irrespective of direction, is not zero, but it is a function of b, i.e. of the length l of the links and the number n of links in the chain.

Another important quantity is the root-mean-square chain length, $\sqrt{\bar{r}^2}$

$$\bar{r}^2 = \int_0^\infty r^2 P(r)\,dr$$

Evaluation gives $\overline{r^2} = nl^2$ or that the root-mean-square length $\sqrt{\bar{r}^2} = l\sqrt{n}$, i.e. it is proportional to the square root of the number of links in the chain. This calculation shows that the Gaussian distribution function holds provided that n is so large that the distance between the ends, r, is much less than the extended chain length, nl.

If $dx\,dy\,dz$ is constant, the number of configurations available to the chain is proportional to the probability $p(x, y, z)$. Thus the entropy of the freely jointed chain, which, according to the Boltzmann relationship, is proportional to the logarithm of the number of configurations, is given by

$$S = c - kb^2r^2 = c - kb^2(x^2 + y^2 + z^2)$$

where c is a constant and k is Boltzmann's constant.

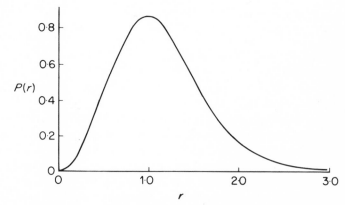

Figure 4.4. The distribution function $P(r) = \text{const } r^2\,e^{-b^2r^2}$.

Elasticity of a Molecular Network

We now wish to calculate the strain-energy function for a molecular network, assuming that this is given by the change in entropy of a network of chains as a function of strain.

The actual network is replaced by an ideal network in which each segment of a molecule between successive points of cross-linkage is considered to be a freely jointed Gaussian chain.

Two additional assumptions are introduced:

(1) In either the strained or unstrained state, each junction point may be regarded as fixed at its mean position.

(2) The effect of the deformation is to change the components of the vector length of each chain in the same ratio as the corresponding dimensions of the bulk material (this is the so-called 'affine' deformation assumption).

As discussed above, we can restrict our discussion to the case of homogeneous pure strain without loss of generality. We again choose principal extension ratios λ_1, λ_2, λ_3 parallel to the three rectangular coordinate axes x, y, z. The affine deformation assumption implies that the relative displacement of the chain ends is defined by the macroscopic deformation. Thus, in Figure 4.5, we take a system of coordinates x, y, z in the undeformed body. In this coordinate system a representative chain PQ

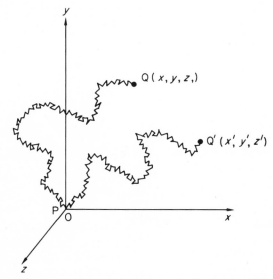

Figure 4.5. The end of the chain Q(x, y, z) is displaced to Q'(x', y', z').

has one end P at the origin. We refer any point in the deformed body to this system of coordinates. Thus the origin, i.e. the end of the chain P, is convected during the deformation. The other end Q (x, y, z) is displaced to the point Q' (x', y', z') and from the affine-deformation assumption we have

$$x' = \lambda_1 x \qquad y' = \lambda_2 y \qquad z' = \lambda_3 z$$

The actual network is formally equivalent to an assembly of freely jointed chains each with one end at the origin. The position of the other end follows the Gaussian distribution function. Thus in the constrained state the number of chains dn with ends in the elemental volume $dx \, dy \, dz$ at the point x, y, z is given by

$$dn = N \frac{b^3}{\pi^{3/2}} e^{-b^2(x^2+y^2+z^2)} \, dx \, dy \, dz.$$

where N is the number of chains per unit volume. Each of these chains has entropy

$$S_n = c - kb^2(x^2 + y^2 + z^2)$$

The total entropy before deformation is given by

$$S = \int S_n \, dn$$

$$= \frac{Nb^3}{\pi^{3/2}} \int_{-\infty}^{\infty} \int_{-\infty}^{\infty} \int_{-\infty}^{\infty} [c - kb^2(x^2 + y^2 + z^2)] \, e^{-b^2(x^2+y^2+z^2)} \, dx \, dy \, dz$$

which can be evaluated to give

$$S = N\left(c - \frac{3k}{2}\right)$$

On deformation the chain ends dn take coordinates $\lambda_1 x, \lambda_2 y, \lambda_3 z$. The entropy is therefore changed to

$$S'_n = c - kb^2(\lambda_1^2 x^2 + \lambda_2^2 y^2 + \lambda_3^2 z^2)$$

The total entropy of the network changes to

$$S' = \int S'_n \, dn$$

i.e. products of the new entropy S'_n and the same set of chains dn are involved. This gives

$$S' = \frac{Nb^3}{\pi^{3/2}} \int\limits_{-\infty}^{\infty} \int\limits_{-\infty}^{\infty} \int\limits_{-\infty}^{\infty} [c - kb^2(\lambda_1{}^2 x^2 + \lambda_2{}^2 y^2 + \lambda_3{}^2 z^2)] e^{-b^2(x^2 + y^2 + z^2)} \, dx \, dy \, dz$$

$$= N[c - \tfrac{1}{2}k(\lambda_1{}^2 + \lambda_2{}^2 + \lambda_3{}^2)]$$

The entropy change on deformation is

$$\Delta S = S' - S = -\tfrac{1}{2}Nk(\lambda_1{}^2 + \lambda_2{}^2 + \lambda_3{}^2 - 3)$$

Assuming no change in internal energy on deformation this gives the change in the Helmholtz Free Energy.

$$\Delta A = -T\Delta S = \tfrac{1}{2}NkT(\lambda_1{}^2 + \lambda_2{}^2 + \lambda_3{}^2 - 3)$$

If we assume that the strain energy U is zero in the undeformed state this gives

$$U = \Delta A = \tfrac{1}{2}NkT(\lambda_1{}^2 + \lambda_2{}^2 + \lambda_3{}^2 - 3)$$

Thus we arrive at the neo-Hookeian form for the strain-energy function U, in which U is a function of the strain invariant $I_1 = \lambda_1{}^2 + \lambda_2{}^2 + \lambda_3{}^2$ only, and

$$U = \tfrac{1}{2}NkT(I_1 - 3)$$

There are a number of refinements and modifications to this simple theory which do not, however, involve a change in the basic principles of the calculation. These will now be discussed in turn.

(1) N has been referred to as the number of chains per unit volume. It is determined by the number of junction points in the network. 'Junction points' can mean either chemical cross-links (as in vulcanized rubber) or physical entanglements (as in an amorphous polymer above its glass-transition temperature). These considerations have led to theoretical attempts to analyse the cross-link situation in more detail[7] to take into account the fact that not all the cross-links are effective. There will be 'loose loops' where a chain folds back on itself, indicated by symbol (a) in Figure 4.6, and 'loose ends' where a chain does not contribute to the network following a cross-link point which is close to the end of a chain molecule, indicated by symbol (b) in Figure 4.6.

(2) We have already hinted at the extension of this treatment to real chains with fixed bond angles and hindered rotation. This leads to the concept of the equivalent freely jointed chain.[8] It can be shown, for example, that for paraffin chains the root-mean-square end-to-end distance is $\sqrt{2}$ times that of a freely jointed chain with bonds of the same length.

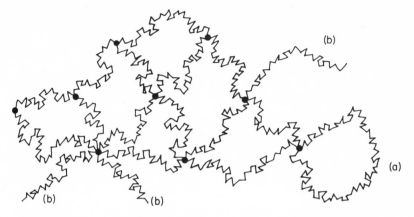

Figure 4.6. Types of network defect: (a) Loose loop; (b) Loose end.

More sophisticated treatments have been undertaken by making computer calculations based on the statistics of random walks[9]. Within the past three years, attempts have been made to take into account the excluded volume, i.e. the fact that a chain cannot bend back to occupy its portion of space twice[10].

(3) One of the basic assumptions of the simple theory is that the junction points are assumed to remain fixed in the material body. It was soon shown by James and Guth[11] that this assumption is unnecessarily restrictive. They found that it was adequate to assume that the cross-links move within the range permitted by the geometry of chemical bonds, i.e. that they fluctuate around their most probable positions. It must be remarked that these refinements of the statistical theory are in many cases highly controversial, and the reader is referred elsewhere for an adequate discussion of the situation[6].

(4) The final extension of the simple Gaussian theory is the so-called 'inverse Langevin approximation' for the probability distribution. The Gaussian distribution is only valid for end-to-end distances which are much less than the extended chain length. It was shown by Kuhn and Grun that removing this restriction (but still maintaining the other assumptions of freely jointed chains) gives a probability distribution $p(r)$ as

$$\ln p(r) = \mathrm{const} - n\left[\frac{r}{nl}\beta + \ln\frac{\beta}{\sinh\beta}\right]$$

In this equation, $\dfrac{r}{nl} = \coth\beta - \dfrac{1}{\beta} = \mathscr{L}(\beta)$

where $\mathscr{L}$ is the Langevin function and $\beta = \mathscr{L}^{-1}(r/nl)$ is the inverse Langevin function. This expression may be expanded to give,

$$\ln p(r) = \text{const} - n\left[\frac{3}{2}\left(\frac{r}{nl}\right)^2 + \frac{9}{20}\left(\frac{r}{nl}\right)^4 + \frac{99}{350}\left(\frac{r}{nl}\right)^6 + \ldots\right]$$

from which it can be seen that the Gaussian distribution is the first term of the series, an adequate approximation for $r \ll nl$.

Treloar[12] later developed a series distribution formula for $p(r)$ which is valid for any number of links, i.e. it does not require $r \ll nl$. Using this formula $p(r)$ was computed for various values of n. Typical results are shown in Figure 4.7.

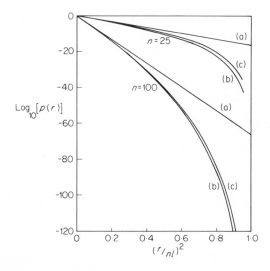

Figure 4.7. Distribution functions for 25 and 100 link random chains (a) Gaussian approximation; (b) Langevin approximation; (c) Treloar (after Treloar).

The final part of the exercise is to reconsider the stress–strain relations using either the inverse Langevin distribution function or Treloar's series-distribution formula. This was done by James and Guth using an analogous development to that for the Gaussian distribution function. It was also undertaken more exactly by Treloar. The results of such calculations are shown in Figure 4.8. These are to be compared with a typical force-extension curve for vulcanized rubber shown in Figure 3.6. It can

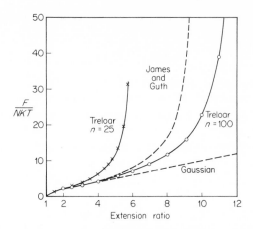

Figure 4.8. Theoretical force–extension curves for networks for non-Gaussian chains (after Treloar).

be seen that there is broad agreement between the form of the force–extension curve predicted by the non-Gaussian statistical theories and that observed in practice.

4.2 THE INTERNAL-ENERGY CONTRIBUTION TO RUBBER ELASTICITY

The simple treatment of rubber elasticity given above makes two assumptions which are in fact interrelated. First, it has been assumed that the internal-energy contribution is negligible, which implies that different molecular conformations of the chains have identical internal energies. Secondly, the thermodynamic formulae which have been derived are, strictly, only applicable to measurements at constant volume, whereas most experimental results are obtained at constant pressure. These two assumptions are interrelated in the sense that the experimental work of Gee (see Section 4.1.1 above) based on the approximation

$$\left(\frac{\partial f}{\partial T}\right)_{P,\lambda} = \left(\frac{\partial f}{\partial T}\right)_{V,l} \qquad \text{(where } \lambda \text{ is the extension ratio)}$$

lead to the conclusion that the internal-energy contribution was zero.

Although Gee's approximation is much better than the assumption that

$$\left(\frac{\partial f}{\partial T}\right)_{P,\lambda} = \left(\frac{\partial f}{\partial T}\right)_{V,l}$$

it has, however, subsequently proved necessary either to develop the theory in more exact form in order to evaluate the internal-energy contribution, or more elegantly to conduct the actual measurements at constant volume[13].

The internal-energy component of the tensile force f is given by

$$f_e = \left(\frac{\partial U}{\partial l}\right)_{T,V} \qquad \text{(Equation 4.2 above)}$$

Volkenstein and Ptitsyn[14] showed that, if the unperturbed dimensions of an isolated chain are temperature dependent, f_e is given by

$$\frac{f_e}{f} = T\frac{\partial(\ln \overline{r_0{}^2})}{\partial T}$$

where $\sqrt{\overline{r_0}{}^2}$ is the root-mean-square length (where the subscript 0 indicates a free chain unconstrained by cross-linkages).

Experimental data on dilute solutions of polymers using light scattering and viscosity measurements show that $\sqrt{\overline{r_0}{}^2}$ depends in general on temperature. This implies that the energy of a chain depends on its conformation and that for a rubber, in general, f_e will differ from zero. Flory and his collaborators[15] have been particularly prominent in performing stress-temperature measurements on polymer networks, together with physico-chemical measurements, to confirm these points and to obtain the energy difference between different conformations, e.g. the *trans* and *gauche* conformations of a polyethylene chain.

The value f_e can be expressed in terms of the tensile-force–temperature relationship by the equation

$$\frac{f_e}{f} = -T\left[\frac{\partial\{\ln f/T\}}{\partial T}\right]_{V,l} \qquad (4.8)$$

which follows by manipulating Equation (4.3) above. The investigations by Allen and his coworkers find f_e/f directly from this equation.

It is more usual however to make measurements at constant pressure, and in this case obtaining f_e/f is then more elaborate.

The procedure adopted by Flory was based on the theory of the Gaussian network. Flory, Ciferri and Hoeve[15] showed that if the rubber network obeys Gaussian statistics, the expression for measurement of simple extension at constant pressure is

$$\left[\frac{\partial\{\ln f/T\}}{\partial T}\right]_{P,l} + \frac{\alpha}{\lambda^3 - 1} = -\frac{d(\ln \overline{r_0^2})}{dT}$$

where α is the coefficient of volume expansion of the rubber at constant pressure, i.e.

$$\alpha = \frac{1}{V}\left(\frac{\partial V}{\partial T}\right)_P$$

Treloar[16,17] has more recently considered the behaviour in torsion and shown that the corresponding relationship is

$$\left[\frac{\partial\left\{\ln \dfrac{M}{T}\right\}}{\partial T}\right]_{P,l,\psi} = \frac{-d(\ln \overline{r_0^2})}{dT} + \alpha$$

where M is the torsional couple and ψ is the torsion (expressed in radians per unit length of the strained axis). M_e, the internal-energy contribution to the couple at constant volume, is given by an equation similar to (4.8).

$$\frac{M_e}{M} = -T\left[\frac{\partial\left\{\ln \dfrac{M}{T}\right\}}{\partial T}\right]_{V,l,\psi} = T\frac{d(\ln \overline{r_0^2})}{dT} \tag{4.9}$$

These and similar equations have been used to measure f_e or M_e and in all cases it has been found that the internal energy makes a significant contribution. In natural rubber f_e is positive and forms approximately 20% of the total force at room temperature[13,18], in polydimethyl silaxane[19] the contribution is positive, whereas in swollen and unswollen polyethylene it is negative[20,21].

Boyce and Treloar[22] showed that M_e/M for natural rubber is about 13%, which was within the range of all reported values of f_e/f, although lower than the 20% values which we have quoted here.

There have been a number of measurements, which, although indicating significant energy contributions, show a definite decrease in f_e with increase in extension ratio[23,24,25]. This decrease does not, of course, agree with the theory in which f_e was shown to be a function of r_0^2 only. Other attempts have been made to formulate a theory which takes into account the energy differences associated with different configurations. One theory does predict a variation of f_e/f with extension ratio, although this variation is not nearly large enough to explain some of the experimental results[26].

4.3 THE MOLECULAR INTERPRETATION OF HIGHER ORDER TERMS IN THE STRAIN-ENERGY FUNCTION

There is considerable variety in the attempts to give a molecular inter-pretation to the C_2 term in the Mooney equation, which is significant at all but the lowest strains.

An extreme view is that the deviation from Gaussian theory at moderate extensions is entirely due to the failure to reach equilibrium. Ciferri and Flory[27] undertook a series of experiments over a wide range of conditions where equilibrium was not necessarily achieved. They claimed that the Mooney equation fitted during increase of strain but did not apply during the retraction. In fact, they showed that the value of C_2 increased with the hysteresis of the straining cycle and differed for different polymers. This conclusion would seem to be at variance with that of Gumbrell and his collaborators[28] who showed that C_2 was almost a constant, independent of the nature of the polymer and the degree of chemical cross-linking, although both sets of workers agreed that C_2 decreases with the swelling of the polymer. It has also been shown that C_2 decreases with the removal of physical entanglements and with the increase of the time of the experiment[29]. The latter workers endorse the opinion of Ciferri and Flory[27] that the deviation from the Gaussian theory in the form of the additional term $C_2(I_2 - 3)$ was entirely due to the failure to reach equilibrium.

A theoretical attempt to explain the C_2 term by Ciferri and Hermans[30] predicts its increase with hysteresis and decrease with swelling, although it does give a much greater reduction of C_2 with increase in time than has been observed in practice. Ciferri and Hermans could only draw the conclusion that a possibility of some deviation from the simple theory cannot, at present, be dismissed. This opinion is held by Roe and Krig-baum[23] who also concluded that the internal energy of the chains makes a significant contribution to C_2, whereas the C_1 term is primarily due to entropy effects. They also point out that using the theory of Krigbaum and Kaneko[26], any non-equilibrium effects if in fact they exist, can only be expected to vanish under equilibrium conditions in the limit of infinite change in chain length between cross-links.

Earlier attempts to explain the C_2 term include those of Isihara and his collaborators[31] who attribute it to a non-Gaussian distribution, Wang and Guth[32], who associate it with the internal energy and Thomas[33] who expresses the free energy of the chain as containing an additional empirical term of the form

$$a(\sqrt{\overline{r_0^2}})^{-n}$$

All these attempts to fit the behaviour of rubbers to the Mooney equation do suffer from the objection that only limited types of deformation were examined. The conclusions must therefore be examined with some reservation.

REFERENCES

1. K. H. Meyer and C. Ferri, *Helv. Chim. Acta*, **18**, 570 (1935).
2. G. Gee, *Trans. Faraday Soc.* **42**, 585 (1946).
3. K. H. Meyer, G. Von Susich and E. Valko, *Kolloidzeitschrift*, **59**, 208 (1932).
4. E. Guth and H. Mark, *Lit. Chem.*, **65**, 93 (1934).
5. W. Kuhn, *Kolloidzeitschrift*, **68**, 2 (1934) and *Kolloidzeitschrift* **76**, 258 (1936).
6. L. R. G. Treloar, *The Physics of Rubber Elasticity*, Clarendon Press, Oxford, 1958.
7. P. J. Flory, *Chem. Rev.*, **35**, 51 (1944).
8. W. Kuhn, *Kolloidzeitschrift*, **76**, 258 (1936); **87**, 3 (1939).
9. F. T. Wall and J. J. Erpenbeck, *J. Chem. Phys.*, **30**, 634 (1959).
10. S. F. Edwards, *Proc. Phys. Soc.*, **91**, 513 (1967); Series **21A**, 15 (1968) and **92**, 9 (1967).
11. H. M. James and E. Guth, *J. Chem. Phys.*, **11**, 455 (1943).
12. L. R. G. Treloar, *Trans. Faraday Soc.*, **42**, 77 (1946).
13. G. Allen, U. Bianchi and C. Price, *Trans. Faraday Soc.*, **59**, 2493 (1963).
14. M. V. Volkenstein and O. B. Ptitsyn, *Dokl. Akad. SSSR*, **91**, 1313 (1953); *Zh. Tekh. Fiz.*, **25**, 649 (1955) and *Zh. Tekh. Fiz.*, **25**, 662 (1955).
15. P. J. Flory, A. Ciferri and C. A. J. Hoeve, *J. Polymer Sci.*, **45**, 235 (1960).
16. L. R. G. Treloar, *Polymer*, **10**, 279 (1969).
17. L. R. G. Treloar, *Polymer*, **10**, 291 (1969).
18. A. Ciferri, *Makromolek. Chem.*, **43**, 152 (1961).
19. A. Ciferri, *Trans. Faraday Soc.*, **57**, 846 (1961).
20. P. J. Flory, C. A. J. Hoeve and A. Ciferri, *J. Polymer Sci.*, **34**, 337 (1959).
21. A. Ciferri, C. A. J. Hoeve and P. J. Flory, *J. Amer. Chem. Soc.*, **83**, 1015 (1961).
22. P. H. Boyce and L. R. G. Treloar, *Polymer*, **11**, 21 (1970).
23. R. J. Roe and W. R. Krigbaum, *J. Polymer Sci.*, **61**, 167 (1962).
24. G. Crespi and U. Flisi, *Makromolek. Chem.*, **60**, 191 (1963).
25. U. Bianchi and E. Pedemonte, *J. Polymer Sci.*, **A2**, 5039 (1964).
26. W. R. Krigbaum and M. Kaneko, *J. Chem. Phys.*, **36**, 99 (1962).
27. A. Ciferri and P. J. Flory, *J. Appl. Phys.*, **30**, 1498 (1959).
28. S. M. Gumbrell, L. Mullins and R. S. Rivlin, *Trans. Faraday Soc.*, **49**, 1495 (1953).
29. G. Kraus and G. A. Moczvgemba, *J. Polymer Sci.*, **A2**, 277 (1964).
30. A. Ciferri and J. J. Hermans, *J. Polymer Sci.*, **B2**, 1089 (1964).
31. A. Isihara, N. Hashitsume and M. Tatibana, *J. Phys. Soc. Japan*, **3**, 289 (1951).
32. M. C. Wang and E. Guth, *J. Chem. Phys.*, **20**, 1144 (1953).
33. A. G. Thomas, *J. Polymer Sci.*, **18**, 177 (1955).

5

Linear Viscoelastic Behaviour

5.1 VISCOELASTIC BEHAVIOUR

In textbooks on properties of matter two particular types of ideal material are discussed, the elastic solid and the viscous liquid. The elastic solid has a definite shape and is deformed by external forces into a new equilibrium shape. On removal of these external forces it reverts exactly to its original form. The solid stores all the energy which it obtains from the work done by the external forces during deformation. This energy is then available to restore the body to its original shape when these forces are removed. A viscous liquid, on the other hand, has no definite shape and flows irreversibly under the action of external forces. Real materials have properties which are intermediate between those of an elastic solid and a viscous liquid. As discussed in Section 2.2 above, one of the most interesting features of high polymers is that a given polymer can display all the intermediate range of properties depending on temperature and the experimentally chosen time scale.

5.1.1 Linear Viscoelastic Behaviour

Newton's law of viscosity defines viscosity η by stating that stress σ is proportional to the velocity gradient in the liquid

$$\sigma = \eta \frac{\partial V}{\partial y}$$

where V is the velocity, and y is the direction of the velocity gradient. For a velocity gradient in the xy plane

$$\sigma_{xy} = \eta \left(\frac{\partial V_x}{\partial y} + \frac{\partial V_y}{\partial x} \right)$$

where $\partial V_x/\partial y$ and $\partial V_y/\partial x$ are the velocity gradients in the y and x directions respectively (see Figure 5.1 for the case where the velocity gradient is in the y direction).

77

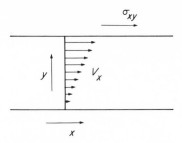

Figure 5.1. The velocity gradient.

Since $V_x = \partial u/\partial t$ and $V_y = \partial v/\partial t$ where u and v are the displacements in the x and y directions respectively, we have

$$\sigma_{xy} = \eta \left[\frac{\partial}{\partial y}\left(\frac{\partial u}{\partial t}\right) + \frac{\partial}{\partial x}\left(\frac{\partial v}{\partial t}\right) \right]$$

$$= \eta \frac{\partial}{\partial t}\left(\frac{\partial u}{\partial y} + \frac{\partial v}{\partial x}\right)$$

$$= \eta \frac{\partial e_{xy}}{\partial t}$$

It can be seen that the shear stress σ_{xy} is directly proportional to the rate of change of shear strain with time. This formulation brings out the analogy between Hooke's Law for elastic solids and Newton's Law for viscous liquids. In the former the stress is linearly related to the *strain*, in the latter the stress is linearly related to the *rate of change of strain* or *strain rate*.

Hooke's Law describes the behaviour of a linear *elastic* solid and Newton's Law that of a linear *viscous* liquid. A simple constitutive relation for the behaviour of a linear viscoelastic solid is obtained by combining these two laws.

For elastic behaviour $(\sigma_{xy})_E = Ge_{xy}$ where G is the shear modulus.
For viscous behaviour $(\sigma_{xy})_V = \eta(\partial e_{xy}/\partial t)$.
A possible formulation of linear viscoelastic behaviour combines these equations; thus

$$\sigma_{xy} = (\sigma_{xy})_E + (\sigma_{xy})_V = Ge_{xy} + \eta \frac{\partial e_{xy}}{\partial t}$$

This makes the simplest possible assumption that the shear stresses related to strain and strain rate are additive. The equation represents one of the simple models for linear viscoelastic behaviour (the Voigt or Kelvin model) and will be discussed in detail in Section 5.2.6 below.

Most of the experimental work on linear viscoelastic behaviour is confined to a single mode of deformation, usually corresponding to a measurement of the Young's modulus or the shear modulus. Our initial discussion of linear viscoelasticity will therefore be confined to the one-dimensional situation, recognizing that greater complexity will be required to describe the viscoelastic behaviour fully. For the simplest case of an isotropic polymer at least two of the situations corresponding to two of the modes of deformation defining two of the quantities E, G and K for an elastic solid must be examined, if the behaviour is to be completely specified.

In defining the constitutive relations for an elastic solid we have assumed that the *strains* are *small* and that there are linear relationships between stress and strain. We now ask how the principle of linearity can be extended to materials where the deformations are time-dependent. The basis of the discussion is the Boltzmann Superposition Principle[1]. This states that in linear viscoelasticity effects are simply additive, as in classical elasticity, the difference being that in linear viscoelasticity it matters at which instant an effect is created. Although the application of stress may now cause a time-dependent deformation, it can still be assumed that each increment of stress makes an independent contribution. From the present discussion it can be seen that the linear viscoelastic theory must also contain the additional assumption that the strains are small. In Chapter 9, we will deal with attempts to extend linear viscoelastic theory either to take into account non-linear effects at small strains or to deal with the situation at large strains.

5.1.2 Creep

It is convenient to introduce the discussion of linear viscoelastic behaviour with the one-dimensional situation of creep under a fixed load. For an elastic solid the following is observed at the two levels of stress σ_0 and $2\sigma_0$ (Figure 5.2a).

The strain follows the pattern of the loading programme exactly and in exact proportionality to the magnitude of the loads applied.

The effect of applying a similar loading programme to a linear visco-elastic solid has several similarities (Figure 5.2b). In the most general case the total strain e is the sum of three separate parts e_1, e_2 and e_3. e_1 and e_2 are often termed the *immediate* elastic deformation and the *delayed* elastic deformation respectively. e_3 is the Newtonian flow,

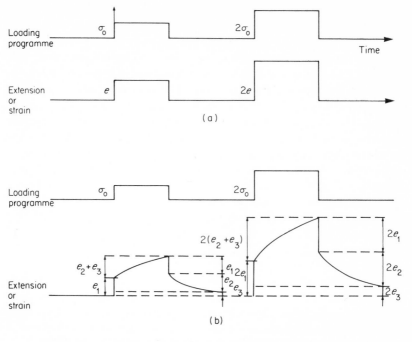

Figure 5.2. (a) Deformation of an elastic solid; (b) Deformation of a linear viscoelastic solid.

i.e. that part of the deformation which is identical with the deformation of a viscous liquid obeying Newton's law of viscosity.

Because the material shows linear behaviour the magnitudes of the strains e_1, e_2 and e_3 are exactly proportional to the magnitude of the applied stress. Thus the simple loading experiment defines a *creep compliance* $J(t)$ which is only a function of time.

$$\frac{e(t)}{\sigma} = J(t) = J_1 + J_2 + J_3$$

where J_1, J_2 and J_3 correspond to e_1, e_2 and e_3 respectively.

The term J_3, which defines the Newtonian flow, can be neglected for rigid polymers at ordinary temperatures, because their flow viscosities are very large. Linear amorphous polymers do show a finite J_3 at temperatures above their glass transitions, but at lower temperatures their behaviour is dominated by J_1 and J_2. Cross-linked polymers do not show a J_3 term, and this is true to a very good approximation for highly crystalline polymers as well.

This leaves J_1 and J_2. At any given temperature the separation of the compliance into terms J_1 and J_2 may involve an arbitrary division, which expresses the fact that at the shortest experimentally accessible times we will observe a limiting compliance J_1. We will assume, however, that there is a real distinction between the elastic and delayed responses. In some texts the immediate elastic response in a creep experiment is called the 'unrelaxed' response to distinguish it from the 'relaxed' response which is observed at times sufficiently long for the various relaxation mechanisms to have occurred. To emphasize that the values of such terms as J_1 are sometimes arbitrary we will enclose them in brackets.

We have already discussed in the introductory chapter how polymers can behave as glassy solids, viscoelastic solids, rubbers or viscous liquids depending on the time scale or on the temperature of the experiment. How does this fit in with our present discussion? Figure 5.3 shows the variation of compliance with time at constant temperature over a very wide time scale for an idealized amorphous polymer with only one relaxation transition. This diagram shows that for short-time experiments the observed compliance is 10^{-10} cm^2/dyn, that for a glassy solid. It is also time independent. At very long times the observed compliance is 10^{-6} cm^2/dyn, that for a rubbery solid, and it is again time independent. At intermediate times the compliance lies between these values and is time dependent; this is the general situation of viscoelastic behaviour.

These considerations suggest that the observed behaviour will depend on the time scale of the experiment relative to some basic time parameter of the polymer. For creep this parameter is called the retardation time τ'

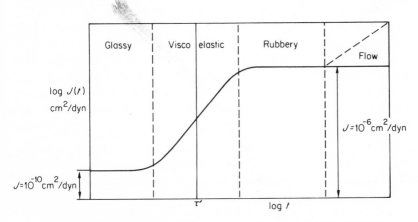

Figure 5.3. The creep compliance $J(t)$ as a function of time t. τ' is the characteristic time (the retardation time).

and falls in the middle range of our time scale as shown in the diagram. The distinction between a rubber and a glassy plastic can then be seen as somewhat artificial in that it depends only on the value of τ' at room temperature for each polymer. Thus for a rubber τ' is very small at room temperature compared with normal experimental times which are greater than say one second, whereas the opposite is true for a glassy plastic. The value of this parameter τ' for a given polymer relates to its molecular constitution, as will be discussed later.

These considerations lead immediately to a qualitative understanding of the influence of temperature on polymer properties. With increasing temperature the frequency of molecular rearrangements is increased, reducing the value of τ'. Thus at very low temperatures a rubber will behave like a glassy solid, as is well known, and equally a glassy plastic will soften at high temperatures to become rubber-like.

In the diagram illustrating creep under constant load the recovery curves are also displayed. We will presently show that the recovery behaviour is basically similar to the creep behaviour if we neglect the quantity e_3, the Newtonian flow. This is a direct consequence of linear viscoelastic behaviour.

5.1.3   Stress Relaxation

The counterpart of creep is stress relaxation, where the sample is subjected to constant strain e, and the decay of stress $\sigma(t)$ is observed. This is illustrated in Figure 5.4.

The assumption of linear behaviour enables us to define the *stress-relaxation modulus* $G(t) = \sigma(t)/e$. In the case of stress relaxation the presence of viscous flow will affect the limiting value of the stress. Where viscous flow occurs the stress can decay to zero at sufficiently long times, but where there is no viscous flow the stress decays to a finite value, and we obtain an equilibrium or relaxed modulus G_r at infinite time. Figure 5.5 is a schematic graph of the stress-relaxation modulus as a function of time. This is to be compared with the corresponding graph for creep (Figure 5.3). The same regions of behaviour, viz. glassy, viscoelastic,

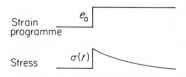

Figure 5.4. Stress relaxation.

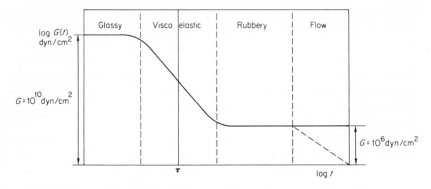

Figure 5.5. The stress-relaxation modulus $G(t)$ as a function of time t.
τ is the characteristic time (the relaxation time).

rubbery and flow, can be identified, and a transition time τ is defined which characterizes the time-scale of the viscoelastic behaviour. We will shortly discuss stress relaxation and creep in detail and show that as customarily defined, the characteristic times τ and τ', although of the same order of magnitude, are not identical. Similar considerations to those discussed for creep apply to the effect of changing temperature on stress relaxation; i.e. changing temperature is equivalent to changing the time scale. Time–temperature equivalence is applicable to all linear viscoelastic behaviour in polymers and is considered fully in Chapter 7. The measurement of G_r may present difficulties as in the case of the elastic response. We will assume that there is a relaxed response to which it relates, but enclose the term involving it in brackets as for those involving the elastic response.

5.2 MATHEMATICAL TREATMENT OF LINEAR VISCOELASTIC BEHAVIOUR

The discussion must now be placed on a quantitative basis. It is a question of personal taste how this is done, as it depends on the relative merits of a formal mathematical treatment with its manipulative advantages, and a less formal treatment which provides more physical insight into viscoelastic behaviour. We will endeavour to strike a balance here by describing the several representations of linear viscoelastic behaviour in turn, and completing the presentation by indicating the formal connections between them.

5.2.1 The Boltzmann Superposition Principle and the Definition of Creep Compliance

The general pattern of creep and stress-relaxation behaviour has been discussed, indicating some of the simpler consequences of assuming linear viscoelastic behaviour. The Boltzmann Superposition Principle[1] is the first mathematical statement of linear viscoelastic behaviour. Boltzmann proposed: (1) that the creep in a specimen is a function of the entire loading history, and (2) that each loading step makes an independent contribution to the final deformation and that the final deformation can be obtained by the simple addition of each contribution.

Consider a several-stage loading programme (Figure 5.6) in which incremental stresses $\Delta\sigma_1, \Delta\sigma_2, \Delta\sigma_3$, etc. are added at times τ_1, τ_2, τ_3, etc., respectively. The total creep at time t is then given by

$$e(t) = \Delta\sigma_1 J(t-\tau_1) + \Delta\sigma_2 J(t-\tau_2) + \Delta\sigma_3 J(t-\tau_3) + \ldots \qquad (5.1)$$

where $J(t-\tau)$ is the *creep-compliance function*. The contribution of each loading step is the product of the incremental stress and a general function of time, the creep-compliance function, which depends only on the interval in time between the instant at which the incremental stress is applied and the instant at which the creep is measured.

Equation (5.1) can be generalized to give the integral

$$e(t) = \int_{-\infty}^{t} J(t-\tau)\, d\sigma(\tau) \qquad (5.2)$$

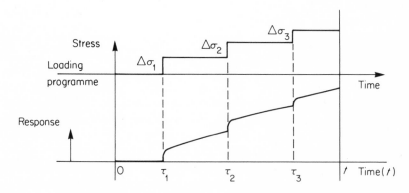

Figure 5.6. The creep behaviour of a linear viscoelastic solid.

and is usually rewritten as

$$e(t) = \left[\frac{\sigma}{G_u}\right] + \int_{-\infty}^{t} J(t-\tau)\frac{\mathrm{d}\sigma(\tau)}{\mathrm{d}\tau}\,\mathrm{d}\tau \qquad (5.3)$$

where the 'immediate elastic' contribution to the compliance is included in terms of an elastic modulus G_u (G_u is the unrelaxed modulus) and the integral term is written in its more correct mathematical form. It is to be noted that the integral is taken from the $-\infty$ to t. This follows from the hypothesis of the Boltzmann principle that *all* previous elements of the loading history were to be taken into account. In all actual experiment a conditioning procedure may be required to destroy the long-time memory of the specimen (see Chapter 6 below).

This integral is called a *Duhamel* integral, and it is a useful illustration of the consequences of the Boltzmann Superposition Principle to evaluate the response for a number of simple loading programmes. Recalling the development which leads to Equation (5.2) it can be seen that the Duhamel integral is most simply evaluated by treating it as the summation of a number of response terms. Consider three specific cases:

(1) Single-step loading of a stress σ_0 at time $\tau = 0$ (Figure 5.7a). For this case

$$J(t-\tau) = J(t) \qquad \text{and} \qquad e(t) = \sigma_0 J(t)$$

(2) Two-step loading, a stress σ_0 at time $\tau = 0$, followed by an *additional* stress σ_0 at time $\tau = t_1$ (Figure 5.7b). For this case

$$e_1 = \sigma_0 J(t), \qquad e_2 = \sigma_0 J(t - t_1)$$

give the creep deformations produced by the two loading steps, and

$$e(t) = e_1 + e_2 = \sigma_0 J(t) + \sigma_0 J(t-t_1)$$

This shows that the 'additional creep' $e'_c(t-t_1)$ produced by the second loading step is given by

$$e'_c(t-t_1) = \sigma_0 J(t) + \sigma_0 J(t-t_1) - \sigma_0 J(t) = \sigma_0 J(t-t_1)$$

This illustrates one consequence of the Boltzmann principle, viz. that the additional creep $e'_c(t-t_1)$ produced by adding the stress σ_0 is identical with the creep which would have occurred had this stress σ_0 been applied without any previous loading at the same instant in time t_1.

(3) Creep and recovery. In this case (Figure 5.7c) the stress σ_0 is applied at time $\tau = 0$ and removed at time $\tau = t_1$. The deformation $e(t)$ at a time t greater than t_1 is given by the addition of two terms

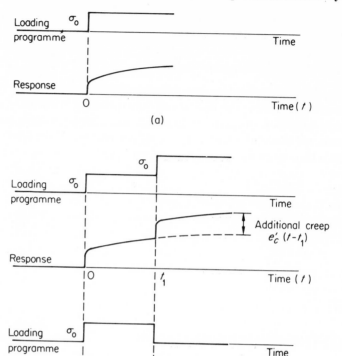

Figure 5.7. Response of a viscoelastic solid to single-step loading (a); two-step loading (b); and loading and unloading (c).

$e_1 = \sigma_0 J(t)$ and $e_2 = -\sigma_0 J(t-t_1)$, which express the application and removal of the stress σ_0 respectively. Thus

$$e(t) = \sigma_0 J(t) - \sigma_0 J(t-t_1)$$

The *recovery* $e_r(t-t_1)$ will be defined as the difference between the anticipated creep under the initial stress and the actual measured response. Thus

$$e_r(t-t_1) = \sigma_0 J(t) - [\sigma_0 J(t) - \sigma_0 J(t-t_1)] = \sigma_0 J(t-t_1)$$

It can be seen that this is identical with the creep response to a stress σ_0 applied at a time t_1. This demonstrates a second consequence of the

Boltzmann Superposition Principle, that the creep and recovery responses are identical in magnitude.

5.2.2   The Stress-Relaxation Modulus

Stress-relaxation behaviour can be represented in an exactly complementary fashion using the Boltzmann Superposition Principle. Consider a stress-relaxation programme in which incremental strains Δe_1, Δe_2, Δe_3, etc., are added at times τ_1, τ_2, τ_3, etc., respectively. The total stress at time t is then given by

$$\sigma(t) = \Delta e_1 G(t - \tau_1) + \Delta e_2 G(t - \tau_2) + \Delta e_3 G(t - \tau_3) + ... \qquad (5.4)$$

where $G(t - \tau)$ is the stress-relaxation modulus. Equation (5.4) may be generalized in an identical manner in which (5.1) leads to (5.2) and (5.3) to give

$$\sigma(t) = [G_r e] + \int_{-\infty}^{t} G(t - \tau) \frac{de(\tau)}{d\tau} dt \qquad (5.5)$$

where G_r is the equilibrium or relaxed modulus.

5.2.3   The Formal Relationship Between Creep and Stress Relaxation

We have already seen that stress relaxation is the converse of creep in a general sense. It is therefore to be expected that they will be formally related through a simple mathematical relationship.

For simplicity consider only the time-dependent terms in Equation (5.3). Then

$$e(t) = \int_{-\infty}^{t} J(t - \tau) \frac{d\sigma(\tau)}{d\tau} d\tau$$

Consider a stress programme starting at time $\tau = 0$ in which the stress decreases exactly as the relaxation function $G(\tau)$. In this case the corresponding strain must remain constant as in a typical stress-relaxation experiment.

Thus if

$$\frac{d\sigma(\tau)}{d\tau} = \frac{dG(\tau)}{d\tau}$$

then

$$\int_{0}^{t} \frac{dG(\tau)}{d\tau} J(t - \tau) \, d\tau = \text{constant} \qquad (5.6)$$

For simplicity we can normalize the definitions of $G(\tau)$ and $J(\tau)$ so that the constant is unity.

We then have

$$\int_0^t \frac{\mathrm{d}G(\tau)}{\mathrm{d}\tau} J(t-\tau)\,\mathrm{d}\tau = 1 \tag{5.7}$$

This expression is sometimes integrated to give

$$\int_0^t G(\tau)J(t-\tau)\,\mathrm{d}\tau = t \tag{5.8}$$

These equations provide a formal connection between the creep and stress-relaxation functions. However, this approach is of greatest interest from a purely theoretical standpoint. In practice the problem of interchangeability of creep and stress-relaxation data is usually dealt with via relaxation or retardation spectra, and by approximate methods.

5.2.4   Mechanical Models, Relaxation and Retardation-Time Spectra

The Boltzmann Superposition Principle is one starting point for a theory of linear viscoelastic behaviour, and is sometimes called the '*integral representation* of linear viscoelasticity', because it defines an integral equation. An equally valid starting point is to relate the stress to the strain by a linear differential equation, which leads to a *differential representation* of linear viscoelasticity. In its most general form the equation has the form

$$P\sigma = Qe$$

where P and Q are linear differential operators with respect to time. This representation has been found of particular value in obtaining solution to specific problems in the deformation of viscoelastic solids[2].

Most generally the differential equation is

$$a_0\sigma + a_1\frac{\mathrm{d}\sigma}{\mathrm{d}t} + a_2\frac{\mathrm{d}^2\sigma}{\mathrm{d}t^2} + \dots = b_0e + b_1\frac{\mathrm{d}e}{\mathrm{d}t} + b_2\frac{\mathrm{d}^2e}{\mathrm{d}t^2} + \dots \tag{5.9}$$

It is often adequate to represent the experimental data obtained over a limited time scale by including only one or two terms on each side of this equation. We will now show that this is equivalent to describing the viscoelastic behaviour by mechanical models constructed of elastic springs which obey Hooke's Law and viscous dashpots which obey Newton's Law of Viscosity.

The simplest models consist of a single spring and a single dashpot either in series or in parallel and these are known as the Maxwell model and the Kelvin or Voigt models respectively.

5.2.5 The Maxwell Model

The Maxwell model consists of a spring and dashpot in series as shown in Figure 5.8.

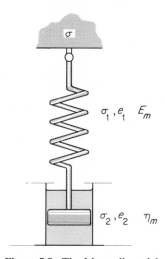

Figure 5.8. The Maxwell model.

The equations for the stress–strain relations are

$$\sigma_1 = E_m e_1 \qquad\qquad (5.10a)$$

relating the stress σ_1 and the strain e_1 in the spring and

$$\sigma_2 = \eta_m \frac{de_2}{dt} \qquad\qquad (5.10b)$$

relating the stress σ_2 and the strain e_2 in the dashpot. Now relate the total stress σ and the total strain e. We have that $\sigma = \sigma_1 = \sigma_2$ since the stress is identical for the spring and dashpot, and $e = e_1 + e_2$ the total strain being the sum of the strain in the spring and the dashpot. Equation (5.10a) can be written as

$$\frac{d\sigma}{dt} = E_m \frac{de_1}{dt}$$

and added to (5.10b) giving

$$\frac{de}{dt} = \frac{1}{E_m}\frac{d\sigma}{dt} + \frac{\sigma}{\eta_m} \tag{5.11}$$

The Maxwell model is of particular value in considering a stress-relaxation experiment.

In this case

$$\frac{de}{dt} = 0 \quad \text{and} \quad \frac{1}{E_m}\frac{d\sigma}{dt} + \frac{\sigma}{\eta_m} = 0$$

Thus

$$\frac{d\sigma}{\sigma} = -\frac{E_m}{\eta_m}\,dt$$

At time $t = 0$, $\sigma = \sigma_0$, the initial stress, and integrating we have

$$\sigma = \sigma_0 \exp\frac{-E_m}{\eta_m}t \tag{5.12}$$

This shows that the stress decays exponentially with a characteristic time constant $\tau = \eta_m/E_m$

$$\sigma = \sigma_0 \exp\frac{-t}{\tau}$$

where τ is called the 'relaxation time'. There are two inadequacies of this simple model which can be understood immediately.

First, under conditions of constant stress, i.e.

$$\frac{d\sigma}{dt} = 0, \qquad \frac{de}{dt} = \frac{\sigma}{\eta_m}$$

and Newtonian flow is observed. This is clearly not generally true for viscoelastic materials where the creep behaviour is more complex.

Secondly, the stress-relaxation behaviour cannot usually be represented by a single exponential decay term, nor does it necessarily decay to zero at infinite time.

5.2.6   The Kelvin or Voigt Model

The Kelvin or Voigt model consists of a spring and dashpot in parallel as shown in Figure 5.9.

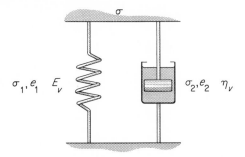

Figure 5.9. The Kelvin or Voigt model.

The stress–strain relations are

$$\sigma_1 = E_v e_1 \tag{5.13}$$

and

$$\sigma_2 = \eta_v \frac{de_2}{dt} \tag{5.14}$$

for the spring and dashpot respectively.

Again it is required to relate the total stress σ to the total strain e. In this case we have

$$e = e_1 = e_2 \quad \text{and} \quad \sigma = \sigma_1 + \sigma_2$$

Thus

$$\sigma = E_v e + \eta_v \frac{de}{dt} \tag{5.15}$$

For stress relaxation, where $de/dt = 0$, the Kelvin model gives $\sigma = E_v e$, i.e. a constant stress, implying that the material behaves as an elastic solid, which is clearly an inadequate representation for general viscoelastic behaviour.

On the other hand, the Kelvin model does represent creep behaviour to a first approximation. For creep under constant load $\sigma = \sigma_0$ it may readily be shown that

$$e = \frac{\sigma_0}{E_v}\left(1 - \exp\frac{-E_v}{\eta_v}t\right) \tag{5.16}$$

In fact, the response of a Kelvin element to constant load conditions is most readily understood by considering the recovery response, where $\sigma = 0$.

Here

$$E_v e + \eta_v \frac{de}{dt} = 0$$

giving a solution for the strain

$$e = e_0 \exp \frac{-t}{\tau'}$$

where $\tau' = \eta_v/E_v$ is a characteristic time constant called the 'retardation time'. All these relationships are exactly analogous to the stress-relaxation behaviour and the relaxation time of the Maxwell model.

5.2.7   The Standard Linear Solid

We have seen that the Maxwell model describes the stress relaxation of a viscoelastic solid to a first approximation, and the Kelvin model the creep

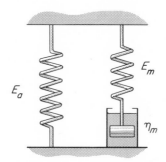

Figure 5.10. The standard linear solid.

behaviour, but that neither model is adequate for the general behaviour of a viscoelastic solid where it is necessary to describe both stress relaxation and creep.

Consider again the general linear differential equation which represents linear viscoelastic behaviour. From the present discussion it follows that to obtain even an approximate description of both stress relaxation and creep, at least the first two terms on each side of Equation (5.9) must be retained, i.e. the simplest equation will be of the form

$$a_0 \sigma + a_1 \frac{d\sigma}{dt} = b_0 e + b_1 \frac{de}{dt} \tag{5.17}$$

This will be adequate to a first approximation for creep (when $d\sigma/dt = 0$) and for stress relaxation (when $de/dt = 0$), giving an exponential response in both cases.

It is very easy to show that the model shown in Figure 5.10 has this form. The stress–strain relationship is

$$\sigma + \tau \frac{d\sigma}{dt} = E_a e + (E_m + E_a)\tau \frac{de}{dt} \quad \text{where} \quad \tau = \frac{\eta_m}{E_m} \qquad (5.18)$$

This model is known as the 'standard linear solid' and is usually attributed to Zener[3]. It provides an approximate representation to the observed behaviour of polymers in their main viscoelastic range. As has been discussed, it predicts an exponential response only. To describe the observed viscoelastic behaviour quantitatively would require the inclusion of many terms in the linear differential Equation (5.9). These more complicated equations are equivalent to either a large number of Maxwell elements in parallel or a large number of Voigt elements in series (Figures 5.11a and b).

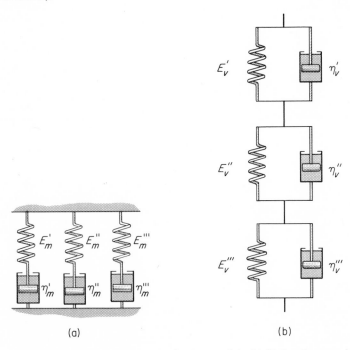

(a)

(b)

Figure 5.11. (a) Maxwell elements in parallel; (b) Voigt elements in series.

5.2.8   Relaxation-Time Spectra and Retardation-Time Spectra

It is next required to obtain a quantitative description of stress relaxation
and creep which will help to form a link with the original mathematical
descriptions in terms of the Boltzmann integrals. It is simple and instructive to do this by development of the Maxwell and Kelvin models.

Consider first stress relaxation, described by

$$\sigma(t) = [G_r e] + \int_{-\infty}^{t} G(t - \tau) \frac{de(\tau)}{d\tau} d\tau \qquad (5.4)$$

where $G(t)$ is the stress-relaxation modulus. For stress relaxation at
constant strain e, equation (5.12) shows that the Maxwell model gives

$$\sigma(t) = E_m e \exp \frac{-t}{\tau}$$

and the stress-relaxation modulus $G(t) = E_m \exp(-t/\tau)$. For a series of
Maxwell elements joined in parallel, again at constant strain e, the stress
is given by

$$\sigma(t) = e \sum^{n} E_n \exp \frac{-t}{\tau_n}$$

where E_n, τ_n are the spring constant and relaxation time respectively of the
nth Maxwell element.

The summation can be written as an integral, giving

$$\sigma(t) = [G_r e] + e \int_{0}^{\infty} f(\tau) \exp \frac{-t}{\tau} d\tau \qquad (5.19)$$

where the spring constant E_n is replaced by the weighting function $f(\tau) d\tau$
which defines the concentration of Maxwell elements with relaxation
times between τ and $\tau + d\tau$.

The stress-relaxation modulus is given by

$$G(t) = [G_r] + \int_{0}^{\infty} f(\tau) \exp \frac{-t}{\tau} d\tau \qquad (5.20)$$

The term $f(\tau)$ is called the 'relaxation-time spectrum'. In practice, it has
been found more convenient to use a logarithmic time-scale. A new
relaxation-time spectrum $H(\tau)$ is now defined, where $H(\tau) d \ln(\tau)$ gives the
contributions to the stress relaxation associated with relaxation times

between $\ln(\tau)$ and $(\ln\tau + \mathrm{d}\ln\tau)$. The stress-relaxation modulus is then given by

$$G(t) = [G_r] + \int_{-\infty}^{\infty} H(\tau)\exp\frac{-t}{\tau}\,\mathrm{d}\ln\tau \qquad (5.21)$$

An exactly analogous treatment using a series of the Kelvin models, leads to a similar expression for the creep compliance $J(t)$:
Thus

$$J(t) = [J_u] + \int_{-\infty}^{\infty} L(\tau)\left(1 - \exp\frac{-t}{\tau}\right)\mathrm{d}\ln(\tau) \qquad (5.22)$$

where J_u is the instantaneous elastic compliance and $L(\tau)$ is the *retardation-time spectrum*, $L(\tau)\,\mathrm{d}(\ln\tau)$ defining the contributions to the creep compliance associated with retardation times between $(\ln\tau)$ and $(\ln\tau + \mathrm{d}\ln\tau)$.

The relaxation-time spectrum can be calculated exactly from the measured stress-relaxation modulus using Fourier or Laplace transform methods, and similar considerations apply to the retardation-time spectrum and the creep compliance. It is more convenient to consider these transformations at a later stage, when the final representation of linear viscoelasticity, that of the complex modulus and complex compliance, has been discussed.

It is important to recognize that the relaxation-time spectrum and the retardation-time spectrum are only mathematical descriptions of the macroscopic behaviour and do not necessarily have a simple interpretation in molecular terms. It is a quite separate exercise to correlate observed patterns in the relaxation behaviour such as a predominant relaxation time, with a specific molecular process. It should also be emphasized, as will be apparent from the further detailed discussion, that qualitative interpretations in general molecular terms can often be obtained from the experimental data directly, without recourse to calculation of the relaxation-time spectrum or the retardation-time spectrum.

5.3 DYNAMICAL MECHANICAL MEASUREMENTS: THE COMPLEX MODULUS AND COMPLEX COMPLIANCE

An alternative experimental procedure to creep and stress relaxation is to subject the specimen to an alternating strain and simultaneously measure the stress. For linear viscoelastic behaviour, when equilibrium is reached, the stress and strain will both vary sinusoidally, but the strain lags behind the stress.

Thus if we write

$$\text{strain } e = e_0 \sin \omega t$$

$$\text{stress } \sigma = \sigma_0 \sin (\omega t + \delta)$$

where ω is the angular frequency, δ is the phase lag.

Expanding $\sigma = \sigma_0 \sin \omega t \cos \delta + \sigma_0 \cos \omega t \sin \delta$ we see that the stress can be considered to consist of two components: (1) of magnitude $(\sigma_0 \cos \delta)$ in phase with the strain; (2) of magnitude $(\sigma_0 \sin \delta)$ 90° out of phase with the strain.

The stress–strain relationship can therefore be defined by a quantity G_1 in phase with the strain and by a quantity G_2 which is 90° out of phase with the strain, i.e.

$$\sigma = e_0 G_1 \sin \omega t + e_0 G_2 \cos \omega t$$

where

$$G_1 = \frac{\sigma_0}{e_0} \cos \delta \quad \text{and} \quad G_2 = \frac{\sigma_0}{e_0} \sin \delta$$

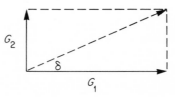

Figure 5.12. The complex modulus $\overset{*}{G} = G_1 + iG_2$ and $\tan \delta = G_2/G_1$.

This immediately suggests a complex representation for the modulus as shown in Figure 5.12.

If we write

$$e = e_0 \exp i\omega t$$

$$\sigma = \sigma_0 \exp i(\omega t + \delta)$$

Then

$$\frac{\sigma}{e} = \overset{*}{G} = \frac{\sigma_0}{e_0} e^{i\delta}$$

$$= \frac{\sigma_0}{e_0} (\cos \delta + i \sin \delta)$$

$$= G_1 + iG_2$$

The real part of the modulus G_1, which is in phase with the strain, is often called the *storage modulus* because it defines the energy stored in the specimen due to the applied strain. The imaginary part of the modulus G_2, which is out of phase with the strain, defines the dissipation of energy and is often called the *loss modulus*. The reason for this is seen by calculating the energy dissipated per cycle, $\Delta\mathscr{E}$.

$$\Delta\mathscr{E} = \int \sigma \, de = \int_0^{2\pi/\omega} \frac{\sigma \, de}{dt} \, dt$$

Substituting for σ and e we have

$$\Delta\mathscr{E} = \omega e_0^2 \int_0^{2\pi/\omega} (G_1 \sin \omega t \cos \omega t + G_2 \cos^2 \omega t) \, dt$$

$$= \pi G_2 e_0^2$$

In most cases G_2 is small compared with G_1. $|\overset{*}{G}|$ is therefore approximately equal to G_1. $|\overset{*}{G}|$ is sometimes loosely referred to as the 'modulus' G. It is customary to define the dynamic mechanical behaviour in terms of the 'modulus' $G \doteqdot G_1$, and the phase angle δ or often $\tan \delta = G_2/G_1$. To a good approximation $\delta = \tan \delta$ when the loss modulus G_2 is small. Typical values of G_1, G_2 and $\tan \delta$ for a polymer would be 10^{10} dyn/cm², 10^8 dyn/cm² and $0 \cdot 01$ respectively.

An exactly complementary treatment can be developed to define a complex compliance

$$\overset{*}{J} = J_1 - iJ_2$$

This is directly related to the complex modulus since

$$\overset{*}{G} = \frac{1}{\overset{*}{J}}$$

We have so far ignored any question of frequency or time dependence. Here there is an exact analogy with creep and stress relaxation and it is necessary to determine G_1 and G_2 (or $\tan \delta$) or J_1 and J_2 (or $\tan \delta$) as a function of frequency if we wish to specify the viscoelastic behaviour completely.

5.3.1   Experimental Patterns for G_1, G_2, etc. as a Function of Frequency

Now consider the complex moduli and compliances as a function of frequency for a typical viscoelastic solid, in a similar manner to the creep and stress relaxation as a function of time.

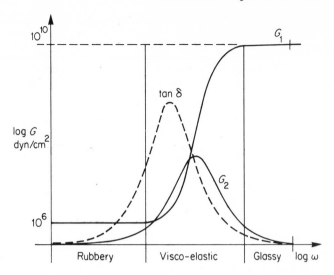

Figure 5.13. The complex modulus $G_1 + iG_2$ as a function of frequency ω.

Figure 5.13 shows the variation of G_1, G_2 and tan δ with frequency for a polymer which shows no flow. At low frequencies the polymer is rubber-like and has a low modulus G_1 of $\sim 10^6$ dyn/cm^2, which is independent of frequency. At high frequencies the polymer is glassy with a modulus of $\sim 10^{10}$ dyn/cm^2, which is again independent of frequency. At intermediate frequencies the polymer behaves as a viscoelastic solid, and its modulus G_1 increases with increasing frequency.

The complementary pattern of behaviour is shown by the loss modulus G_2. At low and high frequencies G_2 is zero, the stress and strain being exactly in phase for the rubbery and glassy states. At intermediate frequencies where the polymer is viscoelastic, G_2 rises to a maximum value, this occurring at a frequency close to that for which the storage modulus is changing most rapidly with frequency. The viscoelastic region is also characterized by a maximum in the loss factor tan δ, but this occurs at a slightly lower frequency than that in G_2, since tan $\delta = G_2/G_1$ and G_1 is also changing rapidly in this frequency range.

An analogous diagram (Figure 5.14) shows the variation of the compliances J_1 and J_2 with frequency.

The next development is to obtain a mathematical representation for the dynamic mechanical behaviour as a function of frequency. As in the

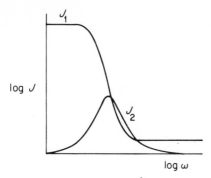

Figure 5.14. The complex compliance $\overset{*}{J} = J_1 - iJ_2$ as a function of frequency ω.

case of stress-relaxation and creep, a very easy starting point for the argument is based on the Maxwell and Voigt models.

Using Equation (5.11) for the Maxwell model,

$$\frac{de}{dt} = \frac{1}{E_m}\frac{d\sigma}{dt} + \frac{\sigma}{\eta_m}$$

and the definition of the relaxation time as $\tau = \eta_m/E_m$ we can write

$$\sigma + \tau\frac{d\sigma}{dt} = E_m\tau\frac{de}{dt}$$

Put

$$\sigma = \sigma_0\,e^{i\omega t} = (G_1 + iG_2)e$$

This gives

$$\sigma_0\,e^{i\omega t} + i\omega\tau\sigma_0\,e^{i\omega t} = \frac{E_m\tau i\omega\sigma_0\,e^{i\omega t}}{G_1 + iG_2}$$

from which it follows that

$$G_1 + iG_2 = E_m\frac{i\omega\tau}{1 + i\omega\tau}$$

i.e.

$$G_1 = E_m\frac{\omega^2\tau^2}{1 + \omega^2\tau^2} \qquad G_2 = \frac{E_m\omega\tau}{1 + \omega^2\tau^2} \quad \text{and} \quad \tan\delta = \frac{1}{\omega\tau} \qquad (5.23)$$

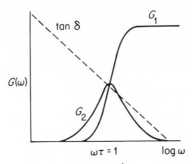

Figure 5.15. The complex modulus $\overset{*}{G} = G_1 + iG_2$ as a function of frequency ω.

This result gives the pattern shown in Figure 5.15 for G_1, G_2 and $\tan \delta$ as a function of frequency (or $\omega\tau$, which is more convenient). It is seen that the qualitative features are correct in the case of G_1 and G_2, but not for $\tan \delta$.

A similar manipulation of the equation representing the Voigt model, introducing the complex compliances, leads to a comparable qualitative picture of J_1, J_2 and $\tan \delta$ as a function of frequency. Again the qualitative features are correct for J_1 and J_2 but not for $\tan \delta$, which in this case is equal to $\omega\tau'$.

The Maxwell and Voigt models are therefore inadequate to describe the dynamic mechanical behaviour of a polymer, as they do not provide an adequate representation of both the creep and stress-relaxation behaviour. A good measure of qualitative improvement could be gained, as in the previous discussion of creep and stress relaxation, by using a three-parameter model, e.g. the standard linear solid, and it is an interesting exercise to show that this model gives a more realistic variation of G_1, G_2 and $\tan \delta$ with frequency.

It is, however, desirable to move directly to the general representation, using the relaxation-time spectrum.

The general representation for the stress-relaxation modulus (Equation 5.21)

$$G(t) = \frac{\sigma(t)}{e} = [G_r] + \int_{-\infty}^{\infty} H(\tau) \exp \frac{-t}{\tau} \, d \ln \tau$$

follows by generalizing the stress-relaxation response from a single Maxwell element where $G(t) = E_m \exp(-t/\tau)$.

The response of a Maxwell element to an alternating strain is defined by the relationships

$$G_1 = E_m \frac{\omega^2 \tau^2}{1+\omega^2\tau^2} \quad \text{and} \quad G_2 = E_m \frac{\omega\tau}{1+\omega^2\tau^2}$$

An identical generalization to the previous one then gives

$$G_1(\omega) = [G_r] + \int_{-\infty}^{\infty} \frac{H(\tau)\omega^2\tau^2}{1+\omega^2\tau^2}\, \mathrm{d}\ln\tau \tag{5.24}$$

and

$$G_2(\omega) = \int_{-\infty}^{\infty} \frac{H(\tau)\omega\tau}{1+\omega^2\tau^2}\, \mathrm{d}\ln\tau \tag{5.25}$$

As previously, the spring constant E_m is replaced by the weighting function $H(\tau)\, \mathrm{d}\ln\tau$ which defines the contribution to the response of elements whose relaxation time is between $\ln\tau$ and $(\ln\tau + \mathrm{d}\ln\tau)$. It is seen that the stress-relaxation modulus $G(t)$ and the real and imaginary parts of the complex compliance G_1 and G_2 can all be directly related to the same relaxation-time spectrum $H(\tau)$.

Similar relationships hold between the creep compliance $J(t)$, the real and imaginary parts of the complex compliance J_1 and J_2 and the retardation-time spectrum $L(\tau)$. These relationships can be readily derived by consideration of the response of a Voigt element to an alternating stress. They will not be derived here, but the results are quoted for completeness.

$$J_1(\omega) = [J_u] + \int_{-\infty}^{\infty} \frac{L(\tau)}{1+\omega^2\tau^2}\, \mathrm{d}\ln\tau \tag{5.26}$$

$$J_2(\omega) = \int_{-\infty}^{\infty} \frac{L(\tau)\omega\tau}{1+\omega^2\tau^2}\, \mathrm{d}\ln(\tau) \tag{5.27}$$

5.4 THE RELATIONSHIPS BETWEEN THE COMPLEX MODULI AND THE STRESS-RELAXATION MODULUS

The exact formal relationships between the various viscoelastic functions are conveniently expressed using Fourier or Laplace transform methods (cf. Section 5.4.2 below).

It is often required to combine dynamic mechanical measurements of complex moduli at high frequencies (or short times) with stress-relaxation modulus measurements at long times (or low frequencies). The most usual approach is to use approximate methods to convert both types of measurement to the determination of the relaxation-time spectrum.

Now

$$G(t) = [G_r] + \int_{-\infty}^{\infty} H(\tau) \exp \frac{-t}{\tau} \, \mathrm{d} \ln \tau \qquad (5.21)$$

A simple approximation is to assume that $e^{-t/\tau} = 0$ up to the time $\tau = t$, and $e^{-t/\tau} = 1$ for $\tau > t$.

Then we can write

$$G(t) = [G_r] + \int_{\ln t}^{\infty} H(\tau) \, \mathrm{d} \ln \tau$$

This gives the relaxation-time spectrum

$$H(\tau) = -\left[\frac{\mathrm{d}G(t)}{\mathrm{d} \ln t}\right]_{t=\tau} \qquad (5.28)$$

which is known as the 'Alfrey approximation'[4].

The relaxation-time spectrum can be expressed to a similar degree of approximation in terms of the real and imaginary parts of the complex modulus

$$H(\tau) = \left[\frac{\mathrm{d}\, G_1(\omega)}{\mathrm{d} \ln \omega}\right]_{1/\omega = \tau} = \frac{2}{\pi}[G_2(\omega)]_{1/\omega = \tau} \qquad (5.29)$$

These relationships are illustrated diagrammatically for the case of a single relaxation transition in Figures 5.16a and b. To obtain the complete relaxation-time spectrum the longer time part of $H(\tau)$ will be found from the stress-relaxation modulus data of Figure 5.16a and the shorter time part from the dynamic mechanical data of Figure 5.16b.

Complementary relationships can be used to obtain the retardation-time spectrum in terms of the complex compliances and the creep compliance.

5.4.1   Formal Representations of the Stress-Relaxation Modulus and the Complex Modulus

A complete exposition of the mathematical structure of linear visco-elasticity has been given by Gross[6]. Here we will only summarize certain parts of his argument to illustrate the use of Laplace and Fourier transforms in establishing the formal connections between various visco-elastic functions:

(1) The stress-relaxation modulus is a continuous, decreasing function which goes to zero at infinite time. In Gross' nomenclature it is repre-

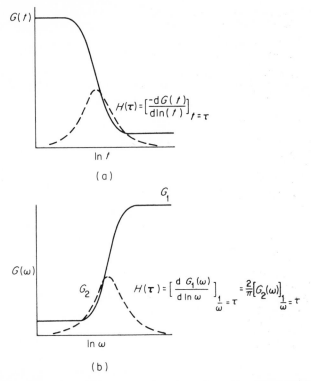

Figure 5.16. The Alfrey approximations for the relaxation-time spectrum $H(\tau)$: (a) from the stress relaxation modulus $G(t)$; (b) from the real and imaginary parts G_1 and G_2 respectively of the complex modulus $G(\omega)$.

sented in integral form as

$$G(t) = [G_r] + \int_0^\infty \bar{\beta}\bar{F}(\tau) \exp\frac{-t}{\tau}\,\mathrm{d}\tau \qquad (5.30)$$

where $\bar{F}(\tau)\,\mathrm{d}\tau$ is the relaxation spectrum, and $\bar{\beta}$ is a normalization factor such that

$$\int_0^\infty \bar{F}(\tau)\,\mathrm{d}\tau = 1, \qquad \bar{\beta} = G(0)$$

This equation, in terms of our models, represents an infinite series of Maxwell elements and it is formally identical to Equation (5.20) above.

It can be transformed into a Laplace integral or Laplace transform by putting $1/\tau = s$, the relaxation frequency, and introducing a frequency function $\bar{N}(s)\,ds$ defined as

$$\bar{N}(s) = \frac{\bar{\beta}\bar{F}(1/s)}{s^2}$$

Thus

$$G(t) = [G_r] + \int_0^\infty \bar{N}(s)\,e^{-ts}\,ds \tag{5.31}$$

The importance of this representation is that when $G(t)$ has been determined, the relaxation-time spectrum can be found in principle by standard methods for the inversion of the Laplace integral. In practice this requires computation methods, as it is not usually possible to find an analytical expression to fit the stress-relaxation modulus.

The Alfrey approximation is now given by putting

$$e^{-ts} = 1 \qquad \text{for } s \leqslant \frac{1}{t}$$

and

$$e^{-ts} = 0 \qquad \text{for } s > \frac{1}{t}$$

and

$$G(t) = [G_r] + \int_0^{1/t} \bar{N}(s)\,ds \tag{5.32}$$

(2) The complex modulus.

The Boltzmann Superposition Principle gives us that

$$\sigma(t) = [G_r e(t)] + \int_{-\infty}^t G(t-\tau)\frac{de(\tau)}{d\tau}\,d\tau$$

Put $e(\tau) = e_0\,e^{i\omega\tau}$. Then

$$\sigma(t) = [G_r e(t)] + i\omega \int_{-\infty}^t G(t-\tau)e_0\,e^{i\omega\tau}\,d\tau \tag{5.33}$$

Put $t-\tau = \mathbf{T}$. Then

$$\sigma(t) = [G_r e(t)] + i\omega \int_0^\infty G(\mathbf{T})\,e^{-i\omega\mathbf{T}}\,d\mathbf{T}e_0\,e^{i\omega t} \tag{5.34}$$

Now $e(t) = e_0\, e^{i\omega t}$. Thus

$$\frac{\sigma(t)}{e(t)} = [G_r] + i\omega \int_0^\infty G(\tau)\, e^{-i\omega\tau}\, d\tau = \overset{*}{G}(\omega)$$

the complex modulus, where we have changed the dummy variable from **T** back to τ.

$$\therefore \quad G_1(\omega) = \omega \int_0^\infty G(\tau) \sin \omega\tau\, d\tau \qquad (5.35)$$

and

$$G_2(\omega) = \omega \int_0^\infty G(\tau) \cos \omega\tau\, d\tau \qquad (5.36)$$

where $\overset{*}{G}(\omega) = (G_r + G_1) + iG_2$. Equations (5.35) and (5.36) are one-sided Fourier transforms.

Inversion gives the stress-relaxation modulus

$$G(t) = \frac{2}{\pi} \int_0^\infty \frac{G_1(\omega)}{\omega} \sin \omega t\, d\omega \qquad (5.37)$$

and

$$G(t) = \frac{2}{\pi} \int_0^\infty \frac{G_2(\omega)}{\omega} \cos \omega t\, d\omega \qquad (5.38)$$

These equations imply a relationship between $G_1(\omega)$ and $G_2(\omega)$, the dispersion or compatibility relations which are the viscoelastic analogue of the Kramers–Krönig relations for optical dispersion and magnetic relaxation.

5.4.2 Formal Representations of the Creep Compliance and the Complex Compliance

Similar relationships hold for the creep compliance and complex compliance to those derived for the stress-relaxation modulus and the complex modulus. The details of the derivations will not be given, but the results are quoted for completeness.

(1) Creep compliance.

In this case the *rate of change* of creep compliance is expressed as a Laplace integral. Thus

$$\frac{dJ(t)}{dt} = \int_0^\infty sN(s)\, e^{-ts}\, ds \qquad (5.39)$$

where

$$N(s) = \frac{F(1/s)}{s^2}, \qquad s = \frac{1}{\tau}$$

and $F(\tau)\,d\tau$ is the distribution of retardation times. Note that $N(s) \neq \bar{N}(s)$, and that $F(\tau) \neq \bar{F}(\tau)$, i.e. the retardation-time spectrum is not identical to the relaxation-time spectrum.

(2) Complex compliance.

Here it is found that

$$J_1(\omega) = \int_0^\infty \frac{dJ(\tau)}{d\tau} \cos \omega\tau \, d\tau \qquad (5.40)$$

and

$$J_2(\omega) = - \int_0^\infty \frac{dJ(\tau)}{d\tau} \sin \omega\tau \, d\tau \qquad (5.41)$$

Again both $J_1(\omega)$ and $J_2(\omega)$ are Fourier transforms, which may be inverted to give the creep compliance in terms of the components of the complex compliance. The inversion formulae both give the creep compliance, implying a relationship between the real and imaginary parts of the complex compliance, as in the case of the complex modulus.

5.4.3   The Formal Structure of Linear Viscoelasticity

Gross[5] has discussed the formal structure of the theory of linear viscoelasticity. A summary of his treatment will be presented here, as a suitable conclusion to our discussion.

There are two groups of experiments:

Group 1: Experiments which take place under a given stress, either fixed or alternating. These define the creep compliance or the complex compliance.

Group 2: Experiments which take place under a given strain, either fixed or alternating. These define the stress-relaxation modulus or the complex modulus.

Within each group the viscoelastic functions exist in three levels:

(a) Top level	Complex Compliance	(Group 1)
	Complex Modulus	(Group 2)
(b) Medium level	Creep Function	(Group 1)
	Relaxation Function	(Group 2)
(c) Bottom level	Retardation Spectrum	(Group 1)
	Relaxation Spectrum	(Group 2)

To go *up* a level, one applies either a Laplace transform or a one-sided complex Fourier transform.

To go *down* a level, one applies either an inverse Laplace transform or an inverse Fourier transform.

The relationships between the groups vary in complexity. At the top level, the complex compliance is merely the inverse of the complex modulus. The relationships between the creep function and the relaxation function and between the retardation spectrum and the relaxation spectrum involve integral equations and integral transforms respectively.

5.5 THE RELAXATION STRENGTH

A concept which is of value in considering the relationship of viscoelastic behaviour to physical and chemical structure is that of 'relaxation strength'. In a stress-relaxation experiment the modulus relaxes from a value G_u at very short times to G_r at very long times (Figure 5.17b).

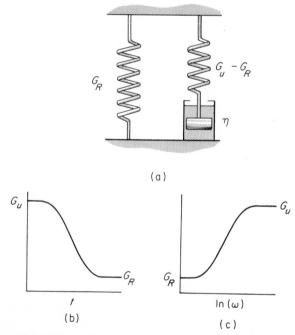

Figure 5.17. The standard linear solid (a) gives a response in a stress relaxation test shown in (b) and in a dynamic mechanical test in (c). G_u is the unrelaxed modulus and G_R the relaxed modulus.

Similarly in a dynamic mechanical experiment the modulus changes from G_r at low frequencies to G_u at very high frequencies. G_u is the unrelaxed modulus and G_r the relaxed modulus (Figure 5.17c).

This behaviour is shown by the standard linear solid of Figure 5.17a. Consider in turn the behaviour in the initial unrelaxed state, and in the final relaxed state.

(1) A total applied stress σ is given in terms of the initial unrelaxed strain e_1 by adding the stresses in both springs. Thus

$$\sigma = G_r e_1 + (G_u - G_r)e_1 = G_u e_1$$

and the initial unrelaxed strain

$$e_1 = \frac{\sigma}{G_u} = \frac{\sigma}{\text{unrelaxed modulus}}$$

(2) The final relaxed strain e_2 is given in terms of an applied stress σ by

$$e_2 = \frac{\sigma}{G_r} = \frac{\sigma}{\text{relaxed modulus}}$$

because the spring $(G_u - G_r)$ is now ineffective. The relaxation strength is conventionally defined as

$$\frac{\text{Final strain} - \text{Initial strain}}{\text{Initial strain}}$$

This is

$$\frac{e_2 - e_1}{e_1} = \left\{ \frac{1}{G_r} - \frac{1}{G_u} \right\} G_u$$

$$= \frac{G_u - G_r}{G_r} \qquad (5.42)$$

The equation for the standard linear solid in Figure 5.17a is

$$\sigma + \tau_1 \frac{d\sigma}{dt} = G_r \left\{ e + \tau_2 \frac{de}{dt} \right\} \qquad (5.43)$$

where

$$\tau_1 = \frac{\eta}{G_u - G_r} \quad \text{and} \quad \tau_2 = \frac{\tau_1 G_u}{G_r}$$

Then for dynamic mechanical measurements it may be shown that

$$G_1(\omega) = \frac{G_r(1+\omega^2\tau_1\tau_2)}{1+\omega^2\tau_1^2} = \frac{G_r+\omega^2\tau^2 G_u}{1+\omega^2\tau^2} \tag{5.44}$$

$$G_2(\omega) = \frac{G_r(\tau_2-\tau_1)\omega}{1+\omega^2\tau_1^2} = \frac{(G_u-G_r)\omega\tau}{1+\omega^2\tau^2} \tag{5.45}$$

and

$$\tan\delta = \frac{(\tau_2-\tau_1)\omega}{1+\omega^2\tau_1\tau_2} = \frac{(G_u-G_r)\omega\tau}{G_r+\omega^2\tau^2 G_u} \tag{5.46}$$

where we have put

$$\tau = \tau_1 = \frac{\eta}{G_u-G_r}$$

The following relationships then hold:

$$\tan\delta_{max}\ (\omega^2\tau^2 = G_r/G_u) = \frac{G_u-G_r}{2\sqrt{G_uG_r}} \tag{5.47}$$

$$G_{2_{max}}(\omega^2\tau^2 = 1) = \frac{G_u - G_r}{2} \tag{5.48}$$

$$\int_{-\infty}^{\infty} G_2(\omega)\,\mathrm{d}\ln(\omega) = \frac{\pi}{2}(G_u-G_r) \tag{5.49}$$

$$\int_{-\infty}^{\infty} \tan\,\mathrm{d}\,(\ln\omega) = \frac{\pi}{2}\frac{(G_u-G_r)}{\sqrt{G_uG_r}} \tag{5.50}$$

All the relationships defined above are proportional to $G_u - G_r$ and hence to the relaxation strength. $\tan\delta_{max}$ and $\int_{-\infty}^{\infty}\tan\delta\,\mathrm{d}\ln(\omega)$ are closest to our original definition in being normalized to a dimensionless quantity. This provides some formal justification for the use of $\tan\delta$ rather than G_2 for estimating the relaxation strength to correlate with structural parameters[6].

REFERENCES

1. L. Boltzmann, *Pogg. Ann. Phys. U. Chem.*, 7, 624 (1876).
2. E. H. Lee, *Proceedings of the First Symposium on Naval Structural Mechanics*, Pergamon Press, Oxford, 1960, p. 456.
3. C. Zener, *Elasticity and Anelasticity of Metals*, Chicago University Press, 1948.
4. T. Alfrey, *Mechanical Behaviour of High Polymers*, Interscience Publishers, New York, 1948.
5. B. Gross, *Mathematical Structure of the Theories of Viscoelasticity*, Hermann, Paris, 1953.
6. R. W. Gray and N. G. McCrum, *J. Polymer Sci.*, B6, 691 (1968).

6

The Measurement of Viscoelastic Behaviour

Experimental studies of viscoelasticity in polymers are extremely extensive, and a very large number of techniques have been developed. In this chapter we will only attempt to indicate the types of methods which are available, together with a few representative examples of actual experimental arrangements. The reader is referred to standard texts on viscoelastic behaviour[1,2] and review articles[3-5] for more detailed expositions of the subject.

To obtain a satisfactory understanding of the viscoelastic behaviour, data are required over a wide range of frequency (or time) and temperature. In Chapter 5, the equivalence of creep, stress-relaxation and dynamic mechanical data has been described. In Chapter 7 the equivalence of time and temperature as variables will be discussed. Although this equivalence can sometimes reduce the required range of experimental data, it is desirable in principle to be able to cover wide ranges of both time and temperature. This can only be done by combining a wide variety of techniques, the approximate time scales of which are shown in Figure 6.1.

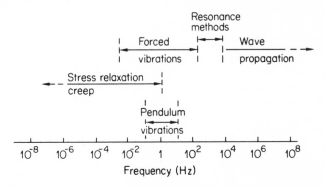

Figure 6.1. Approximate frequency scales for different experimental techniques (after Becker).

The techniques fall into five main classes:
1. Transient measurements: creep and stress relaxation.
2. Low-frequency vibrations: free oscillation methods.
3. High-frequency vibrations: resonance methods.
4. Forced-vibration non-resonance methods.
5. Wave-propagation methods.

These will now be discussed in turn.

6.1 CREEP AND STRESS RELAXATION

The transient methods of measuring viscoelastic behaviour, creep and stress relaxation, cover the frequency range from ~ 1 Hz to very low frequencies. These methods are also most effective in revealing the essential nature of the non-linear viscoelasticity which is typical of most polymers at other than very low strain levels.

6.1.1 Creep: Conditioning

To undertake steady-state quantitative measurements of creep and recovery under various loading conditions (i.e. measurements which are equivalent to dynamic mechanical measurements), a conditioning procedure is essential, as emphasized by Leaderman[6]. The specimen should be subjected to successive creep and recovery cycles, each cycle consisting of application of the maximum load for the maximum period of loading, followed by a recovery period after unloading of about 10 times the loading period. For creep tests over a range of temperatures, this conditioning should be carried out at the highest temperature of measurement.

The conditioning procedure has two major effects on the creep and recovery behaviour. First, subsequent creep and recovery responses under a given load are then identical, i.e. the sample has lost its 'long-time' memory and now only remembers loads applied in its immediate past history. Secondly, after the conditioning procedure the deformation produced by any loading programme is almost completely recoverable provided that the recovery period is about 10 times the period during which loads are applied. For tensile creep measurements over a wide range of temperatures greater elaboration is required.

6.1.2 Extensional Creep

For simplicity of interpretation, creep measurements require a dead-loading procedure. The most elementary procedure is to use a

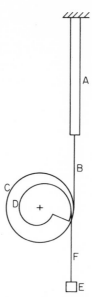

Figure 6.2. Cam arrangement for creep under constant stress (after Leaderman).

cathetometer or travelling microscope to measure the creep between two ink marks on the sample, which eliminates end effects at the clamps.

For accuracy at other than low-strain levels, there should be a device for reducing the load in a manner proportional to the decrease in cross-sectional area, in order to maintain a constant stress. The system adopted by Leaderman[7] is shown in Figure 6.2.

The cylindrical drum C and the specially shaped cam D are attached to a shaft supported in cone bearings. The upper end of the specimen A is fixed, and the lower end attached to the drum by means of a thin flexible steel tape B. One end of a similar tape F is attached to the cam; the other end of this tape supports a constant weight E. A balance arm is used to set the centre of gravity of the rotating system at the axis of the shaft. As the specimen extends under load, the moment arm of the applied weight decreases according to the cam profile.

6.1.3 Extensometers

A more satisfactory technique for high accuracy is to use an extensometer attached directly to the specimen, the strain being converted either into rotation of mirrors or into an electrical signal by a displacement trans-ducer. Dunn, Mills and Turner[8] describe two types of extensometer which

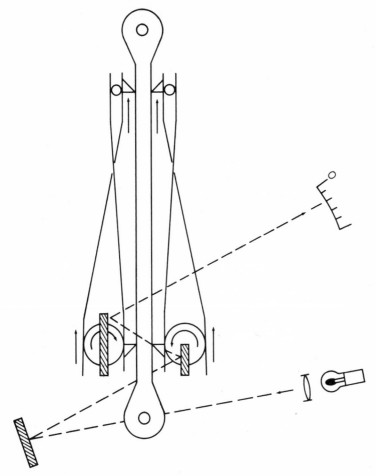

Figure 6.3. Schematic diagram of modified Lamb extensometer (after Dunn, Mills and Turner).

have been used for polymers of high and medium stiffness respectively. For a polymer of high stiffness, the modified Lamb extensometer shown in Figure 6.3 was found satisfactory. A long rectangular prism of polymer (cross-section 0·55 cm × 0·32 cm) is gripped between two pairs of knife edges. The knife edges are attached to four main members and movement takes place by rotation of two steel balls at the top and by rotation of two rollers at the bottom. Mirrors attached to the rollers form part of an optical system so that extension of the specimen is converted by rotation of the rollers into movement of a light beam. A typical deflection of the light beam is 0·26 radians for a 1% strain, i.e. ~13 cm displacement of a light spot on a scale at 50 cm distance.

For specimens of lower stiffness, the Lamb extensometer may distort the specimen at the knife edges. The weight of the extensometer may also cause extension of the specimen. For these reasons an optical lever extensometer was developed, as shown in Figure 6.4 (see Plate II). The two forked arms B possess steel pins which pass through $\frac{3}{64}$-in diameter holes in the specimen. Mirrors attached to these forked arms form part of an optical system so that the extension of the specimen is again converted into the movement of an optical beam. Sensitivities are quoted of ~6 cm deflection of the beam for 1% strain.

6.1.4 High-Temperature Creep

An apparatus designed to measure creep with reasonable accuracy[9] over a wide temperature range (20°c–200°c) is shown in Figure 6.5. A heavy monofilament of 0·100 in diameter is supended from a micrometer head

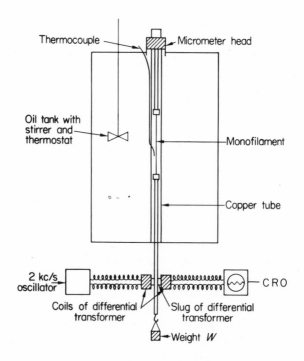

Figure 6.5. High-temperature creep apparatus.

inside a copper tube which is surrounded by an oil bath, the temperature of which is thermostatically controlled to better than $+0.1°C$. The mono-filament is attached to long clamps (~ 12 in) at its extremities so that its entire length is well within the controlled temperature volume. The clamp at the lower end of the monofilament is attached via the slug of a differential transformer to the load W. The total weight of slug and clamp is about 20 g, and at each temperature at least 20 hr are allowed for the sample to creep under this small load. The subsequent creep under this residual load is thus small compared with that due to the larger applied load. The creep is measured by continuously adjusting the micrometer so as to position the slug in the centre of the differential transformer. This gives a minimum output voltage from the transformer (which is observed on the oscilloscope). The input to the transformer comes from a 2 kHz oscillator via suitable matching impedances. The monofilament is typically about 12 inches in length and the micrometer can be read to ± 0.0005 in.

This apparatus was used to undertake creep measurements at low strain levels. There was therefore no attempt to correct for the change in cross-sectional area under load.

6.1.5 Torsional Creep

The measurement of creep in torsion can be made with very great accuracy. This is because the deformation can be used to cause rotation of a mirror directly and thus give large deflection of a light beam. McCrum and his collaborators, for example, have used a torsional creep apparatus of the type shown in Figure 6.6 to make very accurate torsional creep measure-ments on polyethylene[10].

6.1.6 Stress Relaxation

The measurement of stress relaxation can in principle most simply be made by placing the polymer specimen in series with a spring of sufficiently great stiffness to undergo negligible deformation compared with the specimen. The spring may be the elements of a resistance strain-gauge transducer, enabling a direct measurement of stress as in the Instron and Hounsfield tensometer, or it may be incorporated in a differential trans-former so that its displacement again records the stress. For mathe-matical simplicity it is desirable to use a step function of strain, which requires very rapid application of strain if measurements are to be made at short times. An apparatus for stress-relaxation measurements[11] is shown in Figure 6.7.

A more elaborate technique is to maintain the specimen at constant length by continuously adjusting the required force as the stress relaxes. This has been done by Stein and Schaevitz[12] using an automatic servo-mechanism.

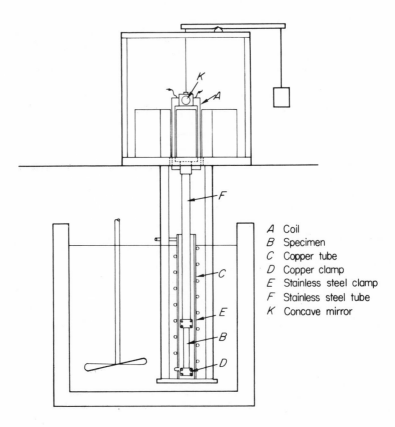

Figure 6.6. Apparatus used for the measurement of creep and stress relaxation (after McCrum and Morris).

It is also worth noting that the stress-relaxation modulus can be obtained from constant rate of strain experiments. This principle has been used by Smith for elastomers[13] and because it involves non-linear viscoelastic behaviour is described in detail in Chapter 9.

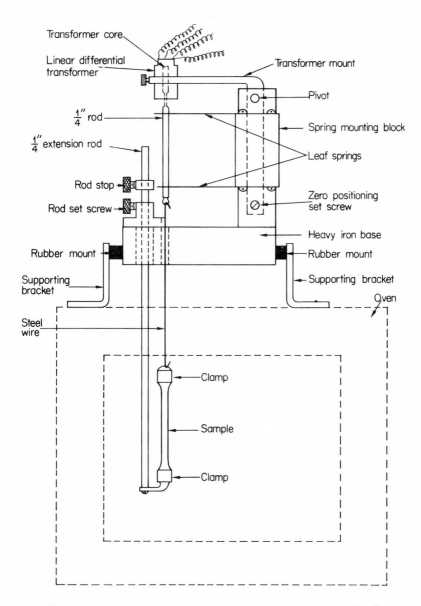

Figure 6.7. Apparatus for measuring stress relaxation in extension by the changes in position of a spring whose stiffness is much greater than that of the sample (after McLoughlin and Ferry).

6.2 DYNAMIC MECHANICAL MEASUREMENTS: THE TORSION PENDULUM

One of the simplest and best-known techniques for making dynamic mechanical measurements is the torsion pendulum. The frequency range of operation is 0·01 Hz to 50 Hz, the upper limit being set by the dimensions of the specimen becoming comparable to the wavelength of the stress waves in the specimen.

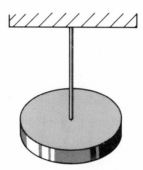

Figure 6.8. The simple torsion pendulum.

A simple torsion pendulum is shown in Figure 6.8. The specimen is a cylindrical rod of polymer, one end of which is rigidly clamped, the other end supporting an inertia disc. The system is set into oscillation and undertakes damped sinusoidal oscillations.

First let us consider the case of an *elastic* rod. The equation of motion is $M\ddot{\theta} + \tau\theta = 0$ where M is the moment of inertia of the disc and τ is the torsional rigidity of the rod, which is related to the torsional modulus G, of the rod by the equation

$$\tau = \frac{G\pi r^4}{2l}$$

(l = length of rod, r = radius of rod)

The system executes simple harmonic motion with a frequency given by

$$\omega = \sqrt{\frac{\tau}{M}} = \sqrt{\frac{\pi r^4 G}{2lM}}$$

Thus G can be found directly from the frequency of oscillation, M being determined separately.

The effect of the viscoelastic behaviour of the polymer is to introduce a damping term proportional to $\dot{\theta}$ into the equation of motion, giving an equation of the form

$$a\ddot{\theta} + b\dot{\theta} + c\theta = 0$$

This has the general solution

$$\theta = A \exp\frac{-b+\sqrt{b^2-4ac}}{2a}t + B \exp\frac{-b-\sqrt{b^2-4ac}}{2a}t$$

For small damping $b^2 < 4ac$ and the motion is oscillatory. A solution is found by taking $\theta = 0$ and $\dot{\theta}$ = constant when $t = 0$.
Then

$$\theta = \theta_0 \exp\frac{-bt}{2a}\sin\omega_1 t \qquad (6.1)$$

where

$$\omega_1 = \frac{b^2-4ac}{2a}$$

This solution represents a damped oscillatory motion. It is customary to describe the damping by what is termed the 'logarithmic decrement' Λ. This is the natural logarithm of the ratio of the amplitude of successive oscillations, and Equation (6.1) is then written as

$$\theta = \theta_0 \exp\left(\frac{-\omega_1\Lambda}{2\pi}t\right)\sin\omega_1 t \qquad (6.2)$$

For a linear viscoelastic solid the torsional modulus is a complex quantity and may be written as $\overset{*}{G} = G_1 + iG_2$. The equation of motion for the torsion pendulum may then be written as

$$M\ddot{\theta} + \frac{\pi r^4}{2l}(G_1 + iG_2)\theta = 0 \qquad (6.3)$$

Assuming a solution of the form

$$\theta = \theta_0 \exp\left(\frac{-\omega\Lambda}{2\pi}t\right)\sin\omega t$$

substituting and equating real and imaginary parts we have

$$\frac{\pi r^4}{2l}\frac{G_1}{M} = \omega^2 \qquad \text{for small damping} \qquad (6.4)$$

and

$$\frac{\pi r^4}{2l}\frac{G_2}{M} = \omega^2 \frac{\Lambda}{\pi} \tag{6.5}$$

or

$$\frac{G_2}{G_1} = \frac{\Lambda}{\pi} = \tan \delta \tag{6.6}$$

Thus the frequency of the oscillation gives the real part of the modulus G_1 and measurement of the logarithmic decrement gives $\tan \delta$ and G_2.

A simple torsion pendulum with an inertia disc mounted directly on the end of the specimen was used by Schmieder and Wolf[14]. In a slightly more elaborate arrangement (Figure 6.9) the specimen is supported by a fine wire or ribbon (or by a counterbalancing weight[15]). This enables measurements to be made in temperature ranges where the weight of the inertia disc would cause extensional creep. The equation of motion for the compound pendulum is a simple extension of Equation (6.3) adding a

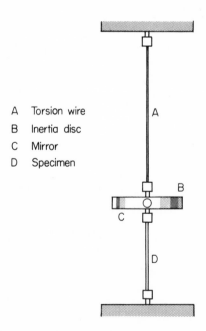

A   Torsion wire
B   Inertia disc
C   Mirror
D   Specimen

Figure 6.9. Apparatus for measuring torsional rigidity at low frequencies.

term ($\tau'\theta = (\pi r'^4/2l')G'\theta$ for a circular wire) from the supporting wire or ribbon. The additional term will not contribute to the damping as the supporting wire can be considered to be elastic.

The polymer specimen is often in the form of a rectangular block or thin sheet. The term $(\pi r^4/2l)\overset{*}{G}$ has then to be replaced by the St. Venant formula for the torsion of rectangular beams. (See Chapter 10.)

Figure 6.9 shows a typical experimental scheme for measuring the behaviour of a fibre monofilament. The frequency and logarithmic decrement are determined by observing the motion of a light beam reflected from the mirror.

Figure 6.10 (see Plate III) shows the apparatus with heating and cooling facilities. The image of the light source falls on the photodyne recorder which plots out the damped sinusoidal oscillations.

6.3   RESONANCE METHODS

At higher frequencies the wavelength of the stress waves decreases until it becomes comparable to the dimensions of the specimen. Assuming a Young's modulus of 10^8 dyn cm^{-2} and a density of 1 g cm^{-3} gives a longitudinal wave velocity of 10^4 cm sec^{-1}. At a frequency of 10^3 Hz this gives a wavelength of 10 cm. Thus at higher frequencies the specimens becoming vibrating systems of standing waves with resonances, and from the variation in amplitude of vibration with frequency we can determine the real and imaginary parts of the complex modulus. The frequency range of these methods is clearly somewhat restricted.

6.3.1   The Vibrating-Reed Technique

The vibrating reed or cantilever is a very popular technique which in many industrial laboratories provides a standard method for polymer evaluation[16,17,18]. The small specimen (Figure 6.11) is clamped at one end in a gramophone cutter head which is driven from a variable-frequency oscillator (at a frequency typically in the range 200–1500 Hz). The light source illuminates a rectangular slit, the image of which is focussed on the tip of the specimen. The motion of the specimen modulates the intensity of the light falling on the photocell and must be arranged so that the specimen is never totally in nor totally out of the illuminated region. The signal from the photocell is amplified and monitored on an oscilloscope for correct alignment of the optics, and its amplitude is measured by a meter. The amplitude of vibration, i.e. the resonance

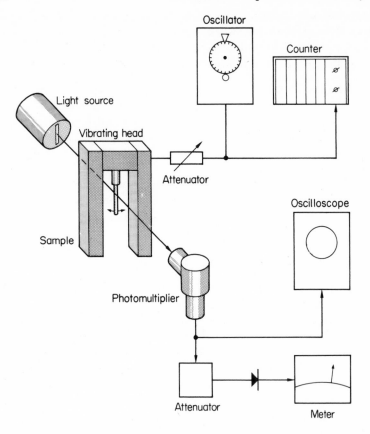

Figure 6.11. Schematic diagram of the vibrating reed apparatus.

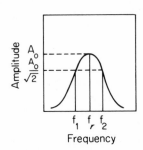

Figure 6.12. Resonance curve for the vibrating reed.
$(f_2 - f_1 = \Delta f$, see text.)

curve, is measured as a function of frequency, at fixed temperatures and a curve of the form shown in Figure 6.12 is obtained.

A complete treatment of the viscoelastic behaviour of the vibrating reed has been given by Bland and Lee[19]. It is, however, satisfactory for our purposes to assume that the losses are small (in practice this means $\tan \delta < 0.2$). The 'modulus' ($|E|$) can then be obtained from the solution to the *elastic* problem and $\tan \delta$ from the resonance curve as follows.

Figure 6.13. The vibrating reed.

The equation of motion of an *elastic* cantilever is (for a detailed discussion see Reference 20),

$$\frac{\partial^2 y}{\partial t^2} + \frac{E}{\rho} k^2 \frac{\partial^4 y}{\partial x^4} = 0 \tag{6.7}$$

where

E = Young's modulus of cantilever
ρ = density

k^2 = radius of gyration = $\dfrac{h^2}{12}$

where h is the thickness of the cantilever, and y is the displacement from its equilibrium position of a point on the reed at a distance x from the fixed end (Figure 6.13).

The general solution to Equation (6.7) is

$$y = \{A \cosh mx + B \sinh mx + C \cos mx + D \sin mx\} \cos \omega t \tag{6.8}$$

where

$$m = \left\{ \frac{\rho \omega^2}{E k^2} \right\}^{1/4}$$

ω = angular frequency of vibration in radians $\sec^{-1}$

The boundary conditions of the vibrating reed are $y = dy/dx = 0$ for all t at the clamped end where $x = 0$ and $d^2y/dx^2 = d^3y/dx^3 = 0$ for all t at the free end. (No couple and no shearing force.)

Substituting these boundary conditions in turn into the general solution (Equation 6.8) gives two simultaneous equations which are compatible if

$$\cosh ml \cos ml = -1 \tag{6.9}$$

where l is the length of the cantilever. The first mode solution of Equation (6.9) is $ml = 1 \cdot 875$. The angular frequency at resonance ω_r is given by

$$\omega_r = \frac{(1 \cdot 875)^2}{l^2} \sqrt{\frac{Ek^2}{\rho}} \tag{6.10}$$

The resonant frequency f_r is then

$$f_r = \frac{1}{2\pi} \frac{(1 \cdot 875)^2}{l^2} \sqrt{\frac{E}{\rho}} \frac{h}{2\sqrt{3}} \tag{6.11}$$

giving

$$E = 38 \cdot 24 \frac{l^4 \rho}{h^2} f_r^2 \tag{6.12}$$

It may also be shown, mostly readily on the basis of the analogy between the viscoelastic reed and an equivalent electrical circuit[16], that

$$\tan \delta = \frac{\Delta f}{f_r} \tag{6.13}$$

where Δf is the band width, or the difference between those frequencies for which the amplitude of vibration has $1/\sqrt{2}$ times its maximum value.

The vibrating-reed method is most often used for very small polymer samples moulded to give an isotropic specimen. It is, however, equally applicable to oriented specimens, for example, by cutting thin strips from oriented sheets.

Figure 6.14 (see Plates IVa and IVb) shows a typical experimental set-up, where the reed is enclosed in a cryostat and measurements can be carried out over a wide temperature range ($-150°$c to $+200°$c).

6.4 FORCED-VIBRATION NON-RESONANCE METHODS

The free-vibration and resonance methods are simplest and most accurate for determining the dynamic mechanical behaviour over a wide range of temperatures. They suffer from the considerable disadvantage that the frequency of measurement depends on the stiffness of the specimen, and as this changes with temperature so does the effective frequency. Thus to determine the frequency and temperature dependence of viscoelastic behaviour, forced-vibration non-resonance methods are preferable.

6.4.1   Measurement of Dynamic Extensional Modulus

The principle of this technique is to apply a sinusoidal extensional strain to the specimen which takes the form of a thin strip of film, a fibre mono-filament or a multifilament yarn, and simultaneously measure the stress. The viscoelastic behaviour is specified from the relative amplitudes of the stress and the strain, and from the phase shift between these (see Section 5.3.1 above).

There are two limitations to design and measurement respectively which are of some importance. First, the length of the sample must be short enough for there to be no appreciable variation of stress along the sample length, i.e. the length of the fibre must be short compared with the wavelength of the stress waves. Assuming the lowest value of the modulus to be measured is 10^8 dyn cm^{-2} and a specimen density of 1 g cm^{-3}, the longitudinal wave velocity is 10^4 cm sec^{-1}. At a frequency of 100 Hz the wavelength of the stress waves is then 100 cm. This suggests an upper limit of approximately 10 cm length on specimens to be measured at 100 Hz.

Secondly, there is the limit imposed by the stress-relaxation time of the material, it being clear that the stress developed in the sample must never vanish.

Several types of apparatus of this general form have been designed[5,21–24].

A block diagram of that due to Takayanagi[22] is shown in Figure 6.15. The specimen, which is usually from 0·1 to 0·3 mm in width and from 4 to 6 cm in length, undergoes an alternating extensional strain due to the electromagnetic vibrator in the frequency range 3·5–100 Hz. The strain and stress are measured by unbonded strain-gauge transducers, the signals from which are fed via additional amplifiers to a phase meter. This provides a direct reading of the relative amplitudes and phase difference of the signals from the two transducers and hence values of the modulus and tan δ. The specimen is enclosed in a thermostatted enclosure so that measurements can be made over the temperature range $-170°$c to $+150°$c.

The range of measurement has been extended to 10^{-3} Hz by an apparatus which is identical in principle but uses a mechanical drive to produce the alternating strain (range 10–10^{-3} Hz) and a Servomex

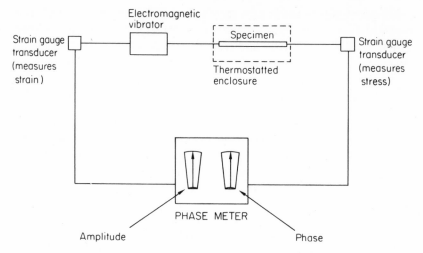

Figure 6.15. Block diagram of forced-vibration apparatus.

Transfer Function Analyser (TFA) to make simultaneous comparison of stress and strain[24]. A schematic diagram of this apparatus is shown in Figure 6.16.

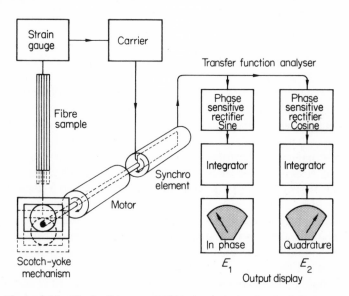

Figure 6.16. Block diagram of TFA dynamic mechanical apparatus (after Pinnock and Ward).

The servo-controlled motor, in addition to providing the sinusoidal strain by driving the Scotch-yoke mechanism, rotates a loop in a high frequency (2 kHz) alternating magnetic field. Since the loop rotates at a very low frequency (ω_m) this produces a carrier wave modulated at the rotational speed, which can be fed to the system under test. In this application the mechanical signal is used, and consequently the return signal presented to the measuring system by the transducer which measures the stress will be of the form $K_1 \sin(\omega_m t + \delta) + N$ where δ is the phase shift (stress leads strain) and N represents unwanted elements such as harmonics, random noise, etc. At this stage a carrier is introduced and the signal is then multiplied simultaneously but separately by $\sin \omega_m t$ and $\cos \omega_m t$. These two signals are subsequently demodulated and integrated over a complete cycle to produce terms which are proportional in magnitude and polarity to the real and imaginary parts of the system output at the fundamental frequency, i.e. to E_1 and E_2.

A photograph of this apparatus is shown in Figure 6.17 (see Plate V). The sample is heated or cooled by means of a flow of air or dry nitrogen respectively. (Temperature range $-150°$c to $+200°$c). To isolate the sample thermally it is connected to the Scotch yoke and the resistance strain gauge at its two extremities by German-silver rods and surrounded by a cylindrical copper can which fits tightly into a thermos tube. The temperature is determined by a series of thermocouples. The air is heated by passing through a hot wire spiral and the nitrogen is cooled by use of a German-silver heat-exchanger cryostat. With these arrangements temperatures can be controlled to about $\pm 1°$c.

A fixed alternating strain of 0·5% ($\pm 0·25$ mm on a sample length of 10 cm) is typical, being close to the highest strain level at which linear viscoelastic behaviour is observed. The sample is also subjected to a static strain greater than 0·25% to prevent it becoming slack.

6.5 WAVE-PROPAGATION METHODS

In the frequency range 10^3–10^4 Hz, the wavelength of the stress waves is the same order of magnitude as the length of sample. A typical stress-wave propagation method[25,26] is shown diagrammatically in Figure 6.18. The polymer is in the form of a monofilament (diameter $\sim 0·010$ in) and is attached to a stiff massive diaphragm (say a loudspeaker) with a quartz piezo-electric crystal which detects the signal amplitude and phase at a variable distance along its length.

Let the displacement of the diaphragm $u_s = A \sin \omega t$ and the displacement of the pick-up be $u_p = B \sin (\omega t + \theta)$.

There are two quantities to measure; first the ratio B/A and secondly the phase difference θ. In the first instance these can be related to the

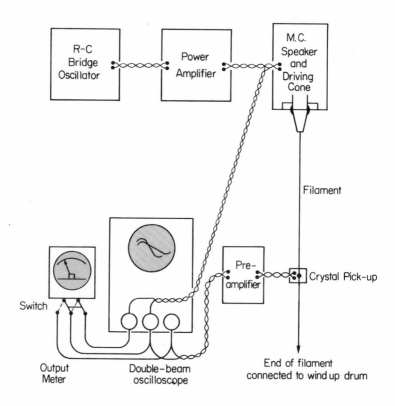

Figure 6.18. The sonic modulus apparatus of Hillier and Kolsky[25].

position of the pick-up l, the reflexion coefficient at the pick-up m, the propagation constant $k = \omega/c$ (c = longitudinal wave velocity along the filament), and the damping or attenuation coefficient.

In the steady state the displacement at any point can be represented by the summation of a single progressive wave travelling down the filament and a single reflected wave from the pick-up travelling back. It is important to note that the filament is assumed to be sufficiently long to ignore reflections from its far end.

The total displacement at any point x is then given by

$$u_T = u_1 + u_2$$

$$= a\,e^{-\alpha x}\sin(\omega t - kx)$$

$$+ ma\,e^{-\alpha(2l-x)}\sin[\omega t - k(2l-x)] \tag{6.14}$$

Substituting $x = 0$ and $x = l$ into this equation gives us equations for θ and A/B in terms of l, m, k and α, from which it can be shown that

$$\tan\theta = \left\{\frac{1 + m\,e^{-2\alpha l}}{1 - m\,e^{-2\alpha l}}\right\}\tan kl \tag{6.15}$$

For large αl, we note that $\theta = kl$. A typical result for low-density polyethylene is shown in Figure 6.19, and it can be seen that a plot of θ against l takes the form of damped oscillations about the line $\theta = kl$, the slope giving the value of k, the propagation constant.

For small α the amplitude–distance relation is of the form shown in Figure 6.20. It can be shown[4] that to a good approximation $V_{max}/V_{min} = \tanh(\alpha l + \beta)$ where V is the amplitude of the signal received. Thus a plot of $\tanh^{-1}(V_{max}/V_{min})$ against l gives us a line of slope α, the attenuation constant.

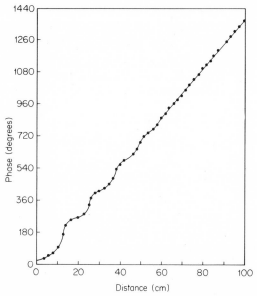

Figure 6.19. The variation of phase angle with distance along a poly-ethylene monofilament for transmission of sound waves at 3000 Hz (after Hillier and Kolsky).

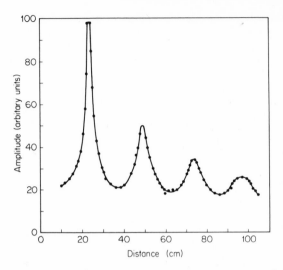

Figure 6.20. The amplitude–distance relationship for transmission of sound waves along a polyethylene monofilament at 3000 Hz (after Hillier and Kolsky).

The second part of the exercise is to relate the propagation constant k and the attenuation coefficient α to E_1, E_2 and $\tan \delta$. It is instructive physically to introduce the phase velocity of propagation c. Then any displacement can be written as

$$u = u_0 \exp \left[-\alpha x + i\omega \left(t - \frac{x}{c} \right) \right] \qquad (6.16)$$

and strain

$$e = \frac{\partial u}{\partial x} = - \left[\alpha + \frac{i\omega}{c} \right] u \qquad (6.17)$$

But stress

$$\sigma = (E_1 + iE_2)e \qquad (6.18)$$

where

$$E_1 = \frac{\sigma_0}{e_0} \cos \delta, \qquad E_2 = \frac{\sigma_0}{e_0} \sin \delta$$

$$\tan \delta = E_2/E_1$$

Newton's Second Law applied to the extensional motion of a rod parallel to the x axis gives

$$\rho \frac{\partial^2 u}{\partial t^2} = \frac{\partial \sigma}{\partial x} \qquad (6.19)$$

Equation (6.18) is substituted into (6.19) and then (6.16) and (6.17) are used. Equating real and imaginary parts and assuming α is small it is found that,

$$E_1 = c^2 \rho = \frac{\omega^2}{k^2} \rho \qquad (6.20)$$

and

$$\alpha = \frac{\omega}{2c} \frac{E_2}{E_1} = \frac{\omega}{2c} \tan \delta \qquad (6.21)$$

or

$$E_2 = \frac{2\alpha c}{\omega} E_1 = \frac{2\alpha c^3 \rho}{\omega} \qquad (6.22)$$

Figure 6.21 shows the propagation and attenuation constants for polyethylene as a function of frequency at room temperature. The velocity

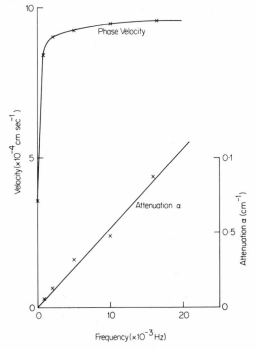

Figure 6.21. Experimental measurements of attenuation and velocity in polyethylene filaments at 10°c (after Kolsky).

is approximately constant; this shows that E_1 is independent of frequency (Equation 6.20). The attenuation, however, rises approximately linearly with frequency; this shows that tan δ is independent of frequency to the same degree of approximation (Equation 6.21).

There have been many other wave propagation experiments, and the reader is referred to review articles 27–29 for further information.

REFERENCES

1. J. D. Ferry, *Viscoelastic Properties of Polymers*, 2nd ed., Wiley, New York, 1970.
2. N. G. McCrum, B. E. Read and G. Williams, *Anelastic and dielectric effects in polymeric solids*, Wiley, New York, 1967.
3. H. Kolsky, *Mechanics and Chemistry of Solid Propellants, Proceedings of the Fourth Symposium on Naval Structural Mechanics*, Pergamon Press, New York, 1966, p. 357.
4. K. W. Hillier, *Progress in Solid Mechanics*, North Holland Publishing Company, Amsterdam, 1961, Chapter 5.
5. G. W. Becker, *Materie Plastiche ed Elastomeri*, **35**, 1387 (1969).
6. H. Leaderman, *Elastic and Creep Properties of Filamentous Materials and Other High Polymers*, Textile Foundation, Washington D.C., 1943.
7. H. Leaderman, *Trans. Soc. Rheol.*, **6**, 361 (1962).
8. C. M. R. Dunn, W. H. Mills and S. Turner, *British Plastics*, **37**, 386, 1964.
9. I. M. Ward, *Polymer*, **5**, 59 (1964).
10. N. G. McCrum and E. L. Morris, *Proc. Roy. Soc.*, **A281**, 258 (1964).
11. J. R. McLoughlin, *Rev. Sci. Instr.*, **23**, 459 (1952).
12. R. S. Stein and H. Schaevitz, *Rev. Sci. Inst.*, **19**, 835 (1948).
13. T. L. Smith, *Trans. Soc. Rheol.*, **6**, 61 (1962).
14. K. Schmieder and K. Wolf, *Kolloidzeitschrift*, **127**, 65 (1952).
15. J. Koppelman, *Kolloidzeitschrift*, **144**, 12 (1955).
16. A. W. Nolle, *J. Appl. Phys.*, **19**, 753 (1948).
17. M. Horio and S. Onogi, *J. Appl. Phys.*, **22**, 977 (1951).
18. D. W. Robinson, *J. Sci. Inst.*, **2**, 32 (1955).
19. D. R. Bland and E. H. Lee, *J. Appl. Phys.*, **26**, 1497 (1955).
20. K. W. Hillier, *Proc. Phys. Soc.*, **B64**, 998 (1951).
21. A. B. Thompson and D. W. Woods, *Trans. Faraday Soc.*, **52**, 1383 (1956).
22. M. Takayanagi, 'Proceedings of Fourth International Congress on Rheology', Part 1, Interscience Publishers, New York, 1965, p. 161.
23. P. R. Pinnock and I. M. Ward, *Proc. Phys. Soc.*, **81**, 261 (1963).
24. P. R. Pinnock and I. M. Ward, *Polymer*, **7**, 255 (1966).
25. K. W. Hillier and H. Kolsky, *Proc. Phys. Soc.*, **B62**, 111 (1949).
26. J. W. Ballou and J. C. Smith, *J. Appl. Phys.*, **20**, 493 (1949).
27. H. Kolsky, *Applied Mechanics Reviews*, Vol. 11, No. 9, 1958.
28. H. Kolsky, *International Symposium on Stress Wave Propagation in Materials*, Interscience Publishers, New York, 1960, p. 59.
29. H. Kolsky, *Structural Mechanics*, Pergamon Press, Oxford, 1960, p. 233.

7

Experimental Studies of the Linear Viscoelastic Behaviour of Polymers

7.1 GENERAL INTRODUCTION

An introduction to the extensive experimental studies of linear visco-elastic behaviour in polymers falls conveniently into three parts, in which amorphous polymers, temperature dependence and crystalline polymers are discussed in turn.

7.1.1 Amorphous Polymers

Most of the earlier investigations of linear viscoelastic behaviour in polymers were confined to studies of amorphous polymers. This is because amorphous polymers show more distinct changes in viscoelastic behaviour with frequency (and as we will shortly discuss, temperature) than crystalline polymers. The most extensive attempt to obtain a complete range of viscoelastic data on an amorphous polymer was in the case of polyisobutylene $\left[-CH_2-\underset{\underset{}{|}}{\overset{(CH_3)^2}{C}}- \right]_n$. R. S. Marvin of the National Bureau of Standards, Washington, collected and analysed results from many laboratories to obtain the complex shear modulus and complex shear compliance of a high-molecular weight sample of this polymer over a very wide range of frequencies[1]. The results are shown in Figures 7.1a and b, where in addition to the experimental points, calculated curves are shown based on the phenomenological theories of Marvin and Oser[2].

These figures show clearly the four characteristic regions for amorphous high polymers, i.e. the glassy, viscoelastic, rubbery and flow regions. It can be seen that the viscoelastic behaviour is not very dissimilar from that of the standard linear solid. The complex modulus at high frequencies is approximately constant at a value of about 10^{10} dyn/cm^2, and decreases through the viscoelastic range to a value of 10^6 dyn/cm^2. In this

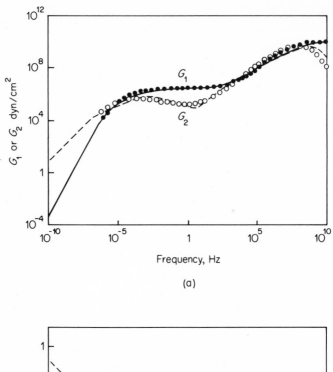

(a)

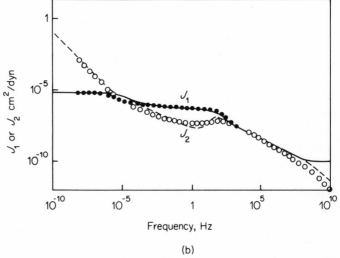

(b)

Figure 7.1. (a) Complex shear modulus, and (b) complex shear compliance for 'standard' polyisobutylene reduced 25°C. Points from averaged experimental measurements, curves from a theoretical model for viscoelastic behaviour (after Marvin and Oser).

high-molecular weight sample it remains approximately constant at this value of 10^6 dyn/cm^2 over the frequency range 10^2–10^{-2} Hz, and molecular flow only sets in appreciably at frequencies below 10^{-5} Hz.

The relaxation spectrum and retardation spectrum for the polymer are shown in Figure 7.2. The relaxation spectrum has the typical 'wedge and

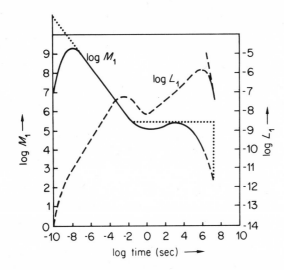

Figure 7.2. Approximate distribution functions of relaxation (M_1) and retardation (L_1) times for polyisobutylene (after Marvin).

box' form to which it approximates in the manner indicated by the dotted lines. This approximation has been found useful for purposes of numerical evaluation[3].

For amorphous polymers large changes in viscoelastic behaviour may be brought about by the presence or absence of chemical cross-links or by changing the molecular weight which controls the degree of molecular entanglement or physical cross-linking.

The influence of chemical or physical cross-links is two-fold. First, chemical cross-links prevent irreversible molecular flow at low frequencies (or, as will be discussed shortly, high temperatures) and thereby produce the rubbery plateau region of modulus or compliance. Physical cross-links due to entanglements will restrict molecular flow by causing the formation of temporary networks. At long times such physical entanglements are usually labile and lead to some irreversible flow.

Secondly, the value of the modulus in the plateau region is directly related to the number of effective cross-links per unit volume; this follows from the molecular theory of rubber elasticity (Section 4.1.2 above).

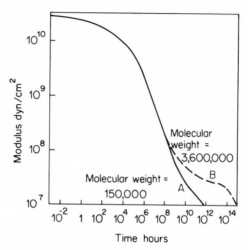

Figure 7.3. Master stress–relaxation curves for low molecular weight (molecular weight = 150,000), curve A, and high molecular weight (molecular weight = 3,600,000), curve B, polymethylmethacrylate (after McLoughlin and Tobolsky).

The influence of molecular entanglements is illustrated by Figure 7.3, which shows the stress-relaxation behaviour for two samples of poly-methylmethacrylate. It is seen that the lower molecular weight sample does not show a plateau region of modulus.

7.1.2   Temperature Dependence of Viscoelastic Behaviour

Previously we have only referred indirectly to the effect of temperature on viscoelastic behaviour. From a practical viewpoint, however, the temperature dependence of polymer properties is of paramount importance because plastics and rubbers show very large changes in properties with changing temperature.

In purely scientific terms, the temperature dependence has two primary points of interest. In the first place, as we have seen in Chapter 6, it is not possible to obtain from a single experimental technique a complete range of measuring frequencies to evaluate the relaxation spectrum at a single temperature. It is therefore a matter of considerable experimental convenience to change the temperature of the experiment, and so bring the relaxation processes of interest within a time scale which is readily available. This procedure, of course, assumes that a simple interrelation exists between time scale and temperature, and we will discuss shortly the extent to which this assumption is justified.

Secondly, there is the question of obtaining a molecular interpretation of the viscoelastic behaviour. In most general terms polymers change from glass-like to rubber-like behaviour as either the temperature is raised or the time scale of the experiment is increased. In the glassy state at low temperatures we would expect the stiffness to relate to changes in the stored elastic energy on deformation which are associated with small displacements of the molecules from their equilibrium positions. In the rubbery state at high temperatures, on the other hand, the molecular chains have considerable flexibility, so that in the undeformed state they can adopt conformations which lead to maximum entropy (or more strictly, minimum free energy). The rubber-like elastic deformations are then associated with changes in the molecular conformations.

The molecular physicist is interested in understanding how this conformational freedom is achieved in terms of molecular motions, e.g. to establish which bonds in the structure become able to rotate as the temperature is raised. One approach, which has proved successful to some degree, has been to compare the viscoelastic behaviour with dielectric relaxation behaviour and more particularly with nuclear magnetic resonance behaviour.

We have tacitly assumed that there is only one viscoelastic transition, corresponding to the change from the glassy low temperature state to the rubbery state. In practice there are several relaxation transitions. For a typical amorphous polymer the situation is summarized in Figure 7.4. At low temperatures there are usually several secondary transitions involving comparatively small changes in modulus. These transitions

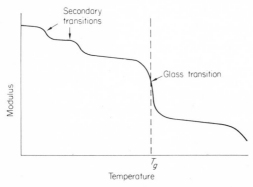

Figure 7.4. Temperature dependence of modulus in a typical polymer.

are attributable to such features as side-group motions, e.g. methyl

$(-CH_3)$ groups in polypropylene $\left[\begin{array}{c} CH_3 \\ | \\ CH_2-CH- \end{array} \right]_n$. In addition, there is

one primary transition which involves a large change in modulus. This is conveniently called the 'glass transition', and the temperature at which it occurs is commonly denoted by T_g.

7.1.3   Crystalline Polymers

The viscoelastic behaviour of crystalline polymers is markedly different from that of amorphous polymers. The four characteristic regions, although they still exist, are not so clearly defined. This is illustrated by the data obtained by Schmieder and Wolf[4] for polychlorotrifluoroethylene $[CClF-CF_2-]_n$ and polyvinylfluoride $[-CH_2CHF-]_n$ which are shown in Figure 7.5.

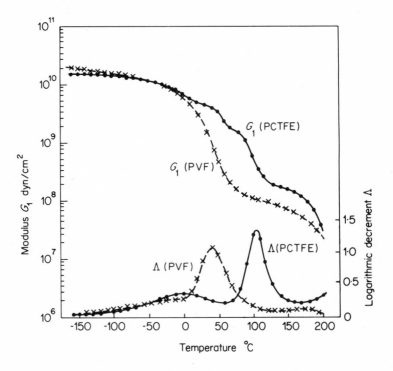

Figure 7.5. Shear modulus G_1 and logarithmic decrement Λ of polychlorotrifluoroethylene (PCTFE) and polyvinyl fluoride (PVF), as a function of temperature at $\sim 3 Hz$ (after Schmieder and Wolf).

For crystalline polymers, the fall in modulus over the glass-transition region is much smaller, generally only involving a fall from 10^{10} dyn cm^{-2} to 10^9 or 10^8 dyn cm^{-2}. Also, the change in modulus or loss factor with frequency or temperature is much more gradual, giving a much broader relaxation-time spectrum.

At high temperatures the molecular mobility is severely curtailed by the crystalline regions so that it is no longer correct to regard the polymer at low frequencies and high temperatures as simply rubber-like. These differences are clearly illustrated by the data for polyethylene terephthalate[5]

$$\left[-O-CH_2-CH_2-O-\overset{\displaystyle O}{\underset{\displaystyle O}{C}}- \bigcirc -\overset{}{C}- \right]_n$$

which can be obtained as an amorphous polymer by quenching rapidly from the melt (Figure 7.6a) or as a semicrystalline polymer by slow cooling or subsequent heat-crystallization treatments (Figure 7.6b).

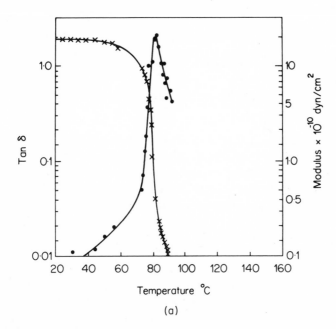

(a)

Figure 7.6a Tensile modulus and loss factor tan δ for unoriented amorphous polyethylene terephthalate as a function of temperature at ~ 1.2 Hz (after Thompson and Woods).

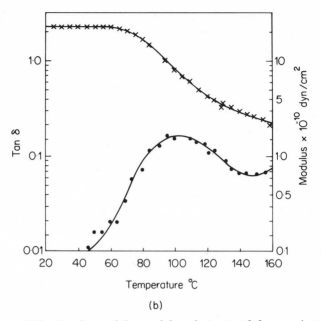

Figure 7.6b Tensile modulus and loss factor tan δ for unoriented crystalline polyethylene terephthalate as a function of temperature at ~ 1.2 Hz (after Thompson and Woods).

7.2 TIME-TEMPERATURE EQUIVALENCE AND SUPERPOSITION

Time-temperature equivalence in its simplest form implies that the visco-elastic behaviour at one temperature can be related to that at another temperature by a change in the time scale only. More sophisticated schemes to be described shortly also allow for changes in the magnitude of the response scale (e.g. compliance in a creep experiment) with temperature.

To fix our ideas, consider the creep-compliance curves of an idealized polymer at two temperatures T_1 and T_2, the time scale being logarithmic (Figure 7.7a). On the simplest scheme for time-temperature equivalence these two creep-compliance curves can be superimposed exactly by a horizontal displacement $\log a_T$. This defines a shift factor a_T.

Consider also the dynamic mechanical behaviour of the same polymer, shown by the determination of tan δ at temperatures T_1 and T_2 (Figure 7.7b). The simplest proposition is that these can also be superimposed by the same shift factor a_T.

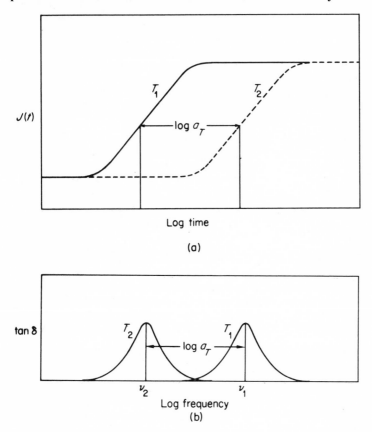

Figure 7.7. Schematic diagrams illustrating the simplest form of time–temperature equivalence for (a) compliance $J(t)$ and (b) loss factor tan δ.

In many studies of the temperature dependence of viscoelastic behaviour in polymers, particularly those involving dynamic mechanical measurements, the shift factors have been calculated on this simple basis. It has, however, been pointed out (notably by McCrum and his coworkers[6]) that it may be necessary to take into account changes in the relaxed and unrelaxed compliances with temperature. The situation is illustrated schematically in Figure 7.8. When we compare the creep-compliance curves at the two temperatures T_1 and T_2 we see that the relaxed and unrelaxed compliances are both changing with temperature. McCrum and Morris[6] propose a scaling procedure for obtaining a modified or 'reduced' compliance curve at the temperature T_1, to give the dashed

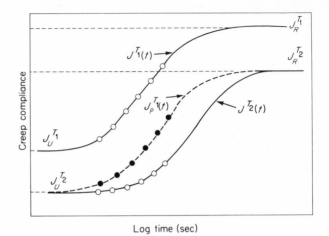

Figure 7.8. Schematic diagram illustrating McCrum's reduction procedure for superposition of creep data. $J_R^{T_1}$ and $J_U^{T_1}$ are the relaxed and unrelaxed compliances respectively at the temperature T_1; $J_R^{T_2}$ and $J_U^{T_2}$ are the corresponding quantities at the temperature T_2.

curve $J_p^{T_1}(t)$ in Figure 7.8. The shift factor is now obtained by a horizontal shift of $J_p^{T_1}(t)$ to superimpose it on $J^{T_2}(t)$.

In many cases it is adequate to assume that $J_R^{T_1}/J_R^{T_2} = J_U^{T_1}/J_U^{T_2}$. This is true for the glass-transition relaxation of an amorphous polymer (see Section 7.4.1 below) and is a very good approximation for some other relaxations. The correction for changes in the relaxed and unrelaxed compliances with temperature is then a vertical shift of the compliance curves (plotted on a logarithmic scale).

7.3 TRANSITION-STATE THEORIES

The simplest theories which attempt to deal with the temperature dependence of viscoelastic behaviour are the transition state or barrier theories. The transition-state theory of time-dependent processes stems from the theory of chemical reactions and is associated with the names of Eyring, Glasstone and others[7]. The basic idea is that for two molecules to react they must first form an *activated complex* or *transition state*, which then decomposes to give the final products of the reaction.

The potential-energy diagram for two reacting molecules is closely analogous to that for the internal rotation of molecules discussed in

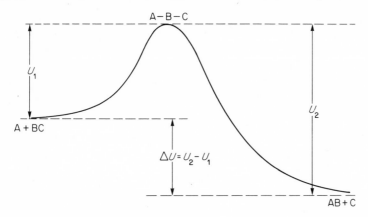

Figure 7.9. Change of potential energy in a chemical reaction (after Glasstone).

Section 1.2.1 above. Figure 7.9 (Reference 8, p. 1092) shows the change in potential energy for the reaction between an atom A and a diatomic molecule BC which results in the formation of a diatomic molecule AB and an atom C. The intermediate step is the formation of the activated complex A–B–C.

The theory of absolute reaction rates now argues as follows. The activated complex can be treated by statistical methods as a normal molecule, except that in addition to having three translational degrees of freedom, it has a fourth degree of freedom of movement along what is termed the reaction coordinate. The reaction coordinate is the direction leading to the lower potential energy of the final reactants. The theory shows that the rate of reaction is the product of two quantities, the probability of forming the activated complex and the effective rate of crossing the energy barrier by the activated complexes. It can be shown that the effective rate of crossing the energy barrier, which is the low frequency vibration of the activated complex in the direction of the reaction coordinate, is equal to kT/h. This is a universal frequency whose value is dependent only on the temperature and is independent of the nature of the reactants and the type of reaction (k is Boltzmann's constant and h is Planck's constant). Because there is equilibrium between the initial reacting species and the transition state, the probability of forming the activated complex is determined in absolute terms by the Boltzmann factor

$$e^{-\Delta G/RT}$$

where ΔG is the free energy difference per mole between the system when the reactants are relatively far from each other and when they form the activated complex. For reactions occurring under constant pressure conditions ΔG is the Gibbs Free Energy difference per mole.

We now argue that, by analogy, the frequency of molecular jumps between two rotational isomeric states of a molecule (Section 1.2.1 above) is given by

$$v = \frac{kT}{h} e^{-\Delta G/RT} \tag{7.1}$$

where ΔG is the Gibbs Free Energy barrier height per mole.

This equation states that the frequency of molecular conformational changes depends on the *barrier height* and not the free-energy difference between the equilibrium sites. Equation (7.1) may be written as

$$v = \frac{kT}{h} e^{\Delta S/R} e^{-\Delta H/RT} = v_0 e^{-\Delta H/RT} \tag{7.2}$$

This form of Equation (7.1) emphasizes the way in which temperature affects v primarily through the activation energy ΔH. To a good approximation the activation energy for the process (actually an enthalpy) is thus given by

$$\Delta H = -R \left[\frac{\partial \ln v}{\partial (1/T)} \right]_P \tag{7.3}$$

Equation (7.2) is known as the 'Arrhenius equation', because it was first shown by Arrhenius[9] that it describes the influence of temperature on the velocity of chemical reactions.

We now take the intuitive step (to be justified below by the Site-Model Theory) that the viscoelastic behaviour can be directly related to a controlling molecular rate process with a constant activation energy.

Consider the $\tan \delta$ curves of Figure 7.7b. At the temperatures T_1 and T_2 the peak value of $\tan \delta$ occurs at frequencies v_1 and v_2 respectively. The assumption is that v_1 and v_2 are related by the equation

$$\frac{v_1}{v_2} = \frac{e^{-\Delta H/RT_1}}{e^{-\Delta H/RT_2}}$$

i.e.

$$\log\frac{v_1}{v_2} = \log a_T = \frac{\Delta H}{R}\left\{\frac{1}{T_2} - \frac{1}{T_1}\right\} \tag{7.4}$$

The activation energy for the process can therefore be obtained from a plot of $\log a_T$ against the reciprocal of the absolute temperature. For large values of ΔH, changes in temperature give very large changes in frequency. Dynamic mechanical data on polymers are often dealt with in terms of the Arrhenius equation and a constant activation energy. In some cases this can be regarded as only an approximate treatment due to the limited range of experimental frequencies available. In general it has been found that the temperature dependence of the glass-transition relaxation behaviour of amorphous and crystalline polymers does not fit a constant activation energy, in contrast to more localized molecular relaxations.

7.3.1   The Site-Model Theory

The site-model theory is based on transition-state theory, and although first developed to explain the dielectric behaviour of crystalline solids[10,11] has also been applied to mechanical relaxations in polymers.[12]

In its simplest form there are two sites, separated by an equilibrium free-energy difference $\Delta G_1 - \Delta G_2$, the barrier heights being ΔG_1 and ΔG_2 per mole, respectively (Figure 7.10).

The transition probability for a jump from site 1 to site 2 is given by

$$\omega_{12}^0 = A' \, e^{-\Delta G_1/RT} \tag{7.5}$$

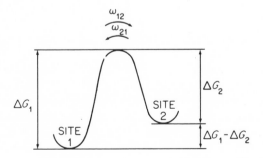

Figure 7.10. The two-site model.

and for a jump from site 2 to site 1 by

$$\omega_{21}^0 = A' e^{-\Delta G_2/RT} \tag{7.6}$$

where A' is a constant.

(In some treatments the change in molecular conformation is imagined to be a simple rotation of 180° around one bond. It is then considered that the transition probability is $2\omega_{12}^0$ where ω_{12}^0 is the probability for a jump in either a clockwise or anticlockwise direction.)

To give rise to a mechanical relaxation process, the energy difference between the two sites must be changed by the application of the applied stress. There is then a change in the populations of site 1 and site 2, and it is assumed that this relates directly to the strain. It is not difficult to imagine how this might arise at a molecular level if, for example, the uncoiling of a molecular chain involved internal rotations. Locally, the chain conformations could be changing from crumpled *gauche* conformations to extended *trans* conformations (see Section 1.2.1).

Assume that the applied stress σ causes a small linear shift in the free energies of the sites such that

$$\delta G'_1 = \lambda_1 \sigma \tag{7.7}$$

and

$$\delta G'_2 = \lambda_2 \sigma \tag{7.8}$$

for sites 1 and 2 respectively, where λ_1 and λ_2 are constants with the dimensions of volume. The transition probabilities ω_{12} and ω_{21} in the presence of the applied stress are then given by

$$\omega_{12} \simeq \omega_{12}^0 \left[1 - \frac{\delta G'_1}{RT} \right] = \omega_{12}^0 \left[1 - \frac{\lambda_1 \sigma}{RT} \right] \tag{7.9}$$

where ω_{12}^0 is the transition probability in the absence of the stress.
Similarly

$$\omega_{21} \simeq \omega_{21}^0 \left[1 - \frac{\lambda_2 \sigma}{RT} \right] \tag{7.10}$$

The rate equations for sites 1 and 2 are then

$$\frac{dN_1}{dt} = -N_1 \omega_{12} + N_2 \omega_{21} \tag{7.11}$$

$$\frac{dN_2}{dt} = -N_2\omega_{21} + N_1\omega_{12} \tag{7.12}$$

where we can write the occupation number N_1 of state 1 as $N_1 = N_1{}^0 + n$ and similarly $N_2 = N_2{}^0 - n$ where $N_1{}^0$ and $N_2{}^0$ are the occupation numbers at zero stress, $N_1{}^0 + N_2{}^0 = N_1 + N_2 = N$.

Combining these equations and making suitable approximations gives a rate equation

$$\frac{dn}{dt} + n(\omega_{12}^0 + \omega_{21}^0) = N_1{}^0\omega_{12}^0\left[\frac{\lambda_1 - \lambda_2}{RT}\right]\sigma \tag{7.13}$$

which describes the change in the site population n as a function of time. Assuming that this change in site population is directly related to the observed strain e, e is given by

$$e = e_u + n\bar{e} \tag{7.14}$$

In this equation e_u is the instantaneous or unrelaxed elastic deformation and it is considered that each change in site population produces a proportionate change in strain by an amount $\bar{e}$.

Equation (7.13) can then be seen to have the form

$$\frac{de}{dt} + Be = C$$

where B and C are constants. This is formally identical to the equation of a Voigt element, with a characteristic retardation time given by $\tau' = 1/B$ which is

$$\tau' = \frac{1}{(\omega_{12}^0 + \omega_{21}^0)} = \frac{e^{\Delta G_2/RT}}{A'\left[\exp\dfrac{-(\Delta G_1 - \Delta G_2)}{RT} + 1\right]} \tag{7.15}$$

Since RT is usually small compared with the equilibrium free-energy difference we may approximate to

$$\tau' = \frac{1}{A'}e^{\Delta G_2/RT} \tag{7.16}$$

Equation (7.16) is formally equivalent to Equation (7.1).

It follows that the time–temperature behaviour of the relaxation process is governed by the unperturbed transition probabilities, and to a good approximation by ΔG_2, a free energy of activation. The *magnitude* of the relaxation[12,13], on the other hand is proportional to

$$p\left[\frac{\exp\left[-(\Delta G_1 - \Delta G_2)/RT\right]}{(1 + \exp\left[-(\Delta G_1 - \Delta G_2)/RT\right])^2}\right]\frac{(\lambda_1 - \lambda_2)^2}{RT}$$

where p is the number of species per unit volume. Thus the intensity of the relaxation on this model is low at both high and low temperatures and passes through a maximum when the free-energy difference $(\Delta G_1 - \Delta G_2)$ and RT are of the same order of magnitude.

The site model is applicable to relaxation processes showing a constant activation energy, e.g. those associated with localized motions in the crystalline regions of semicrystalline polymers.

7.4 THE TIME–TEMPERATURE EQUIVALENCE OF THE GLASS-TRANSITION VISCOELASTIC BEHAVIOUR IN AMORPHOUS POLYMERS AND THE WILLIAMS, LANDEL AND FERRY (WLF) EQUATION

In considering time–temperature equivalence of the glass-transition behaviour in amorphous polymers, we will follow a treatment very close to that given by Ferry[14]. To fix our ideas, consider the storage compliance J_1 of an amorphous polymer (poly n-octyl methacrylate) as a function of temperature and frequency (Figure 7.11). It can be seen that there is an overall change in the shape of the compliance–frequency curve as the temperature changes. At high temperatures there is an approximately constant high compliance, the rubbery compliance. At low temperatures the compliance is again approximately constant but at a low value, the glassy compliance. At intermediate temperatures there is the frequency-dependent viscoelastic compliance.

The simplest way of applying time–temperature equivalence is to produce a 'master compliance curve' by choosing one particular temperature and applying only a horizontal shift on a logarithmic time scale to make the compliance curves for other temperatures join as smoothly as possible onto the curve at this particular temperature. This simple procedure is very nearly, but not quite, the procedure adopted by Ferry and his coworkers. The molecular theories of viscoelasticity suggest that there should be an additional small vertical shift factor $T_0\rho_0/T\rho$ in changing from the actual temperature $T°K$ (at a density ρ) to the reference temperature $T_0°K$ (at a density ρ_0). The physical meaning of this vertical correction factor is that the molecular theories suggest that the equilibrium

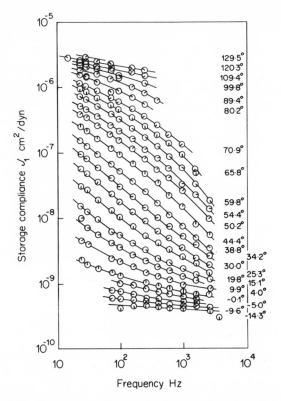

Figure 7.11. Storage compliance of poly-*n*-octyl methacrylate in the glass transition region plotted against frequency at 24 temperatures as indicated (after Ferry).

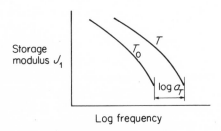

Figure 7.12. Diagram illustrating shift factor $\log a_T$ for change in temperature T to T_0.

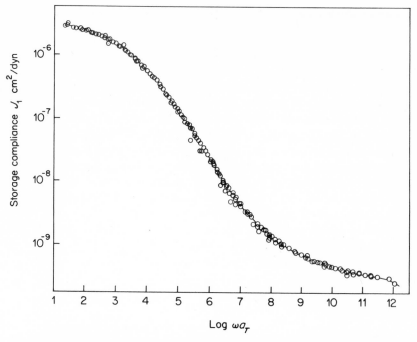

Figure 7.13. Composite curve obtained by plotting the data of Fig. 7.11 with suitable shift factors, giving the behaviours over an extended frequency scale at temperature T_0 (after Ferry).

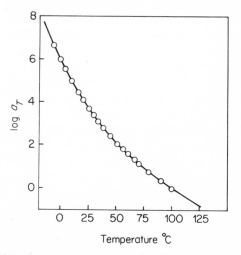

Figure 7.14. Temperature dependence of the shift factor a_T used in plotting Figure 7.13. Points, chosen empirically; curve is WLF equations with a suitable choice of T_g (or T_s) (after Ferry).

modulus changes with temperature in the transition range in a manner according to the theory of rubber elasticity (see Chapter 4). This is quite distinct from changes in the molecular-relaxation times, which affect the measured modulus at a given time or frequency due to affecting the visco-elastic behaviour. In practice, the correction factor has a very small effect in the viscoelastic range of temperatures compared with the large changes in the viscoelastic behaviour. Thus it is usually adequate to apply a simple horizontal shift on the time scale only (see Figure 7.12).

This procedure gives the storage compliance as a function of frequency over a very wide range of frequencies, as shown in Figure 7.13. Thus it is now possible to calculate the retardation-time spectrum, and compare this with any theoretical models which may be proposed.

We may also consider the significance of the horizontal shift on the logarithmic time scale, as shown in Figure 7.14.

The remarkable observation, which was established largely by the work of Williams, Landel and Ferry, is that for all amorphous polymers this shift-factor–temperature relationship is *approximately* identical.

It was found that the relationship

$$\log a_T = \frac{C_1(T - T_s)}{C_2 + (T - T_s)}$$

where C_1 and C_2 are constants, and T_s is a reference temperature peculiar to a particular polymer, holds extremely well over the temperature range $T = T_s \pm 50°c$ for all amorphous polymers. This equation, known as the 'WLF equation'[15] (we shall see that there are other forms for the WLF equation), was originally considered to be only an empirical equation and the constants C_1 and C_2 were originally determined by arbitrarily choosing $T_s = 243°\kappa$ for polyisobutylene.

Following this empirical discovery, there was naturally some specula-tion as to whether the WLF equation has a more fundamental interpreta-tion. This brings us to considerations of the dilatometric glass transition and to discussion of the use of the concept of free volume.

The glass transition can be defined on the basis of dilatometric measure-ments. As shown in Figure 7.15 if the specific volume of the polymer is measured against temperature, a change of slope is observed at a charac-teristic temperature, which we may call T_g. In the first place this change in slope may be somewhat less sharp than this diagram suggests. Secondly, it is known that if dilatometric measurements are carried out at very slow rates of temperature change, one approaches a roughly constant value for the glass-transition temperature T_g. The value of T_g will vary by only 2–3°c when the heating rate is decreased from 1°c per minute to

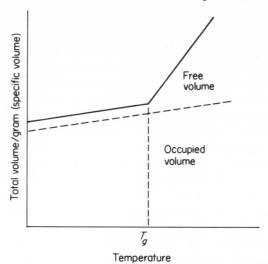

Figure 7.15. The volume–temperature relationship for a typical amorphous polymer.

1°C per day. Thus it appears possible to define a rate-independent value of T_g to at least a very good approximation.

It has subsequently been shown that the original WLF equation can be rewritten in terms of this dilatometric transition temperature such that

$$\log a_T = \frac{C_1{}^g(T - T_g)}{C_2{}^g + (T - T_g)}$$

where $C_1{}^g$ and $C_2{}^g$ are new constants and $T_g = T_s - 50°C$.

Moreover it is now possible to give a plausible theoretical basis to the WLF equation in terms of the concept of free volume[15].

In liquids, the concept of free volume has proved useful in discussing transport properties such as viscosity and diffusion. These properties are considered to relate to the difference $v_f = v - v_0$, where v is the total macroscopic volume, v_0 is the actual molecular volume of the liquid molecules, the 'occupied volume', and v_f is the proportion of holes or voids, the 'free volume'.

Figure 7.15 shows the schematic division of the total volume of the polymer into both occupied and free volumes. It is argued that the occupied volume increases uniformly with temperature. The discontinuity in the expansion coefficient at T_g then corresponds to a sudden onset of expansion in the free volume. This suggests that certain molecular processes which control the viscoelastic behaviour commence at T_g, and not merely that T_g is the temperature when their time scale becomes comparable with that of the measuring time scale. This would seem to imply that T_g is a genuine thermodynamic temperature. This point is not, however, completely resolved, and it has been shown by Kovacs[16] that the T_g measured dilatometrically is still sensibly dependent on the time scale, i.e. the rate of heating. However, as already mentioned, this time dependence is small. Thus to a good approximation it can be assumed that the free volume is constant up to T_g and then increases linearly with increasing temperature.

The fractional free volume $f = v_f/v$ can therefore be written as

$$f = f_g + \alpha_f(T - T_g) \tag{7.17}$$

where f_g is the fractional free volume at the glass transition T_g and α_f is the coefficient of expansion of the free volume.

The WLF equation can now be obtained in a simple manner. The model representations of linear viscoelastic behaviour all show that the relaxation times are given by expressions of the form $\tau = \eta/E$ (see the Maxwell model in Section 5.2.5 above), where η is the viscosity of a dashpot and E the modulus of a spring.

If we ignore the changes in the modulus E with temperature compared with changes in the viscosity η, this suggests that the shift factor a_T for changing temperature from T_g to T will be given by

$$a_T = \frac{\eta_T}{\eta_{T_g}} \tag{7.18}$$

At this juncture, we introduce Doolittle's viscosity equation[17], which relates the viscosity to the free volume. This equation is based on experimental data for monomeric liquids and gives

$$\eta = a \exp\left(\frac{bv}{v_f}\right) \tag{7.19}$$

where a and b are constants. Using (7.18) and (7.19) it can be shown that the Doolittle equation becomes

$$\ln a_T = b\left\{\frac{1}{f} - \frac{1}{f_g}\right\} \tag{7.20}$$

Substituting $f = f_g + \alpha_f(T - T_g)$ we have

$$\log a_T = -\frac{(b/2{\cdot}303 f_g)(T - T_g)}{f_g/\alpha_f + T - T_g} \tag{7.21}$$

which is the WLF equation.

Ferry and his coworkers have given further consideration to the exact form of the WLF equation. It can be shown that a better fit to data for different polymers can be obtained by changing the constants $C_1{}^g$ and $C_2{}^g$; and that the actual values obtained for $C_1{}^g$ and $C_2{}^g$ yield values for f_g and α_f which are plausible on physical grounds. The reader is referred to Ferry's book[14] for detailed discussion of these points. We will, however, note here that the free volume at the glass-transition temperature f_g is $0{\cdot}025 \pm 0{\cdot}003$ for most amorphous polymers. The thermal coefficient of expansion of free volume α_f is a more variable quantity, but has the physically reasonable 'universal' average value of $4{\cdot}8 \times 10^{-4}\,\mathrm{deg}^{-1}$.

It is of some interest to complete our discussion of the WLF equation by indicating the lines of its derivation by Bueche[18] using a transition-state model.

It is possible to develop the transition-state theory on the basis of free volume by expressing the frequency v of the controlling molecular process by the equation

$$v = A \int_{f_c}^{\infty} \phi(f)\,\mathrm{d}f$$

It is assumed that the required unit of structure can move when the local fractional free volume f exceeds some critical value f_c.

Bueche evaluated $\phi(f)$, and showed that with some approximations

$$v = v_g \exp\left\{-Nf_c\left[\frac{1}{f} - \frac{1}{f_g}\right]\right\} \tag{7.22}$$

where v_g is the frequency at T_g. If $f = f_g + \alpha_f(T - T_g)$ it may be shown that

$$\ln\frac{v}{v_g} = \frac{\left(N\dfrac{f_c}{f_g}\right)(T - T_g)}{T - T_g + f_g/f} \tag{7.23}$$

Assuming that there is a direct link between the shift factor a_T and the ratio of the frequencies of the controlling molecular process, Equation (7.23) is identical in form to Equation (7.21). Note also that Equation (7.22) is Bueche's analogy to the Doolittle equation, Equation (7.20).

In conclusion we observe that for time–temperature equivalence to be exact, a necessary simplicity is implied. At a molecular level, the individual relaxation times for molecular processes must shift uniformly with temperature. In phenomenological terms the spectrum of relaxation times must shift as a unit on a logarithmic time scale to shorter times with increasing temperature.

Staverman and Schwarzl[19] call these materials thermorheologically simple, and Lee and his collaborators[20] have worked out the theoretical consequences of this assumption, so that complex problems concerning the deformation of viscoelastic solids in variable-temperature situations can be solved.

7.4.1   The Williams, Landel and Ferry Equation, the Free-Volume Theory and other related theories

The WLF equation gives the shift factor for time–temperature super-position as

$$\log a_T = \frac{C_1{}^g(T - T_g)}{C_2{}^g + (T - T_g)}$$

We have seen that this relationship can be regarded as describing the change in the internal viscosity of the polymer as we change the temperature from the glass-transition temperature T_g to the test temperature T (Equation 7.18).

We may therefore write the WLF equation as:

$$\log \eta_T = \log \eta_{T_g} + \frac{C_1{}^g(T - T_g)}{C_2{}^g + (T - T_g)}$$

where η_T, η_{T_g} is the viscosity of the polymer at temperatures T, T_g respectively. In this form the equation implies that at a temperature $T = T_g - C_2{}^g$ (i.e. $T = T_g - 51 \cdot 6$ for the WLF equation in its universal form) the viscosity of the polymer is infinite.

This has led to the view that the WLF equation should be related at molecular level to the temperature $T = T_g - 51 \cdot 6$, which we will call T_2, rather than to the dilatometric glass transition T_g.

There have been two basic approaches along these lines:

(1) The free-volume theory is modified so that the changes in free volume with temperature relate to a discontinuity which occurs at T_2 rather than T_g. This is discussed in Section 7.4.2.

(2) It is considered that T_2 represents a true thermodynamic transition temperature. A modified transition-state theory is developed in which the frequency of molecular jumps relates to the cooperative movement of a group of segments of the chain. The number of segments acting cooperatively is then calculated from statistical thermodynamic considerations. This is the theory of Adam and Gibbs[23], which is described in Section 7.4.3 below.

7.4.2   The Free-Volume theory of Cohen and Turnbull

Cohen and Turnbull[24] have proposed that the free volume v_f corresponds to that part of the excess volume $v - v_0$ (v = total measured specific volume, v_0 is occupied volume as in Section 7.4) which can be redistributed without a change in energy. It is then assumed on the basis of arguments concerning the nature of the 'cage' formed round a molecule by its neighbours that the redistribution can take place without a change in energy at temperatures above a critical temperature, which is to be identified with T_2, where the cage reaches a critical size.

Thus

$$v_f = 0 \qquad \text{for } T < T_2$$

and

$$v_f = \alpha \bar{v}_m (T - T_2) \qquad \text{for } T \geqslant T_2$$

where α is the average expansion coefficient and $\bar{v}_m$ the average value of the molecular volume v_0 in the temperature range T_2 to T.

For a viscosity equation of the form:

$$\eta = a \exp \frac{bv}{v_f} \qquad \text{(Equation 7.19)}$$

this gives

$$\eta = a \exp \frac{B'}{T - T_2}$$

where B' is a constant and correspondingly for the average relaxation time:

$$\tau = \tau_0 \exp \frac{B'}{T - T_2}$$

There is much experimental evidence from dielectric relaxation for the validity of this equation for amorphous polymers. As we have discussed, putting $T_2 = T_g - 51 \cdot 6$ gives us the WLF equation and the relaxation time will become infinitely long as we approach T_2 due to the disappearance of free volume.

7.4.3   The Statistical Thermodynamic theory of Adam and Gibbs

Gibbs and Di Marzio[21,22] proposed that the dilatometric T_g is a manifestation of a true equilibrium second-order transition at the temperature T_2. In a further development, Adams and Gibbs[23] have shown how the WLF equation can then be derived. On their theory the frequency of molecular jumps is given by

$$v_c = A \exp -\frac{n\Delta \overset{*}{G}}{kT} \tag{7.24}$$

where A is a constant $[A = kT/h$ on the transition-state theory], $\Delta \overset{*}{G}$ is the free-energy difference hindering rearrangement *per segment* (the barrier height) and n is the number of segments acting cooperatively as a unit to make a configurational rearrangement.

The essence of the Adam and Gibbs theory is that n can be calculated on thermodynamic-equilibrium grounds as follows:

If S is the configurational entropy of the system, i.e. the entropy for a mole of segments,

$$S = \frac{N_A}{n} s_n \tag{7.25}$$

where N_A is Avogadro's number and s_n is the entropy of a unit of n segments. Thus

$$n = \frac{N_A s_n}{S} \quad \text{and} \quad v_c = A \exp \frac{-N_A s_n \Delta \overset{*}{G}}{SkT} \tag{7.26}$$

It is assumed that s_n is independent of temperature and that S, the configurational entropy of the system, can be calculated directly for any temperature from the specific heat at constant pressure.

A further assumption is that $S = 0$ at the thermodynamic transition temperature T_2. In molecular terms n becomes infinite and there are no configurations available into which the system may rearrange. We may note that although the entropy S is assumed to be zero at T_2 this is not necessarily (or ever, in practice) a state of complete order.

This gives the entropy $S(T)$ at a temperature T as

$$S(T) = \Delta C_p \ln \frac{T}{T_2} \tag{7.27}$$

where ΔC_p is the difference in specific heat between the supercooled liquid and the glass at T_g and is assumed to be constant over the temperature range considered.

Substituting for $S(T)$ and approximating somewhat we find that

$$v_c = A \exp -\frac{N_A \Delta \overset{*}{G} s_n}{k \Delta C_p (T - T_2)} \tag{7.28}$$

This gives a relaxation-time equation of the form

$$\tau = \tau_0 \exp \left\{ \frac{B}{(T - T_2)} \right\}$$

which as we have seen reduces to the WLF equation if we put

$$T_2 = T_g - 51 \cdot 6.$$

7.4.4   An Objection to Free-Volume Theories

Hoffman, Williams and Passaglia[12] have raised a serious objection to the free-volume ideas. Williams[25] showed that the β-relaxations of polymethylacrylate and polypropylene oxide behave somewhat similarly under constant pressure and constant-volume conditions. It would be expected, however, on the free-volume concept that because the occupied volume v_0 would increase with temperature the results for these two conditions would be very different. Williams concluded that the dielectric relaxation time was not a unique function of volume. He suggested that this implied that the free volume did not remain constant for constant *total* volume while temperature and pressure are varied. An alternative view is that relationships of the form

$$\eta = \eta_0 \exp U_\eta / R(T - T_2)$$

and

$$\tau = C \exp U_\tau / R(T - T_2)$$

should indeed be based on concepts which are more fundamental than those of free volume.

7.5   NORMAL-MODE THEORIES BASED ON MOTION OF ISOLATED FLEXIBLE CHAINS

We have so far discussed two types of theories, those based on the site model, and those based on the WLF equation and its ramifications, which deal with time–temperature equivalence. The site-model theories predict constant-activation energies and are more applicable to relaxation transitions originating from localized chain motions, whereas the WLF equation theories deal with the glass-transition behaviour in amorphous polymers.

In the introductory section on amorphous polymers (Section 7.1.1) we considered the relaxation spectrum of amorphous polymers and noted that it was quite complex. The normal-mode theories, now to be discussed, attempt to predict the relaxation spectrum for amorphous polymers, as well as the time–temperature equivalence.

These theories are associated with the names of Rouse, Bueche and

Zimm[27,28,29] and are based on the idea of representing the motion of polymer chains in a viscous liquid by a series of linear differential equations. They are essentially *dilute solution* theories, but we shall see that rather unexpectedly perhaps, they can be extended to predict the behaviour of the pure polymer. Because of its simplicity we will give an account of the theory due to Rouse[27].

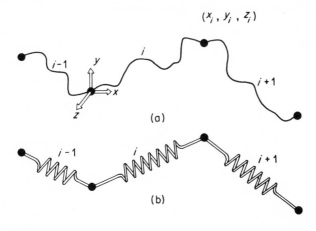

Figure 7.16. The Rouse model. (a) The network of chains. (b) The representation of the network as a combination of springs and beads.

Each polymer chain is considered to consist of a number of submolecules. This is similar to the composition of a rubber network where molecular chains join the cross-link points (see Figure 7.16a). We can then represent the polymer molecules as a system of beads connected by springs whose behaviour is that of a freely jointed chain on the Gaussian theory of rubber elasticity (Figure 7.16b). The molecular chains between the beads are all of equal length, this portion of the polymer chain being long enough for the separation of its ends to approximate to a Gaussian probability distribution. It is assumed that only the beads interact directly with the solvent molecules. If a bead is displaced from its equilibrium position there are two types of forces acting on it; first, the forces due to this viscous interaction with the solvent molecules and secondly, the forces due to the tendency of the molecular chains to return to a state of maximum entropy by Brownian diffusional movements.

Consider the motion of the bead situated at the point $(x_i y_i z_i)$ between the i and $i+1$ submolecule. The origin of coordinates is the bead between the $i-1$ and i submolecule, i.e. at the other end of the ith submolecule.

For a Gaussian distribution of links in the submolecule the probability that this bead will lie at the point $x_i y_i z_i$ in the volume element $dx_i \, dy_i \, dz_i$ is

$$p_i(x_i y_i z_i) \, dx_i \, dy_i \, dz_i = \frac{b^3}{\pi^{3/2}} \exp -b^2(x_i^2 + y_i^2 + z_i^2) \, dx_i \, dy_i \, dz_i$$

where

$$b^2 = \frac{3}{2zl^2}$$

with

$$l = \text{length of each link}$$

$$n = \text{total number of links in molecular chain}$$

$$m = \text{number of submolecules}$$

giving $z = n/m = $ number of links in a submolecule.

The conformational probability of the entire chain can be represented by a point in $3m$ dimensional space. The probability that this point lies at the point $x_1 y_1 \dots z_m$ in the volume element $dx_1 \, dy_1 \dots dz_m$ is given by

$$P_m \, dx_1 \dots dz_m = \prod_{i=1}^{m} p_i(x_i y_i z_i) \, dx_i \, dy_i \, dz_i$$

$$= \left(\frac{b^3}{\pi^{3/2}}\right)^m \exp -b^2 \left[\sum_{i=1}^{m} x_i^2 + y_i^2 + z_i^2\right] dx_1 \dots dz_m$$

At equilibrium, the most probable values of the x_i, y_i and z_i coordinates are zero, i.e. each submolecule is in a coiled-up configuration. Any change from the equilibrium position will result in a decrease of entropy ΔS, or an increase in Helmholtz free energy $\Delta A = -T\Delta S$. (All conformations are assumed to have the same internal energy.)

Consider the change of x_i due to displacements of the ith submolecule from the equilibrium situation $(0, 0, 0)$, i.e. the change of x_i is referred to a private coordinate system with its origin at the bead between the $i-1$ and ith submolecule.

There will be a restoring force

$$T\left(\frac{\partial S_m}{\partial x_i} - \frac{\partial S_m}{\partial x_{i-1}}\right)$$

due to displacements of the bead between the submolecules $i-1$ and i and a restoring force

$$T\left(\frac{\partial S_m}{\partial x_{i+1}} - \frac{\partial S_m}{\partial x_i}\right)$$

due to displacements of the bead between the submolecules i and $i+1$.

The total equation of motion is

$$\eta \dot{x}_i = T\left(2\frac{\partial S_m}{\partial x_i} - \frac{\partial S_m}{\partial x_{i-1}} - \frac{\partial S_m}{\partial x_{i+1}}\right) \tag{7.29}$$

η is the coefficient of friction defining the viscous interaction between the beads and the solvent. S_m is the entropy of a molecule of conformation $x_1 y_1 \dots z_m$ and is given by

$$S_m = k \ln P_m \tag{7.30}$$

Combining Equations (7.29) and (7.30) we have

$$\eta \dot{x}_i + \frac{3kT}{zl^2}(2x_i - x_{i-1} - x_{i+1}) = 0 \tag{7.31}$$

with $3m$ equations for coordinates $x_1 y_1 \dots z_m$. If we make the intuitive connection between displacement and strain, we can see that these equations of motion for the chain molecules are directly equivalent to the equation of a Voigt element which has the form $\eta \dot{e} + Ee = 0$.

It therefore follows that these equations can be regarded as defining a set of creep compliances and stress-relaxation moduli, or complex compliances and moduli.

The mathematical problem is to uncouple the $3m$ equations using a normal coordinate transformation. This involves obtaining the eigenfunctions which are linear combinations of the positions of the submolecules. Each eigenfunction then describes a configuration which decays with a time constant given by an associated eigenvalue, i.e. a single viscoelastic element with characteristic time-dependent properties.

For stress relaxation and dynamic mechanical experiments respectively it can be shown that the stress-relaxation modulus $G(t)$ and the real part of the complex modulus $G_1(\omega)$ are given by

$$G(t) = NkT \sum_{p=1}^{m} e^{-t/\tau_p} \tag{7.32}$$

and

$$G_1(\omega) = NkT \sum_{p=1}^{m} \frac{\omega^2 \tau_p^2}{1 + \omega^2 \tau_p^2} \tag{7.33}$$

where N is the number of molecules/cm^3 and τ_p, the relaxation time of the pth mode, is given by

$$\tau_p = zl^2\eta[24kT\sin^2\{p\pi/2(m+1)\}]^{-1} \qquad (7.34)$$

$$p = 1, 2, \ldots m$$

These equations predict that $G(t)$ and $G_1(\omega)$ are determined by a discrete spectrum of relaxation times, each of which characterizes a given normal mode of motion. These normal modes are shown schematically in Figure 7.17. In the first mode, corresponding to $p = 1$, the ends of the molecule move whilst the centre of the molecule remains stationary. In the second mode, there are two nodes in the molecule. The general case of the pth mode has p nodes, with motion of the molecule occurring in $p+1$ segments.

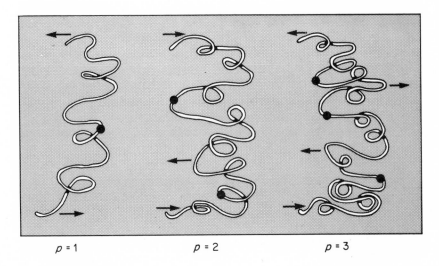

$p = 1$ $p = 2$ $p = 3$

Figure 7.17. Illustration of the first three normal modes of a chain molecule.

On this model the submolecule is the shortest length of chain which can undergo relaxation and the motion of segments *within* the submolecules are ignored. But such motions contribute to the relaxation spectrum for values of m less than about five. Thus we would only expect the Rouse theory to be applicable for $m \gg 1$ where the equation for τ_p

reduces to

$$\tau_p = \frac{m^2 z l^2 \eta}{6\pi^2 p^2 kT} = \frac{n^2 l^2 \eta_0}{6\pi^2 p^2 kT} \qquad (7.35)$$

where $\eta_0 = \eta/z$ is the friction-coefficient per random link. The relaxation times depend on temperature directly through the factor $1/T$, through the quantity nl^2 which defines the equilibrium mean square separation of the chain ends and may change due to differences in the energy of different chain conformations, and through changes in the friction coefficient η_0. η_0 changes rapidly with temperature and is primarily responsible for changes in τ_p. The fact that each τ_p has the same temperature dependence on this molecular theory, shows that it satisfies the requirements of thermorheological simplicity and gives theoretical justification for time–temperature equivalence.

Rouse's theory is the simplest molecular theory of polymer relaxation. A later theory of Zimm[29] does not assume that the velocity of the liquid solvent is unaffected by the movement of the polymer molecules (the 'free-draining' approximation). The hydrodynamic interaction between the moving submolecules is taken into account and this gives a modified relaxation spectrum.

The Rouse, Zimm and Bueche theories are satisfactory for the longer relaxation times which involve movement of submolecules. This has been confirmed for dilute polymer solutions, where the theory would be expected to be most appropriate[30,31]. More remarkably, it also holds for solid amorphous polymers, (Reference 14, Chapter 12), provided that the friction coefficient is suitably modified.

Ferry has shown that if the three longest relaxation times are ignored the distribution of relaxation times $H(\ln \tau)$, is given by

$$H(\ln \tau)\, \mathrm{d} \ln \tau = -NkT\left(\frac{\mathrm{d}p}{\mathrm{d}\tau}\right)\mathrm{d}\tau \qquad (7.36)$$

and from Equation (7.35)

$$H(\ln \tau) = \left(\frac{Nnl}{2\pi}\right)\left(\frac{kT\eta_0}{6}\right)^{1/2}\tau^{-1/2}$$

This equation predicts that the plot of log $H(\ln \tau)$ against log τ should have a slope of $-\frac{1}{2}$. The results for five methacrylate polymers summarized in Figure 7.18 confirm this prediction for long relaxation times. The Zimm theory predicts a slope of $-\frac{2}{3}$ and is perhaps a better fit at shorter relaxation times. At very short relaxation times the theory fails

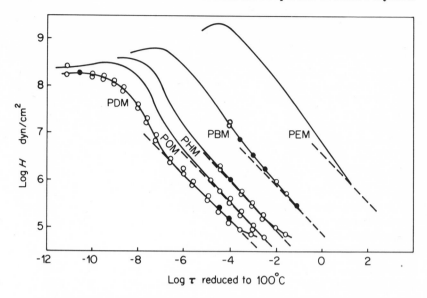

Figure 7.18. Relaxation time spectra $H(\ln\tau)$ for poly-*n*-dodecyl methacrylate (PDM), poly-*n*-octyl methacrylate (POM), poly-*n*-hexyl methacrylate (PHM), poly-*n*-butyl methacrylate (PBM), polyethyl methacrylate (PEM). Dashed lines are a slope of $-\frac{1}{2}$ predicted by the Rouse theory (after Ferry).

completely, as we have anticipated, because the movement of short segments is involved. Another way of looking at this, suggested by Williams[32], is that a theory based essentially on the Gaussian statistics of polymer chains can only hold for low values of the 'modulus', i.e. for values less than $\sim 10^8$ dyn/cm^2.

REFERENCES

1. R. S. Marvin, *Proceedings of the Second International Congress of Rheology*, Butterworth, London, 1954.
2. R. S. Marvin and H. Oser, *J. Res. Nat. Bur. Stand.*, **B66**, 171 (1962).
3. R. S. Marvin and J. T. Berger, *Viscoelasticity: Phenomenological Aspects*, Academic Press, New York, 1960, p. 27.
4. K. Schmieder and K. Wolf, *Kolloidzeitschrift*, **134**, 149 (1953).
5. A. B. Thompson and D. W. Woods, *Trans. Faraday Soc.*, **52**, 1383 (1956).
6. N. G. McCrum and E. L. Morris, *Proc. Roy. Soc.*, **A281**, 258 (1964).
7. S. Glasstone, K. J. Laidler and H. Eyring, *The Theory of Rate Processes*, McGraw-Hill, New York, 1941.

8. S. Glasstone, *Textbook of Physical Chemistry*, 2nd ed., Macmillan, London, 1953.
9. Z. Arrhenius, *J. Phys. Chem.*, **4**, 226 (1889).
10. P. Debye, *Polar Molecules*, Dover Publications, New York, 1945.
11. H. Fröhlich, *Theory of Dielectrics*, Oxford University Press, 1949.
12. J. D. Hoffman, G. Williams and E. Passaglia, *J. Polymer Sci.*, **C14**, 173 (1966).
13. J. B. Wachtman, *Phys. Rev.*, **131**, 517 (1963).
14. J. D. Ferry, *Viscoelastic Properties of Polymers*, Wiley, New York, 1961, Chapter 11.
15. M. L. Williams, R. F. Landel and J. D. Ferry, *J. Amer. Chem. Soc.*, **77**, 3701 (1955).
16. A. Kovacs, *J. Polymer Sci.*, **30**, 131 (1958).
17. A. K. Doolittle, *J. Applied Phys.*, **22**, 1471 (1951).
18. F. Bueche, *J. Chem. Phys.*, **21**, 1850 (1953).
19. A. J. Staverman and F. Schwarzl, *Die Physik der Hochpolymeren*, Springer–Verlag, Berlin, 1956, Chapter 1.
20. E. H. Lee, *Proceedings of the First Symposium on Naval Structural Mechanics* (*London*), Pergamon Press, Oxford, 1960, p. 456.
21. J. H. Gibbs and E. A. di Marzio, *J. Chem. Phys.*, **28**, 373 (1958).
22. J. H. Gibbs and E. A. di Marzio, *J. Chem. Phys.*, **28**, 807 (1958).
23. G. Adam and J. H. Gibbs, *J. Chem. Phys.*, **43**, 139 (1965).
24. M. H. Cohen and D. Turnbull, *J. Chem. Phys.*, **31**, 1164, (1959).
25. G. Williams, *Trans. Faraday Soc.*, **60**, 1556 (1964).
26. G. Williams, *Trans. Faraday Soc.*, **61**, 1564 (1965).
27. P. E. Rouse, *J. Chem. Phys.*, **21**, 1272 (1953).
28. F. Bueche, *J. Chem. Phys.*, **22**, 603, (1953).
29. B. H. Zimm, *J. Chem. Phys.*, **24**, 269 (1956).
30. N. W. Tschoegl and J. D. Ferry, *Kolloidzeitschrift*, **189**, 37 (1963).
31. J. Lamb and A. J. Matheson, *Proc. Roy. Soc.*, **A281**, 207 (1964).
32. G. Williams, *J. Polymer Sci.*, **62**, 87 (1962).

8

Relaxation Transitions and their Relationship to Molecular Structure

In Chapter 7 we considered the main features of viscoelastic behaviour, with particular emphasis on time–temperature equivalence. The treatment was a general one and the ideas of rate-dependent processes, free volume and the dynamics of long-chain molecules are applicable to all polymers. We will now be much more specific and discuss the assignment of the viscoelastic relaxations in a molecular sense to different chemical groups in the molecule, and in a physical sense to different parts of the structure, e.g. to the motion of molecules in the crystalline or amorphous regions.

Because there are fewer structural features in amorphous polymers, discussions of the viscoelastic behaviour are less extensive and will precede those for crystalline polymers.

8.1 RELAXATION TRANSITIONS IN AMORPHOUS POLYMERS: SOME GENERAL FEATURES

Polymethylmethacrylate

$$\left[\begin{array}{c} H_3C-O-\overset{\displaystyle |}{C}=O \\ -\overset{\displaystyle |}{\underset{\displaystyle |}{C}}-CH_2- \\ CH_3 \end{array} \right]_n$$

shows four relaxation transitions as a function of temperature (Figure 8.1). It is customary to label relaxation transitions in polymers in alphabetical order α, β, γ, δ, etc. with decreasing temperature, irrespective of their molecular origin. The highest temperature relaxation, the α-relaxation, is the glass transition and is associated with a large change in modulus. The β-relaxation has been shown by a combination of comparative studies on similar polymers, NMR and dielectric measurements[1-5], to be associated with side-chain motions of the ester group. The γ- and δ-

166

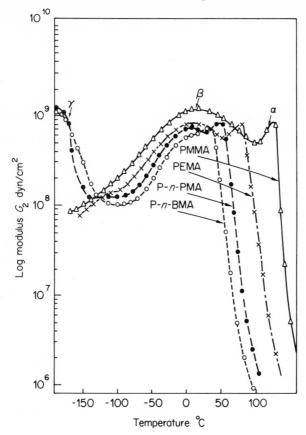

Figure 8.1. Temperature dependence of loss modulus G_2 for poly-methylmethacrylate (PMMA), polyethylmethacrylate (PEMA), poly-n-propylmethacrylate (P-n-PMA) and poly-n-butylmethacrylate (P-n-BMA) (after Heijboer).

relaxations involve motion of the methyl groups attached to the main chain and to the side chain, respectively.

The viscoelastic relaxations of similar polymers (such as polyethylmethacrylate) can be identified with molecular motions in an analogous manner.

The principles described for polymethylmethacrylate thus form a good basis for the interpretation of the relaxation transitions in a number of amorphous polymers. There is a high-temperature transition, the glass transition, associated with the onset of main-chain segmental motion, and secondary transitions which can be assigned to:

(1) Motion of side groups, which is usually well authenticated.

(2) Restricted motion of the main chain or end-group motions, which are usually less well authenticated.

It should be emphasized that in many cases the assignment of a relaxation process is tentative and by no means as straightforward as in the methacrylate and acrylate polymers. For example, in amorphous polyethylene terephthalate two major relaxation transitions are observed, the high-temperature transition being the glass transition and associated with general motion of the molecular chains. The secondary relaxation has been assigned only tentatively either to motion of chain ends or to some restricted motion of the main chain[5-8]. In this case, there are no side groups to provide an alternative assignment.

8.2 GLASS TRANSITIONS IN AMORPHOUS POLYMERS: DETAILED DISCUSSION

We will now discuss various factors which influence glass transitions in amorphous polymers. There are two principal approaches to the interpretation of the influence of main-chain structure, side groups and plasticizers, etc. The first approach is to consider that these produce changes in the molecular flexibility, modifying the ease with which conformational changes can take place. The second approach is to relate all these effects to changes in free volume, and in particular to the temperature at which this reaches its critical value which is, on this scheme, the glass transition.

8.2.1 Effect of Chemical Structure

The effect of chemical structure on the glass transition is a feature of polymer behaviour which has been intensively studied because of its importance in influencing the choice of polymers for useful applications. Much of our knowledge is of an empirical nature, due primarily to the difficulty of separating the intramolecular and intermolecular effects. In spite of this some generalities can be cited:

(1) *Main-chain Structure*

The presence of flexible groups such as an ether link will make the main chain more flexible and reduce the glass-transition temperature, whereas the introduction of an inflexible group, e.g. a terephthalate residue, will increase the glass-transition temperature.

(2) *Influence of Side-groups*[10]

It is generally true that bulky, inflexible side groups increase the temperature of the glass transition. This is illustrated by Table 8.1 for a series of substituted poly α-olefines $\left[-CH_2-\underset{\underset{R}{|}}{CH}-\right]_n$

Table 8.1. Glass transition of some vinyl polymers[10].

Polymer	R	Transition temperature in °C at ~1 Hz
Polypropylene	CH_3	0
Polystyrene	C_6H_5	116
Poly-N-vinylcarbazole	(N-carbazolyl group)	211

There is a difference between the effect of rigid and flexible side groups, which is shown by the series of polyvinyl butyl ethers $\left[-CH_2-\underset{\underset{OR_1}{|}}{CH}-\right]_n$ (Table 8.2). All these polymers have the same atoms in the side group R_1 (R_1 represents the butyl isomeric form) but the more compact arrangements reduce the flexibility of the molecule and the transition temperature is markedly increased.

Table 8.2. Glass transition of some isomeric polyvinyl butyl ethers[10].

Polymer	R_1	Transition temperature in °C at ~1 Hz
Polyvinyl *n*-butyl ether	$CH_2{\cdot}CH_2{\cdot}CH_2{\cdot}CH_3$	−32
Polyvinyl isobutyl ether	$CH_2{\cdot}CH(CH_3)_2$	−1
Polyvinyl *t*-butyl ether	$C(CH_3)_3$	+83

In an analogous manner, increasing the length of the flexible side groups reduces the temperature of the main transition. This can also be understood as increasing the free volume at any temperature. Data showing this are given in Table 8.3 for a series of polyvinyl n-alkyl ethers,

$$\left[-CH_2-\underset{\underset{OR_2}{|}}{CH}- \right]_n$$ where R_2 represents the n-alkyl group.

Table 8.3. Glass transition of some polyvinyl n-alkyl ethers[10].

Polymer	R_2	Transition temperature in °C at ~1 Hz
Polyvinyl methyl ether	CH_3	-10
Polyvinyl ethyl ether	$CH_2 \cdot CH_3$	-17
Polyvinyl n-propyl ether	$CH_2 \cdot CH_2 \cdot CH_3$	-27
Polyvinyl n-butyl ether	$CH_2 \cdot CH_2 \cdot CH_2 \cdot CH_3$	-32

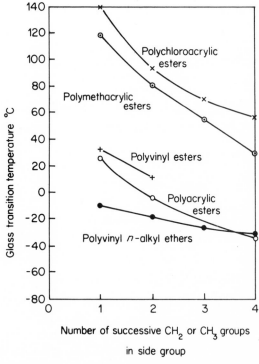

Figure 8.2. The effect of polarity on the position of the glass-transition temperature for five polymer series (after Vincent).

(3) *Effect of Main-Chain Polarity*

In Figure 8.2, the temperature of the glass transition is plotted against the number of successive $-CH_2$ or $-CH_3$ groups in the side groups, for five polymer series. Each series consists of polymers of similar main-chain composition, and it can be seen that the glass-transition temperature increases with increasing main-chain polarity. The associated reduction in the main-chain mobility is presumed to be due to the increase in intermolecular forces. In particular it is suggested that the higher curve for the polychloracrylic esters is due to the increased valence forces associated with the chlorine molecules.

An even more striking case is the effect of chlorinating polyethylene[11] which is shown in Figure 8.3.

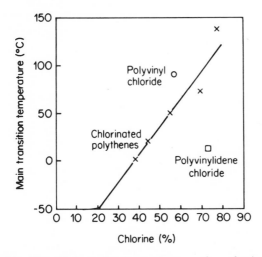

Figure 8.3. The effect of chlorine content on the softening point of polymers with varying degrees of chlorination (after Schmieder and Wolf).

8.2.2 Effect of Molecular Weight and Cross-linking

Molecular weight does not affect the dynamic mechanical properties of polymers in the glassy low-temperature state, although at low molecular weights, the glass-transition temperature T_g is affected by molecular weight. This is usually explained[12,13] on the basis that the chain ends, by reducing the closeness of molecular packing, introduce extra free volume, and hence lower T_g.

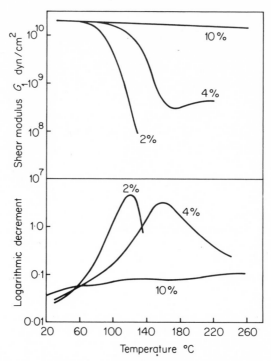

Figure 8.4. Shear modulus G_1 and logarithmic decrement of a phenol-formaldehyde resin cross-linked with hexamethylene tetramine at stated concentrations (after Nielsen).

As has already been discussed (Section 7.1.1 above), molecular weight has a large effect in the glass-transition range, transforming the behaviour from viscous flow to a plateau range of rubber-like behaviour with increasing molecular weight. This is shown in Figure 7.3. The explanation of this behaviour is that chain entanglements prevent irreversible flow. The effects of introducing chemical cross-links are shown in Figure 8.4, where phenolformaldehyde resin has been cross-linked with hexamethylene tetramine to concentrations of 2%, 4% and 10% respectively. Chemical cross-linking raises the temperature of the glass transition and broadens the transition region[14]. It can be seen that in very highly cross-linked materials there is no glass transition.

This behaviour can again be interpreted on the basis of changes in free volume. Chemical cross-linking, by bringing adjacent chains close together, reduces the free volume and hence raises T_g.

8.2.3 Blends, Grafts and Copolymers

The mechanical properties of blends and graft polymers are determined primarily by the mutual solubility of the two homopolymers. If two polymers are completely soluble in one another the properties of the mixture are nearly the same as those of a random copolymer of the same composition. Figure 8.5 illustrates this point by showing that a 50/50 mixture of polyvinyl acetate and polymethylacrylate has very similar properties to those of a copolymer of vinyl acetate and methyl acrylate[14]. It is to be noted that the damping peak for the mixture and the copolymer occurs at 30°c while the peaks for polymethylacrylate and polyvinyl acetate occur at about 15°c and 45°c respectively.

A theoretical interpretation of the glass transition temperatures of copolymers has been given on the basis of the ideas of free volume. We have seen that the glass transition can be considered to occur at a constant value of the free volume. Gordon and Taylor[15] assume that in an ideal

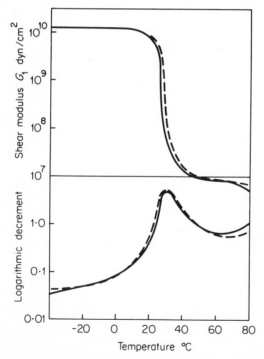

Figure 8.5. Shear modulus G_1 and logarithmic decrement for a miscible blend of polyvinyl acetate and polymethyl acrylate —— and a copolymer of vinyl acetate and methyl acrylate ――― (after Nielsen).

copolymer the partial specific volumes of the two components are constant and equal to the specific volumes of the two homopolymers. It is further assumed that the specific volume–temperature coefficients for the two components in the rubbery and glassy states remain the same in the copolymers as in the homopolymers, and are independent of temperature. It can then be shown that the glass-transition temperature T_g for the copolymer is given by[16,17]

$$\frac{1}{T_g} = \frac{1}{(w_1 + Bw_2)}\left[\frac{w_1}{T_{g_1}} + \frac{Bw_2}{T_{g_2}}\right]$$

where w_1 and w_2 are the weight fractions of the two monomers whose homopolymers have transitions at temperatures T_{g_1} and T_{g_2} respectively and B is a constant which is close to unity.

If the two polymers in a mixture are insoluble they exist as two separate phases and two glass transitions are observed instead of one. This is

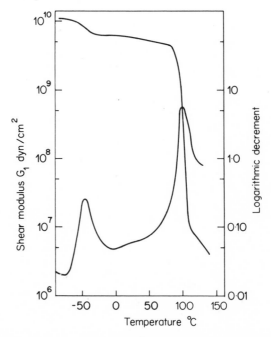

Figure 8.6. Shear modulus G_1 and logarithmic decrement for an immiscible polyblend of polystyrene and a styrene–butadiene copolymer (after Nielsen).

illustrated by the results shown in Figure 8.6 for a polyblend of polystyrene and styrene-butadiene rubber[14]. Two loss peaks are observed
which are very close to those in pure polystyrene and pure styrene-
butadiene rubber.

8.2.4   Effect of Plasticizers

Plasticizers are low-molecular weight materials which are added to rigid
polymers to soften them. Plasticizers must be soluble in the polymer and
usually they dissolve it completely at high temperatures. Figure 8.7 shows
the change in the loss peak associated with the glass transition of polyvinylchloride, when plasticized with various amounts of di(ethyl-hexyl)
phthalate[18]. The major effect of plasticizer is to lower the temperature
of the glass transition; essentially plasticizers make it easier for changes
in molecular conformation to occur.

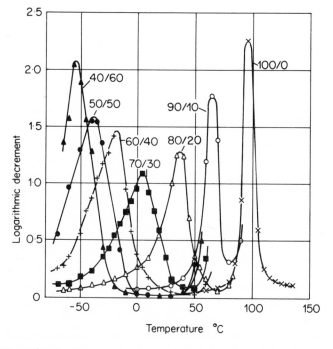

Figure 8.7. The logarithmic decrement of polyvinyl chloride plasticized
with various amounts of di(ethylhexyl)phthalate (after Wolf).

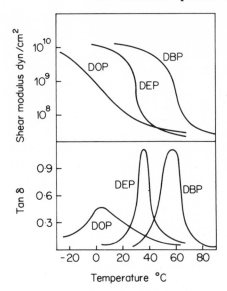

Figure 8.8. Shear modulus and loss factor tan δ for polyvinyl chloride plasticized with diethyl phthalate (DEP), dibutyl phthalate (DBP) and *n*-dioctyl phthalate (DOP) (after Nielsen).

Plasticizers also broaden the loss peak, and the degree of broadening depends on the nature of the interaction between the polymer and the plasticizer. If the plasticizer has a limited solubility in the polymer, or if the plasticizer tends to associate in the presence of the polymer, a broad damping peak is found. Thus the width of the damping peak increases as the plasticizer becomes a poorer solvent. This is shown in Figure 8.8 for plasticized polyvinyl chloride[14]. Diethyl phthalate is a relatively good solvent, dibutyl phthalate is a poorer solvent, and dioctyl phthalate is a very poor solvent.

8.3 THE CRANKSHAFT MECHANISM FOR SECONDARY RELAXATIONS

In a review of relaxation transitions Willbourn[19] suggested that the γ-relaxation in both amorphous and crystalline polymers could in many cases be attributed to a restricted motion of the main chain which required at least four $-CH_2$ groups in succession on a linear part of the chain. This proposal has led to the so-called 'crankshaft' mechanisms of Shatzki[20] and Boyer[21] (Figure 8.9). Shatzki's mechanism involves the simultaneous

PLATE I

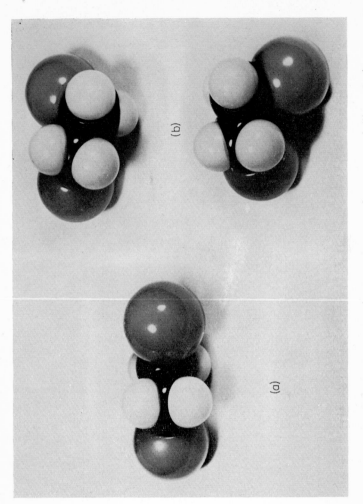

(b)

(a)

Figure 1.6. The *trans* (a) and *gauche* (b) conformations of dibromoethane.

PLATE II

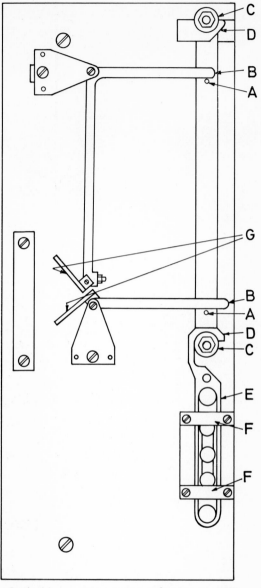

Figure 6.4. Optical lever extensometer (after Dunn, Mills and Turner).

A. steel pins.

B. forked arms.

C. coaxial discs.

D. slotted hooks

E. guide bar.

F. guide bar support slot.

G. mirrors.

PLATE III

Figure 6.10. The free-vibration torsion pendulum.

PLATE IV

(a)

(b)

Figure 6.14. The vibrating reed apparatus.

PLATE V

Figure 6.17. Apparatus for measurement of dynamic extensional modulus.

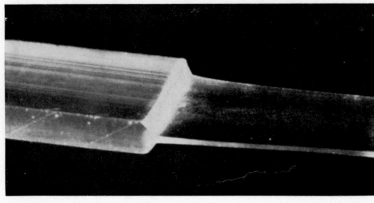

Figure 11.1. Photograph of a neck formed in the redrawing of oriented polyethylene.

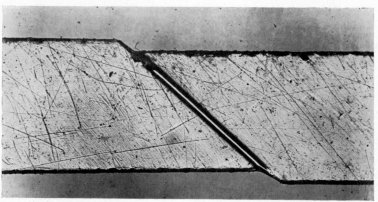

Figure 11.21. Photograph of a deformation band in an oriented sheet of polyethylene terephthalate.

PLATE VII

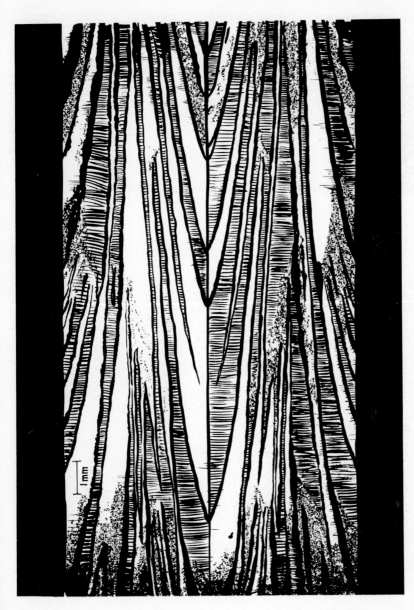

Figure 12.14.(a) Fracture surfaces of a cleavage sample of polymethylmethacrylate showing colour alternation (green filter) (after Berry).

Figure 12.14(b). Craze formation in polystyrene.

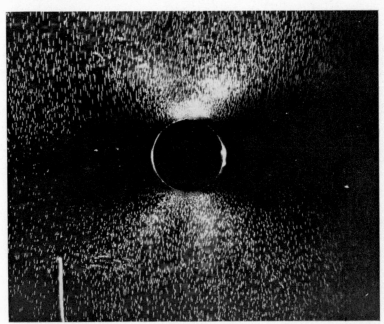

Figure 12.15(a). Craze formation in the vicinity of a hole in a strip of polymethylmethacrylate loaded in tension (result obtained by L. S. A. Smith).

PLATE IX

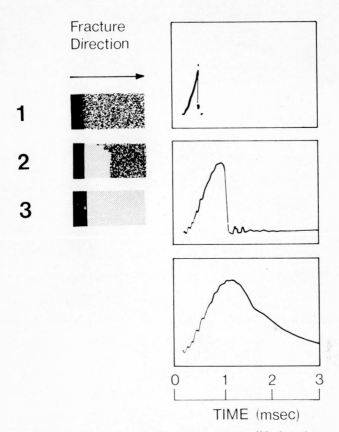

Figure 12.19(a). Fracture surfaces of modified poly-styrene notched Izod impact specimens; (top) broken at −70°C, type I fracture; (centre) broken at +10°C, type II fracture; (bottom) broken at +150°C, type III fracture (after Bucknell).

PLATE X

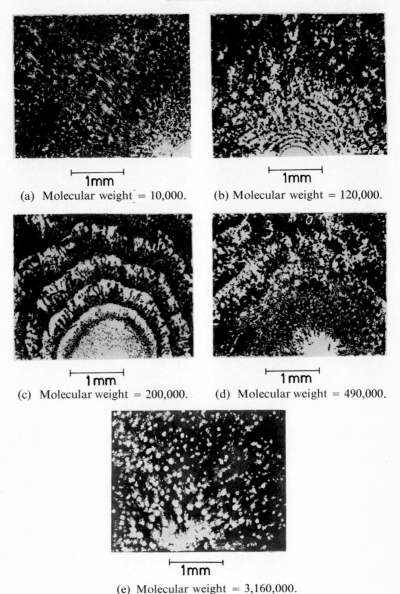

(a) Molecular weight = 10,000.

(b) Molecular weight = 120,000.

(c) Molecular weight = 200,000.

(d) Molecular weight = 490,000.

(e) Molecular weight = 3,160,000.

Figure 12.22. Fracture surfaces of tensile specimen of polymethyl-methacrylate of various molecular weights (after Wolock, Kies and Newman).

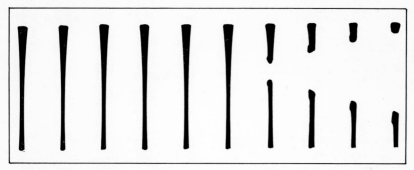

Figure 12.24(a). Fracture of silicone rubber (16,300 frames per second) (after Bueche and White).

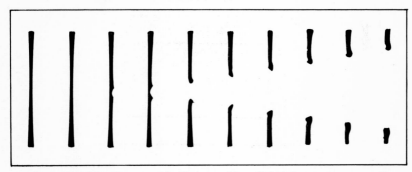

Figure 12.24(b). Fracture of silicone rubber (9,100 frames per second) (after Bueche and White).

PLATE XII

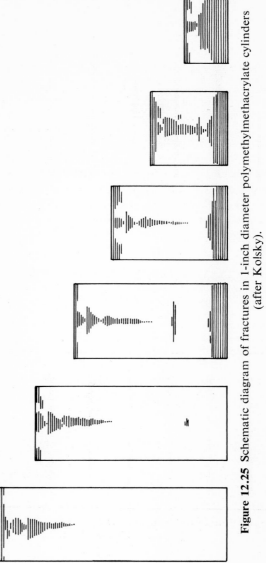

Figure 12.25 Schematic diagram of fractures in 1-inch diameter polymethylmethacrylate cylinders (after Kolsky).

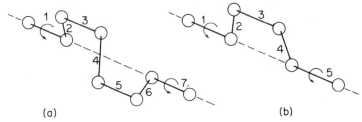

Figure 8.9. The crankshaft mechanisms of (a) Shatzki and (b) Boyer
(after McCrum, Read and Williams).

rotation about the two bonds 1 and 7, such that the intervening carbon
bonds move as a crankshaft. It is to be noted that bonds 1 and 7 are
collinear, which means that bonds on either side may remain unaffected
so that only a relatively small volume is required for the movement.

It has been proposed that this mechanism is relevant to the γ-relaxation
in polyethylene, polyamides, polyesters, polyoxymethylene and polypropylene oxide.

8.4 RELAXATION TRANSITIONS IN CRYSTALLINE POLYMERS

8.4.1 General Discussion

The interpretation of the viscoelastic behaviour of crystalline polymers
is still at a very speculative stage with regard to a detailed understanding.
Attempts at specific interpretations are very much influenced by the
opinions of the individual investigator on the structure of crystalline
polymers.

The simplest starting point for discussions of viscoelastic relaxations in
crystalline polymers is the two-phase fringed-micelle model (see Section
1.2.2 above). On this model we would expect to identify some transitions
with the crystalline regions and some with the amorphous regions. This
is supported by empirical correlations between the magnitude of loss
processes in crystalline polymers and the crystalline–amorphous ratio as
determined by X-ray or density methods.

Two good examples of this come from the work of McCrum[22] on
polytetrafluorethylene (PTFE) and that of Illers and Breuer[9] on polyethylene terephthalate (PET).

In this discussion we will tacitly assume that tan δ can be used as a
measure of the relaxation strength. Although this is plausible in the light

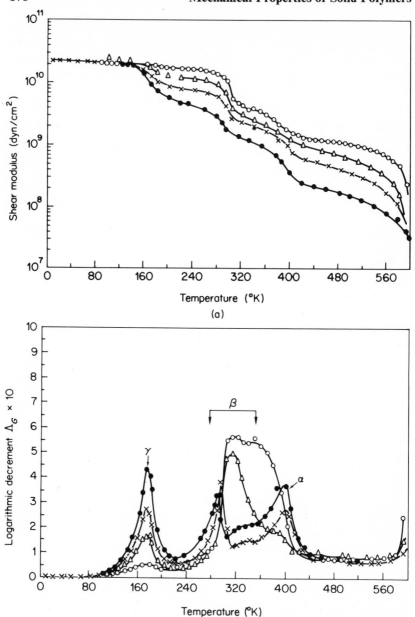

Figure 8.10. Temperature dependence of (a) shear modulus and (b) the logarithmic decrement Λ_G at ~ 1 c/s for PTFE samples of 92% (0), 76% ($\triangle$), 64% ($\times$) and 48% ($\bullet$) crystallinity (after McCrum).

of previous considerations (Section 5.5 above), it is essentially at the hypothesis stage and complete justification for the use of tan δ or some other measure will only be obtained when more sophisticated structural models have been given a quantitative treatment (for further discussion see Reference 5, p. 139).

Figure 8.10 shows the temperature dependence of the logarithmic decrement for three transitions in PTFE as a function of the degree of crystallinity. The lowest temperature relaxation decreases in magnitude with increasing crystallinity in a very clear manner and on the two-phase model is identified with a transition in the amorphous regions of the polymer. The β-relaxation, on the other hand, increases in magnitude with increasing crystallinity and is therefore associated with the crystalline regions. The analysis of the α-relaxation is somewhat dependent on the method of resolving the loss peaks and determining the strength of a relaxation process, but the consensus of opinion (see Reference 5, Chapter 11) would seem to confirm that it decreases in magnitude with increasing crystallinity and is therefore associated with the amorphous regions.

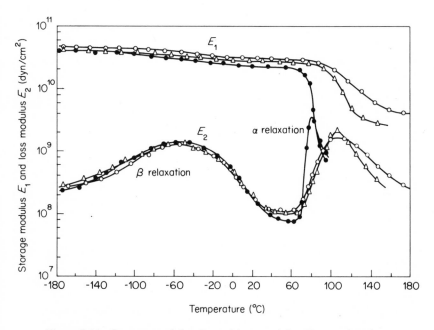

Figure 8.11. Storage modulus E_1 and loss modulus E_2 as a function of temperature at 138 Hz for PET samples of differing degrees of crystallinity ($\bullet$ 5% $\triangle$ 34% $\bigcirc$ 50%) (after Takayanagi).

In PET the effect of crystallinity on the β-relaxation is very small and has led to a very complex interpretation in terms of this loss peak being composed of several relaxation processes. No clear distinction can be drawn between relaxation processes occurring in the crystalline and the amorphous regions.

The effect of crystallinity on the α-relaxation is very striking (Figure 8.11) and more complex than in the case of the γ-relaxation of PTFE. The height of the loss peak decreases with increasing crystallinity, the peak broadens, becomes very asymmetrical and moves to higher temperatures. This suggests not only that the α-relaxation occurs in the amorphous regions of the polymer but also that the presence of the crystallites now imposes a variety of considerable constraints on the amorphous regions, thus influencing the molecular movements associated with this relaxation process. These ideas have been confirmed by NMR measurements of molecular mobility in a series of corresponding samples[23].

The adequacy of the two-phase fringed-micelle model was seriously called into question following the discovery of polyethylene single crystals, and must be severely modified to allow for chain-folding in the bulk, as illustrated in Figure 1.14. The implications of this new model are that we must now identify the following types of relaxations.

(1) Relaxations which occur in the amorphous phase. These will include the glass transition, an example of which is the α-relaxation in polyethylene terephthalate. Another example of an amorphous transition is the γ-relaxation in PTFE.

(2) Relaxations which occur in the crystalline phase. These may be of two types: (a) those involving cooperative motions of the molecular chains along the length of the crystallite—these relaxations will be related to the lamellar thickness, and (b) those associated with defects such as end groups in the crystal—these relaxations could also involve chain movement.

(3) Relaxations which occur in both the crystalline and amorphous phases, with perhaps some detailed differences depending on which phase is involved. Such relaxations will involve movement of restricted length of the chain, e.g. four CH_2 segments in a hydrocarbon chain, examples of which are the previously discussed crank-shaft motions proposed by Shatzki and Boyer. The γ-relaxation in polyethylene terephthalate may be of this nature.

(4) Relaxations associated with details of the morphology. An example of this would be motion of a chain fold. It is also possible that some relaxations are associated with cooperative motion of large morphological elements, e.g. interlamellar shear or interfibrillar shear.

8.4.2   Relaxation Processes in Polyethylene

For an intensive discussion of relaxation processes in crystalline polymers it is clearly most satisfactory to choose a case where the structure has been investigated in great detail. Polyethylene is therefore an obvious choice. Let us begin by looking at the temperature dependence of $\tan \delta$ for high-density and low-density polyethylene measured at about 1 c/s (Figure 8.12). Low-density polyethylene, although basically consisting of long chains of $-CH_2$ groups, contains a significant number of short side branches (of the order of three per 100 carbon atoms in a typical commercial polymer) together with a few long branches (about one per molecule). High density polyethylene, on the other hand, is very much closer to being the pure $(CH_2)_n$ polymer, and the number of branches is often less than five per 1000 carbon atoms.

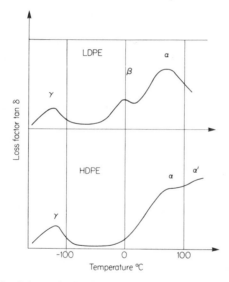

Figure 8.12. Schematic diagram showing α, α', β and γ relaxation processes in low-density polyethylene (LDPE) and high-density polyethylene (HDPE).

The low-density polymer shows three clearly distinguishable loss peaks; these are conventionally labelled α, β and γ. In the high-density polymer the γ-relaxation is very similar to that in the low-density polymer; the β-relaxation is almost absent and the α-relaxation has been considerably modified. There is some controversy regarding the changes in the α-relaxation with structure. This arises partly because the analysis depends on whether tan δ or G_2 is plotted as a function of temperature, and partly because the α-process appears to be a composite process consisting of at least two relaxation processes (usually called α and α') with different activation energies. These facts together mean that the observed shape of the loss peaks in the α-relaxation region can vary greatly according to the different workers.

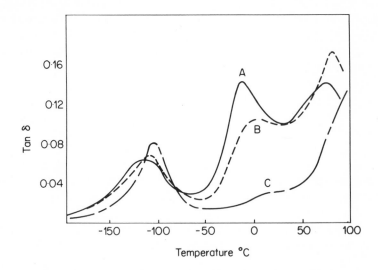

Figure 8.13. Loss factor tan δ of polyethylene as a function of branching. Type A: $3 \cdot 2$ CH$_3$'s/100 CH$_2$; Type B: $1 \cdot 6$ CH$_3$'s/100 CH$_2$; Type C: $<0 \cdot 1$ CH$_3$'s/100 CH$_2$ (after Kline, Sauer and Woodward).

The assignment of these three relaxations was first attempted in a general sense, by comparing the intensities of the loss peaks in different polyethylenes. The comparison shown in Figure 8.13 suggests that the β-relaxation is associated with the relaxation of branch-points. This was demonstrated quantitatively by Kline, Sauer and Woodward[24] who examined a series of polyethylenes of widely varying densities due to varying branch-point content.

The assignment of the α-relaxation to a crystalline relaxation was shown by the reduction in the intensity of the α-loss peak on chlorination of polyethylene[11]. This loss peak disappears at the degree of chlorination required to remove any X-ray diffraction evidence of crystallinity in the polymer.

The γ-relaxation peak was found to decrease in intensity as the crystallinity increased. It was therefore assigned to an amorphous relaxation.

These 'assignments' of the relaxation processes do not extend to a specific understanding of the molecular processes involved. The latter can only come from more detailed knowledge of the structure of polyethylene.

In this respect, our understanding has been much advanced by the comparison of the relaxations in the bulk high-density polymer with those observed in mats of solution crystallized high-density polyethylene. These mats consist of an aggregate of polyethylene crystals in which the molecular chain axes are approximately perpendicular to the plane of the mat. Figure 8.14 shows results obtained by Takayanagi[25], in which the dynamical mechanical properties were determined with the alternating stress applied in the plane of the film. The bulk-crystallized sample shows, as expected from previous work, what appear at first sight to be two principal relaxations, at $+70°$C and $-120°$C respectively. These are labelled α (and α') and γ in the illustration, and it is interesting to note the difference in shape of the α- and α'-relaxations due to plotting E_2 rather than tan δ. Takayanagi's results show that change of frequency affects the high-temperature relaxation in the bulk polymer in an asymmetric fashion, confirming that this is a composite of the two overlapping α- and α'-relaxations.

The crystal-mat results are shown at the top of Figure 8.14. There are two principal points to note concerning the α- and γ-relaxations. First, there is now a simple shift of the α-relaxation with frequency, i.e., the α'-transition is absent. Secondly, the γ-relaxation is present in both samples but is of somewhat greater intensity in the bulk sample.

Takayanagi made the following interpretation of this data. First he suggested that the γ-relaxation is associated with the non-crystalline phase and defects in the crystalline phase, the reduction in intensity in the crystal mats being attributed to their greater degree of perfection. It was tentatively proposed that the molecular motion was a 'local twisting of the molecular chains'.

Secondly, Takayanagi attributed the α-relaxation to a similar molecular motion within the crystalline phase. Finally, he noted that the α-relaxation is absent in the viscoelastic pattern of the crystal mats. He attributed this relaxation to the crystalline phase, and suggested that its absence in the crystal mats was due to the orientation of the applied stress within the crystal mat being unfavourable for the particular molecular process which he proposed. This is a translational motion of chain segments

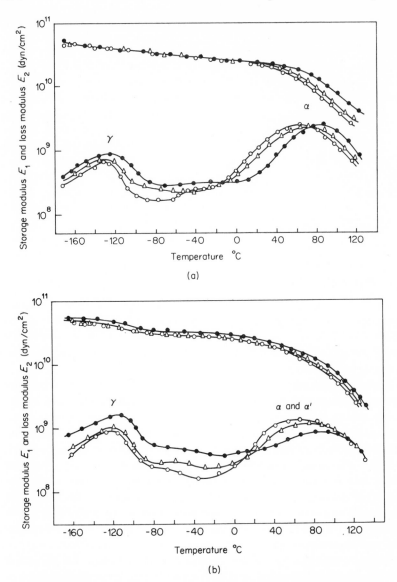

Figure 8.14. (a) Temperature dependence of storage modulus E_1 and loss modulus E_2 at $\bigcirc$ 3·5, $\triangle$ 11 and $\bullet$ 110 Hz measured along the direction perpendicular to the chain axis (c-axis) for a single crystal mat of high-density polyethylene; (b) Temperature dependence of E_1 and E_2 for a bulk-crystallized specimen of high-density polyethylene (after Takayanagi).

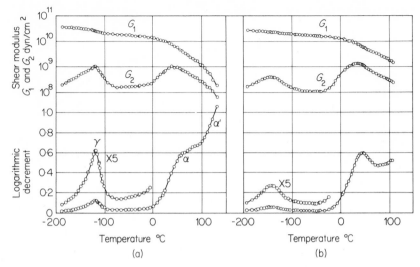

Figure 8.15. Temperature dependence of shear moduli G_1 and G_2 and logarithmic decrement for high-density polyethylene, (a) bulk polymer and (b) a single crystal mat (after Sinnot).

along the chain axis within the crystal lattice (akin to the Fisher and Schmidt process for the annealing of polyethylene single crystals[26]).

Sinnot[27] has also examined the behaviour of bulk and solution-crystallized high-density polyethylene, and his results are summarized in Figures 8.15 and 8.16. Figure 8.15 shows a comparison between the behaviour of the bulk polymer and a crystal mat. Although it must be noted that the measurements are made in torsion rather than extension, and that the logarithmic decrement (equivalent to tan δ) is measured rather than E_2, the results are very similar to those of Takayanagi. Sinnot also made detailed studies of the influence of annealing on the intensity of the α- and γ-relaxations in the crystal mats, and the results are shown in Figure 8.16. It can be seen that the α-relaxation decreases in magnitude and the γ-relaxation increases in magnitude with progressively higher annealing temperatures and that both relaxations move to higher temperatures. Sinnot concluded that because both these relaxations are present in the crystal mats they must be associated with the lamellae. He proposed that the α-relaxation is due to the reorientation of the folds at the surfaces of the lamellae and the γ-relaxation to the stress-induced reorientation of defects within the lamellae. These assignments are consistent with the observed changes in magnitude of the relaxations on

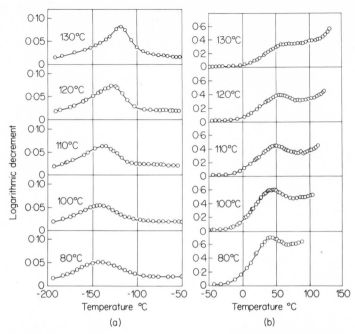

Figure 8.16. Influence of annealing on (a) γ transition and (b) α transition of solution crystallized polyethylene mat (after Sinnot).

annealing. The magnitude of the α-relaxation was found to be inversely proportional to the lamellar thickness; it was therefore proposed that its decrease in intensity on annealing is therefore directly proportional to the decrease in the number of chain folds as the lamellae thicken. The increase in magnitude of the γ-relaxation on annealing is attributed to the generation of defects within the lamellae, perhaps due to the motion of dislocations and point defects.

Further information of value comes from studies of the α- and α'-relaxations by McCrum and Morris[28]. Figure 8.17 compares the logarithmic decrement observed in a torsional pendulum experiment for two slow-cooled bulk-crystallized polyethylenes, one of which has had its lateral surface removed by milling. The consequent reduction in the α'-relaxation was explained by proposing that this relaxation is extremely sensitive to the orientation of the lamellae with respect to the direction of applied stress. Milling removes an oriented layer with the preferred orientation for maximum energy loss. It is a result consistent with that obtained by Takayanagi and Sinnot. McCrum and Morris proposed

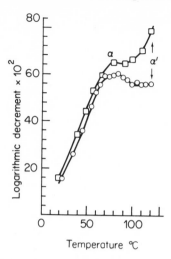

Figure 8.17. Temperature dependence of logarithmic decrement at 0.67 c/s for a polyethylene specimen crystallized by slow cooling from the melt (□). The points (○) are for the same specimen with the oriented surface layers removed by milling (after McCrum and Morris).

that the α'-relaxation can be interpreted as slip at the lamellar boundary. They envisage that the lamellae bend under the applied stress like an elastic beam in a viscous liquid. To explain the fact that their observed creep is totally recoverable, they assume that the lamellae are pinned along their length. This is equivalent to assuming a mechanical model where an elastic spring is now placed in parallel with the original spring and dashpot, i.e. the standard linear solid.

McCrum and Morris found that large variations in spherulitic structure have no influence on the α-relaxation. They agreed with Takayanagi in associating this relaxation with molecular motion within the crystallites. Both Sinnot and McCrum and Morris observed the effect of electron irradiation on the relaxations. The α- and α'-peaks are much affected by irradiation, whereas the γ-peak is unaffected. Sinnot concludes that this supports his assignment of the α-relaxation to motion of chain folds, since these are preferentially cross-linked. McCrum and Morris also claim support for their hypothesis, because the α'-relaxation is affected in the manner expected for an increase in the internal viscosity of the bulk polymer due to cross-linking, i.e. for measurements at constant frequency it appears to move to higher temperatures.

8.4.3 Application of the Site Model

The above investigations leave us with a largely empirical knowledge of the γ-relaxation is associated with the non-crystalline regions and defects with the crystalline regions (which have a lamellar texture); the β-relaxation is associated with branch points in the amorphous regions, the γ-relaxation is associated with the non-crystalline regions and defects in the crystalline regions (i.e. the lamellae). A review article by Hoffman, Williams and Passaglia[29] has considered most of the available experimental data, and attempted a theoretical analysis based on the site model for mechanical relaxations. Experimental data of Illers[30] on n-paraffins was considered together with that for bulk polymer and single-crystal mats. Illers found that the α- and γ-relaxations were observed in crystals of n-alkanes, and moreover that the temperature of maximum loss (at fixed frequency of measurement) increased with increasing numbers of $-CH_2$ groups per molecule.

On the site model (Chapter 7, Section 7.3.1) the relaxation time τ is given by

$$\tau = \frac{1}{A'} e^{\Delta G_2 / RT}$$

where ΔG_2 is the free energy of activation. This can be written in terms of an enthalpy ΔH_2 and entropy ΔS_2 of activation so that

$$\tau = \frac{1}{A'} e^{-\Delta S_2 / R} e^{\Delta H_2 / RT}$$

We will consider that the chain-relaxation process in a linear paraffin involves the rotation of the chain through 180° together with its simultaneous *translation* along its axis to reach a new equilibrium position. If we assume that the chain rotation is identical to the rotation of a rigid rod

$$\Delta H_2(n) = 2\Delta H_{end} + (n-2)\Delta H_{CH_2}$$

where $\Delta H_2(n)$ now refers to a chain of n segments and ΔH_{end} and ΔH_{CH_2} refer to the enthalpy of each end group and that of each segment respectively.

Similarly

$$\Delta S_2(n) = 2\Delta S_{end} + (n-2)\Delta S_{CH_2}$$

where $\Delta S_2(n)$, ΔS_{end} and ΔS_{CH_2} are the corresponding entropies.

This gives

$$\Delta G_2(n) = 2(\Delta H_{end} - T\Delta S_{end}) + (n-2)(\Delta H_{CH_2} - T\Delta S_{CH_2})$$

T_{max}, the temperature at which a loss peak occurs in a plot of loss versus T at a fixed measuring frequency f_m, where $2\pi f_m \tau_m = 1$, is given by

$$\ln \tau_m = \ln \frac{1}{A'} + \frac{(2\Delta H_{end} + (n-2)\Delta H_{CH_2})}{R T_{max}} - \frac{(2\Delta S_{end} + (n-2)\Delta S_{CH_2})}{R}$$

or

$$T_{max} = T_0 \frac{(a+n)}{(b+n)}$$

where

$$T_0 = \frac{\Delta H_{CH_2}}{\Delta S_{CH_2}}$$

is the value of T_{max} for very long chains ($n \to \infty$) and

$$a = \frac{2(\Delta H_{end} - \Delta H_{CH_2})}{\Delta H_{CH_2}}$$

$$b = \frac{[R \ln (A'\tau_m) + 2(\Delta S_{end} - \Delta S_{CH_2})]}{\Delta S_{CH_2}}$$

This gives a value of T_{max} which rises rapidly for small n and then levels off to a value T_0 as n becomes large.

There is one necessary modification to this simple theory. The rigid-rod approximation cannot hold for long chains, because here the rotation of one end would be expected to leave the other end unaffected. Thus $\ln \tau_m$ and T_{max} will fall below the rigid rod curves at large values of n, and approach an asymptote.

Hoffman, Williams and Passaglia distinguish three different models for the relaxation.

(1) $\alpha_c - A$ Model: Chain-folded crystal with folds and interior chain coupled. The theoretical treatment here is similar to that given above for the n-paraffins, with

$$\Delta H_2(n) = \Delta H_{fold} + n\Delta H_{CH_2}$$

and

$$\Delta S_2(n) = \Delta S_{fold} + n\Delta S_{CH_2}$$

replacing the values previously given for $\Delta H(n)$ and $\Delta S(n)$.

(2) $\alpha_c - B$ Model: n-paraffins and extended-chain crystals. This is the model for which the theoretical treatment has been given.

(3) $\alpha_c - C$ Model: Chain-folded crystals with independent chain fold and interior-chain relaxations. The interior-chain relaxation is similar

to the $\alpha_c - A$ and $\alpha_c - B$ models, with suitable modification of the theory. The independent chain fold relaxation will now be different in character, only involving the chain fold and hence having a much lower activation energy.

There are therefore three relaxation processes $\alpha_c - A$, $\alpha_c - B$ and the interior-chain relaxation $\alpha_c - C$ which is of a different character. On a

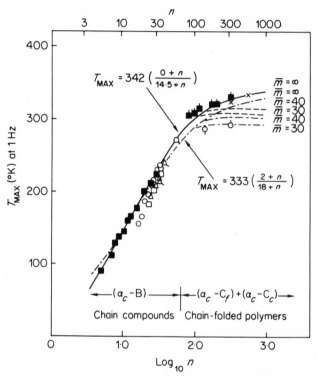

Figure 8.18. The temperature of maximum loss at 1 Hz plotted against $\log n$ (α_c processes). *Mechanical:* (■) single crystal mats polyethylene; (×) bulk linear polyethylene; (■) *n*-paraffins. *Dielectric:* (□) Esters; (△) ethers; (○) and (△) solid solutions of ketones and ethers, respectively, in *n*-paraffins; (Q) solid solutions of ethers in low density polyethylene; (⌀) and (Q) oxidized bulk polyethylene and single crystal mat polyethylene respectively. The upper calculated curve (——) corresponds to the relation $T_{max} = 342(0+n)/(14.5+n)$. The dashed lines ($\bar{m} = 40$ and $\bar{m} = 30$) falling off this curve correspond to the modification of this rigid-rod approximation to include chain twisting. Similarly, the lower calculated curve (—·—) corresponds to $T_{max} = 333(2+n)/(18+n)$ and the lines $\bar{m} = 40$ and $\bar{m} = 30$ take into account twisting of the chains (after Hoffman, Passaglia and Williams).

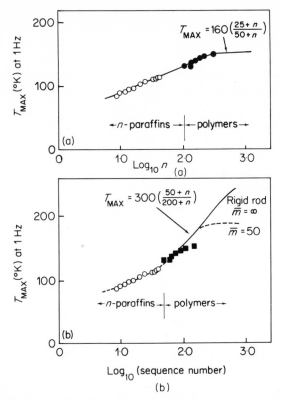

Figure 8.19. The temperature T_{max} of maximum mechanical loss for the γ_c process in single crystal mats of polyethylene. Data for low temperature process observed for n-paraffins in polystyrene matrix shown for comparison: (a) T_{max} versus $\log_{10} n$ where n is taken to be the number of —C— atoms for the n-paraffins, and the number of —C— atoms in a fold period for the single crystals; (b) T_{max} versus $\log_{10}$ (sequence number). The sequence number is taken to be n for the n-paraffins and $(n/2)$ for the single crystals. Dashed line on right-hand side shows effect of chain twisting for case $\bar{m} = 50$ (after Hoffman, Passaglia and Williams).

plot of T_{max} against n, using data from a wide variety of sources (Figure 8.18) it appears that all the processes are indistinguishable, and that a single characteristic law

$$T_{max} = T_0 \frac{(a+n)}{(b+n)}$$

can be obtained for suitably chosen a and b.

The $\alpha_c - C$ process corresponding to the fold motion is distinguishable in other ways, from its different activation energy (Takayanagi) and its variation in intensity with lamellar thickness in single crystals (Sinnot).

A similar correlation was shown to hold for the γ-relaxation in the n-paraffins and single crystals (Figure 8.19). Hoffman et al emphasize, however, that it is difficult to accept that the molecular processes are identical in all these materials. The γ-relaxation in bulk polyethylene and in single-crystal mats has been attributed to a relaxation of defects. If this relaxation is coupled to the main-chain motion, a theoretical model similar to that described here will be applicable, and n will relate to the number of participating segments, which will in turn relate to the lamellar thickness. It seems unlikely that there is a sufficient concentration of such defects in the n-paraffin crystals, for the γ-relaxation observed in that case to be of a similar nature.

It therefore seems likely that the γ-relaxation is a composite relaxation, and that it involves not only defects in the crystals, but also molecules in the amorphous regions.

8.4.4   Use of Mechanical Anisotropy

The work of Takayanagi and McCrum–Morris indicated that the mechanical relaxations in an oriented sample can be dependent on the direction of the applied stress. Stachurski and Ward[31–33] have obtained experimental data on oriented sheets of low-density and high-density poly-

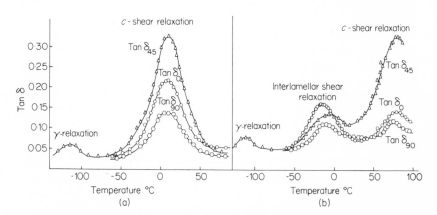

Figure 8.20. Temperature dependence of tan δ in three directions in (a) cold-drawn, and (b) cold-drawn and annealed low-density polyethylene sheets at approximately 500 Hz (after Stachurski and Ward).

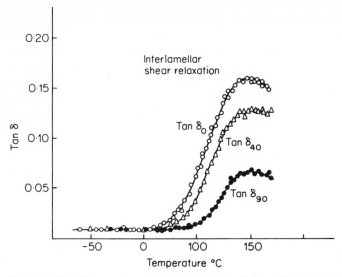

Figure 8.21. Temperature dependence of $\tan\delta$ in cold-drawn and annealed high-density polyethylene sheet in different directions at 50 Hz (after Stachurski and Ward).

ethylene which confirm this expectation, and give further information concerning the origins of the α- and β-relaxations.

The key results are summarized in Figures 8.20 and 8.21. The dynamic mechanical measurements were taken on strips cut at chosen angles to the original draw direction. In the cold-drawn low-density polyethylene sheet (Figure 8.20a) the relaxation process at about 0°C shows striking anisotropy with maximum loss for the 45° sheet. This is the anisotropy to be expected for a relaxation which involves shear parallel to the draw direction in a plane containing the draw direction. Comparison with the cold-drawn and annealed low-density polyethylene sheet (Figure 8.20b) where this relaxation has moved to about 70°C, and with other specially oriented sheets, suggests that the c-axis orientation of the crystalline regions is the significant factor. For this reason the relaxation has been termed the 'c-shear relaxation', i.e. shear in the c-axis direction in planes containing the c-axis. The cold-drawn and annealed low-density sheet also shows a second relaxation at about 0°C with a different anisotropy. The losses are now greatest when the stress is applied along the 0° direction, and a similar anisotropy has been observed for high-density sheet. (Figure 8.21). It was proposed that this relaxation process, which corresponds to the β-relaxation in low-density polyethylene and to the α-relaxation in high-density polyethylene, is an interlamellar shear process.

This interpretation, which owes much to the examination of specially oriented orthorhombic sheets of low-density polyethylene, is based on proposals that the lamellae in such sheets are arranged so that the lamellar planes make an acute angle of about 40° with the initial draw direction[34,35]. Applying the stress along the initial draw direction then gives the maximum resolved shear stress parallel to the lamellar planes.

These results are consistent with McCrum and Morris' interpretation of the α-relaxation in high-density polyethylene and produce the rather surprising conclusion that this relaxation is similar mechanically to the β-relaxation in low-density polymer. This does not mean that both relaxations are associated with identical *molecular* processes. Indeed it is likely that the α-relaxation in the high-density polymer requires mobility of the fold surfaces (as proposed by Hoffman, Williams and Passaglia) whereas the β-relaxation in the low density polymer requires mobility of chains close to branch points. Because the anisotropy of the α-relaxation in the low-density polymer is related to the orientation of the crystalline regions it must be concluded that the chains taking part in the relaxation process thread these regions. It seems probable that the chains involved form interlamellar ties so that the stresses are transmitted throughout the bulk of the polymer.

These results emphasize that it is difficult to assign the mechanical relaxations to specific mechanisms on the basis of analogous behaviour in different polymers. In particular, the labelling of the relaxations α, β, γ, etc. in order of decreasing temperature may be misleading when it comes to comparisons between polymers. We have seen that in polyethylene the low-density polymer shows α, β and γ relaxations in the isotropic state. Cold-drawn low-density polymer shows only two relaxations in the same temperature range because the α-relaxation swamps the β-relaxation so that the latter is not discernable. Cold-drawn and annealed low-density polyethylene does show α, β and γ relaxations and from the mechanical anisotropy the α-relaxation is identified as the c-shear relaxation and the β-relaxation as an interlamellar shear process.

Isotropic high-density polyethylene shows α, α' and γ relaxations. From the orientation studies the α and α' relaxations show anisotropy which can be explained on the basis of interlamellar shear. It is therefore proposed that the nature of the α-relaxation in low-density and high-density polyethylene are quite different.

8.4.5 Conclusion

We can conclude that appreciable progress is being made in unravelling the complex relaxation processes in polyethylene and it is hoped that this account will serve as a guide-line for the interpretation of the relaxations in other crystalline polymers. Certainly there are no comparable data for other polymers, and this can be attributed to the comparatively smaller amount of structural information.

REFERENCES

1. K. Deutsch, E. A. W. Hoff and W. Reddish, *J. Polymer Sci.*, **13**, 365 (1954).
2. J. G. Powles, B. I. Hunt and D. J. H. Sandiford, *Polymer* **5**, 505 (1964).
3. K. M. Sinnot, *J. Polymer Sci.*, **42**, 3 (1960).
4. J. Heijboer, *Physics of Non-Crystalline Solids*, North Holland, Amsterdam, 1965, p. 231.
5. N. G. McCrum, B. E. Read and G. Williams, *Anelastic and Dielectric Effects in Polymer Solids*, Wiley, London, 1967.
6. A. B. Thompson and D. W. Woods, *Trans. Faraday Soc.*, **52**, 1383 (1956).
7. W. Reddish, *Trans. Faraday Soc.*, **46**, 459 (1950).
8. G. Farrow, J. McIntosh and I. M. Ward, *Makromolek. Chem.*, **38**, 147 (1960).
9. K. H. Illers and H. Breuer, *J. Colloid Sci.*, **18**, 1 (1963).
10. P. I. Vincent, *Physics of Plastics*, Iliffe Books, Ltd., London, 1965.
11. K. Schmieder and K. Wolf, *Kolloidzeitschrift*, **134**, 149 (1953).
12. T. G. Fox and P. J. Flory, *J. Appl. Phys.*, **21**, 581 (1950).
13. T. G. Fox and P. J. Flory, *J. Polymer Sci.*, **14**, 315 (1954).
14. L. E. Nielson, *Mechanical Properties of Polymers*, Reinhold, New York, 1962.
15. M. Gordon and J. S. Taylor, *J. Appl. Chem.*, **2**, 493 (1952).
16. L. Mandelkern, G. M. Martin and F. A. Quinn, *J. Res. Nat. Bur. Stand.*, **58**, 137 (1959).
17. T. G. Fox and S. Loshaek, *J. Polymer Sci.*, **15**, 371 (1955).
18. K. Wolf, *Kunststoffe*, **41**, 89 (1951).
19. A. H. Willbourn, *Trans. Faraday Soc.*, **54**, 717 (1958).
20. T. F. Shatzki, *J. Polymer Sci.*, **57**, 496, (1962).
21. R. F. Boyer, *Rubber Rev.*, **34**, 1303 (1963).
22. N. G. McCrum, *J. Polymer Sci.*, **34**, 355 (1959).
23. I. M. Ward, *Trans. Faraday Soc.*, **56**, 648 (1960).
24. D. E. Kline, J. A. Sauer and A. E. Woodward, *J. Polymer Sci.*, **22**, 455 (1956).
25. M. Takayanagi, *Proceedings of the Fourth International Congress of Rheology*, Part 1, Interscience Publishers, New York, 1965, p. 161.
26. E. W. Fischer and G. F. Schmidt, *Z. Angew. Chem.*, **74**, 551 (1962).
27. K. M. Sinnot, *J. Appl. Phys.*, **37**, 3385 (1966).
28. N. G. McCrum and E. L. Morris, *Proc. Roy. Soc.*, **A292**, 506 (1966).
29. J. D. Hoffman, G. Williams and E. A. Passaglia, *J. Polymer Sci.*, **C14**, 173 (1966).
30. K. H. Illers, *Rheol. Acta.*, **3**, 194 (1964).
31. Z. H. Stachurski and I. M. Ward, *J. Polymer Sci.*, **A-2**, **6**, 1083 (1968).
32. Z. H. Stachurski and I. M. Ward, *J. Polymer Sci.*, **A-2**, **6**, 1817 (1968).
33. Z. H. Stachurski and I. M. Ward, *J. Macromol. Sci. (Phys.)*, **B3(3)**, 445 (1969).
34. I. L. Hay and A. Keller, IUPAC Symposium, Prague, 1965, Preprint No. P.325, *J. Materials Sci.*, **2**, 538 (1967).
35. T. Seto and T. Hara, *Rept. Prog. Polymer Phys. (Japan)*, **7**, 63 (1967).

9

Non-linear Viscoelastic Behaviour

9.1 GENERAL INTRODUCTION

In many practical applications of plastics, although the ultimate strains produced are recoverable, the viscoelastic behaviour does not satisfy the tests of linearity required by the Boltzmann Superposition Principle. This can occur for several reasons. In the first instance, there is the restriction of this representation to small strains* because the definition of strain and superposition of strains does not apply at large strains. This limitation applies in particular to studies on synthetic textile fibres where one may be interested in strains of at least 10% or in elastomers where the strains may be as high as 100%.

Secondly, although the experiments are restricted to small strains, it may still be that linear viscoelastic behaviour is not obtained. In this connection it is quite usual to observe linear viscoelastic behaviour at short times at given stress levels, but for the behaviour to be markedly non-linear for long times at the same stress levels.

There is not at present a representation of non-linear viscoelasticity which gives an adequate description of the behaviour and provides some physical insight into the origins of this behaviour. This is a subject where the divergence of the experimentalist and the theoretician is most marked. Faced with non-linear viscoelastic behaviour the experimentalist makes a number of measurements, necessarily finite, and then reduces his data empirically to a series of equations relating stress, strain and time. Although these equations can be extremely valuable in reducing the experimental data to manageable proportions they often do not reveal anything of the essential *nature* of the non-linearity, and may even be misleading in this respect.

The theoretician, on the other hand, will attempt to form a constitutive relation of a most general nature and examine how the form of this relation is determined by such features as 'short-time' memory, material

---

* By 'small strains' we mean that the quadratic terms in the displacement gradients can be neglected (Chapter 3, Section 3.1).

symmetry and invariance under rigid-body rotation. The disadvantage of this approach is that in many cases it is too general. The experimentalist may well conclude that it is of no relevance to his particular problem, particularly if it does not appear to provide any physical insight into the situation.

As the subject of non-linear viscoelastic behaviour cannot be provided with an approach which satisfies all these requirements the various attempts to deal with the situation will be considered under three headings:

(1) The engineering approach. The design engineer requires the ability to predict behaviour exactly for a proposed situation in terms of as few initial experiments as possible. Empirical relations which describe the performance are adequate, and these need not have any physical significance.

(2) The molecular approach. The non-linearity in the stress–strain relationships is suggested to be a consequence of the molecular mechanisms for viscoelasticity.

(3) The rigorous-continuum approach. Attempts are made to extend the formal descriptions of linear viscoelastic behaviour to non-linear behaviour. For example, modifications of the Boltzmann Superposition Principle are examined.

It must be admitted that at the present stage none of these approaches is entirely satisfactory. With some reservations therefore they will be considered in turn, noting their shortcomings where appropriate.

9.2 THE DESIGN ENGINEER'S APPROACH TO NON-LINEAR VISCOELASTICITY

9.2.1 Use of the Isochronous Stress–strain Curves

For a linear viscoelastic solid the creep behaviour is completely specified at a given temperature by a measurement of the response to a constant stress over the required period of time. For a non-linear viscoelastic solid the behaviour over the range of stress required must be mapped out in detail over the required period of time. We will also see that because the Boltzmann Superposition Principle does not hold it is necessary to carry out systematic programmes of loading and unloading. The behaviour for any loading programme is not defined by the data obtained from a single step loading which gives a creep curve, or even from a two-step loading and unloading which gives creep and recovery.

In spite of these pitfalls, the design engineer starts with the stress–strain–time relationship obtained for creep under a constant stress. This produces

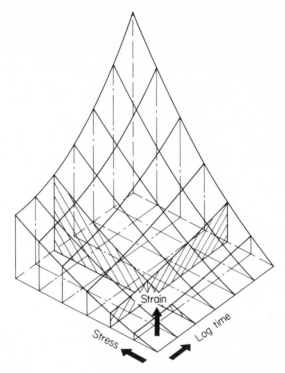

Figure 9.1. The stress–strain–time relationship obtained from creep (after Turner). [NN] Constant time section: isochronous stress–strain curve. [///] Constant stress section: creep curve.

the three-dimensional surface shown in Figure 9.1. It has been proposed by Turner[1] that this surface is in many practical cases defined to a sufficient degree of accuracy by a combination of two types of measurements.

(1) The relationship between stress and strain for a fixed time of measurement. This is the section in Figure 9.1 normal to the log time axis and is called the isochronous stress–strain curve. It is obtained by making a series of single-step-loading tests at different levels of stress and measuring the creep after a fixed time in each case.

(2) At least two creep curves at different stress levels, over a suitable time range, at the same temperature as the stress–strain curve.

Turner has presented results for creep of polypropylene which show how one isochronous stress–strain curve and two creep curves can be combined to allow a complete mapping of the creep behaviour (Figure 9.2). Although this procedure is very economical of experimental effort, data obtained

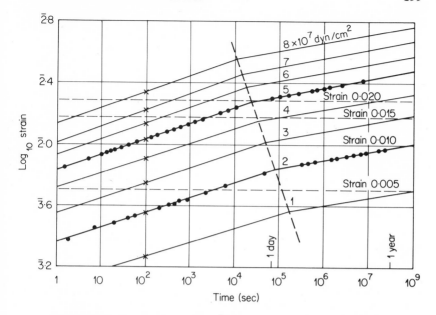

Figure 9.2. Tensile creep of polypropylene at 60°c. The stress and time dependence are approximately separable and therefore creep curves at intermediate stresses can be interpolated from a knowledge of two creep curves ● and the isochronous stress–strain relationship × (after Turner).

from single step loading tests have severe limitations when we attempt to use them to predict behaviour in more complex tests. The simplest of these is the recovery following the removal of load.

Consider the creep and recovery loading programme shown in Figure 5.7c. The recovery at time t, $e_r(t - t_1)$ is defined as the *difference* between (i) the strain at time t under continuous application of the initial stress and (ii) the strain at time t due to the application of the initial stress at zero time followed by its removal at time t_1.

For a *linear* viscoelastic solid the arguments presented in Section 5.2.1 above show that

$$e_r(t - t_1) = e_c(t) - [e_c(t) - e_c(t - t_1)] = e_c(t - t_1) \qquad (9.1)$$

where $e_c(t)$ is the creep under the applied stress for a time t and $e_c(t - t_1)$ is the creep under the applied stress for a time $(t - t_1)$.

The recovery behaviour of polypropylene under typical conditions is shown in Figure 9.3, together with the predicted behaviour on the basis of Equation 9.1. There is a very appreciable divergence. A rigorous

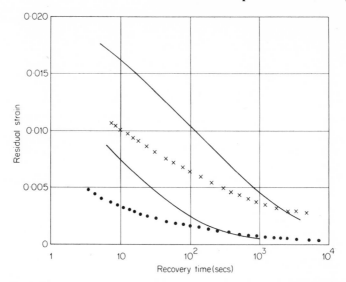

Figure 9.3. Recovery of polypropylene polymer at 20°C after creep under a stress of 2×10^8 dyn/cm^2. × after creep for 1000 secs; ● after creep for 100 secs; — recovery predicted for linear viscoelastic behaviour (after Turner).

treatment, to be given later, describes this divergence in terms of more complex memory functions than those provided by the Boltzmann Superposition Principle. At this stage in the discussion we shall develop an empirical description of the behaviour.

Turner[1] has proposed that two new quantities be introduced. The first one is called the 'Fractional Recovery' (FR) and is defined as:

$$\text{FR} = \frac{\text{strain recovered}}{\text{maximum creep strain}} = \frac{e_c(t_1) - e'_r(t)}{e_c(t_1)}$$

where $e_c(t_1)$ is the creep strain at $t = t_1$ i.e. the time at which the load is removed, and $e'_r(t)$ is a new quantity called the residual strain. $e'_r(t)$ is the strain at a time t in a creep and recovery programme such as that shown in Figure 5.7c.
i.e. for a linear viscoelastic material

$$e'_r(t) = e_c(t) - e_c(t - t_1)$$

When FR is used as the parameter the family of diverse recovery curves obtained for fixed t_1 under different stresses can be brought into approximate coincidence. This is shown in Figures 9.4a and b.

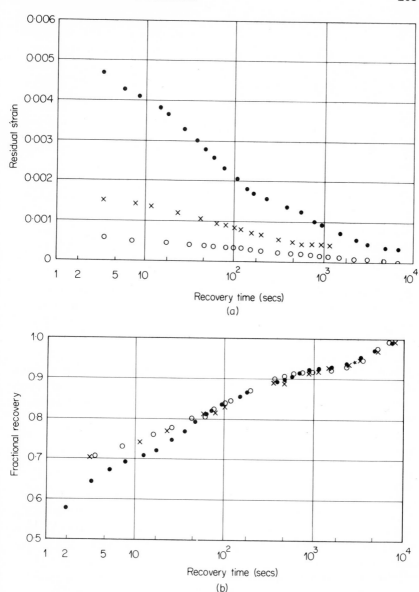

Figure 9.4. Recovery of polypropylene polymer after tensile creep for 1000 seconds at three stress levels: ● $1{\cdot}0 \times 10^8$ dyn/cm^2, × $5{\cdot}0 \times 10^7$ dyn/cm^2, ○ $2{\cdot}0 \times 10^7$ dyn/cm^2 (after Turner). The ordinate is either residual strain (a) or fractional recovery (b).

A second generalization is achieved when a second quantity, 'reduced time', is used. If we plot the FR versus log recovery time for one value of initial load, but for different times of application of this initial load (i.e. different t_1), a family of curves is obtained, as shown in Figure 9.5a. Defining a reduced time t_R as:

$$t_R = \frac{\text{recovery time}}{\text{creep time}} = \frac{t-t_1}{t_1}$$

and replotting the recovery now as a function of t_R, gives approximate coincidence (Figure 9.5b).

Turner points out that this recovery behaviour, in terms of FR and reduced time, is to be expected if linear superposition can be applied to a material which obeys a power law relationship. The assumption is that although the creep under a variety of loads is non-linear, superposition still applies. We will see some further experimental justification for this hypothesis in later work (Section 9.5.2 below).

It is assumed that creep is determined by a power law of the form $e_c(t) = At^n$ where A is a (non-linear) function of stress.

The creep under load for a time $t = t_1$ is then $e_c(t_1) = At_1{}^n$ and the residual strain is given by:

$$e'_r(t) = At^n - A(t-t_1)^n$$

The FR is therefore

$$\frac{e_c(t_1) - e'_r(t)}{e_c(t_1)} = \frac{At_1{}^n - [At^n - A(t-t_1)^n]}{At_1{}^n}$$

$$= 1 - \left(\frac{t}{t_1}\right)^n + \left(\frac{t}{t_1} - 1\right)^n$$

Since

$$t_R = \frac{t}{t_1} - 1$$

we have

$$\text{FR} = 1 + t_R{}^n - (t_R + 1)^n$$

i.e. FR is a unique function of the reduced time t_R.

These two simplifications proposed by Turner mark the present limit of what has been achieved without recourse to a more sophisticated representation. The essential idea behind these simplifications is that the non-linearity in stress and time can be separated.

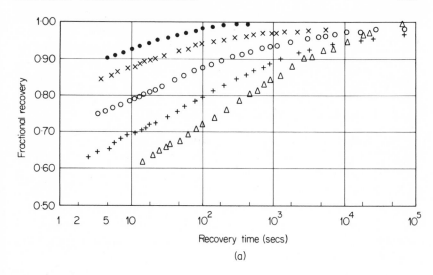

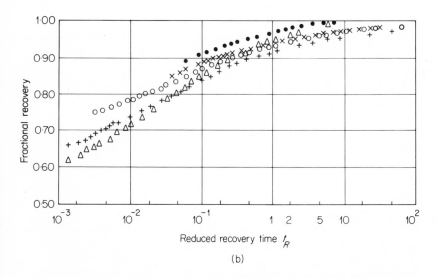

Figure 9.5. Fractional recovery of polypropylene polymer after tensile creep under a stress of 1×10^8 dyn/cm^2 for various times (after Turner). Creep time ● 50 sec, × 100 sec, ○ 1000 sec, + 3000 sec, △ 10,000 sec. (a) Fractional recovery as a function of recovery time; (b) Fractional recovery as a function of reduced recovery time.

We will discuss shortly how more general representations of non-linear viscoelasticity require a reformulation of the Boltzmann Superposition Principle. This leads to a considerable increase in complexity and it can well be argued that for practical purposes this is unprofitable unless it leads to a greater physical understanding.

9.2.2 Power Laws for Non-linear Viscoelasticity

Findley and his collaborators[2] have attempted to fit the creep of many plastics and plastic laminates to analytical relationships similar to those suggested for metals[3].

It was found that the creep strain e_c and time t could be related by an equation of the form

$$e_c(t) = e_0 + mt^n$$

where e_0 and m are functions of stress for a given material and n is a material constant.

Further work suggested that the results could be represented by

$$e_c(\sigma, t) = e'_0 \sinh \frac{\sigma}{\sigma_e} + m't^n \sinh \frac{\sigma}{\sigma_m}$$

where m', σ_e and σ_m are constants for the material.

This equation was a good fit to creep data obtained from single-step loading tests only, i.e. it is exactly equivalent to Turner's stress–strain–time surface.

A similar relationship was found for the creep of nitrocellulose by Van Holde.[4] He proposed that:

$$e_c(t) = e_0 + m't^{1/3} \sin \alpha\sigma$$

where α is a constant. This relationship reduces at low stresses to the Andrade creep law for metals[5].

$$e_c(t) = e_0 + \beta't^{1/3}$$

where β' is a constant.

The latter equation is consistent with *linear* viscoelastic behaviour. Plazek and his collaborators[6] have suggested that the Andrade creep law holds for several polymers and gels, although there is a divergence from linear behaviour at long times.

Findley's empirical equations are very useful to the design engineer for constant stress loading conditions as he can predict the creep for a given material if he is given the constants e_0, m, and n or e'_0, m', n, σ_e and σ_m.

The empirical approaches suggested so far have two principal limitations.

(1) They do not provide a general representation for creep, recovery and behaviour under complicated loading programmes.

(2) Creep data in these formulations cannot be simply related to stress relaxation and dynamic mechanical data.

9.3   THE MOLECULAR APPROACH

A simple starting point for the molecular approach to non-linear visco-elastic behaviour is closely allied to the attempts made to gain a molecular understanding of solution viscosities on the basis of the theory of thermally activated rate processes. This approach is due to Eyring and his co-workers[7].

It is assumed that deformation of the polymer involves the motion of chain molecules or parts of a chain molecule over potential-energy barriers.

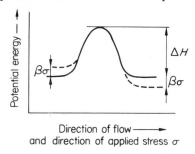

Figure 9.6. The Eyring model for creep.

The basic molecular process could be either intermolecular (e.g. chain-sliding) or intramolecular (e.g. a change in the conformation of the chain). The situation is illustrated schematically in Figure 9.6. With no stress acting, a dynamic equilibrium exists, chain segments moving with a frequency v over the potential barrier in each direction where $v = v_0 e^{-\Delta H/RT}$.

This equation is identical to Equation 7.2 above, describing the frequency of a molecular event. ΔH is an activation energy and v_0 involves the fundamental vibration frequency and the entropy contribution to the free energy.

It is assumed that the applied stress σ produces linear shifts $\beta\sigma$ of the energy barriers in a symmetrical fashion.

We then have a flow

$$v_1 = v_0 \exp\left(-\frac{(\Delta H - \beta\sigma)}{RT}\right)$$

in the forward direction (i.e. the direction of application of the stress) and

$$v_2 = v_0 \exp\left(-\frac{(\Delta H + \beta\sigma)}{RT}\right)$$

in the backward direction. This gives a net flow

$$v^1 = v_1 - v_2 = v_0\, e^{-\Delta H/RT}\{e^{\beta\sigma/RT} - e^{-\beta\sigma/RT}\}$$

in the forward direction.

If we assume that the net flow in the forward direction is directly related to the rate of change of strain we have that

$$\frac{de}{dt} = A \sinh \alpha\sigma$$

where A is a new constant of suitable dimensions and $\alpha = \beta/RT$.

This equation defines an 'activated' non-Newtonian viscosity. Eyring developed his ideas around 1940. It was therefore natural that he should attempt to incorporate them into the spring and dashpot models of viscoelasticity which were in vogue at that time.

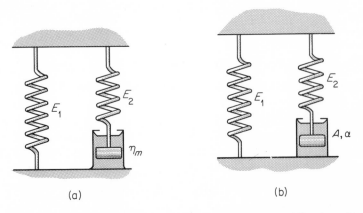

Figure 9.7. (a) The standard linear solid; (b) Eyring's modification of the standard linear solid with the activated dashpot.

Consider the standard linear solid (Figure 9.7a) and replace the dashpot with viscosity η_m by the activated dashpot defined by constants A, α (Figure 9.7b). This leads to a more complicated relationship between stress and strain than that for the standard linear solid, giving non-linear viscoelastic behaviour.

Eyring and his collaborators took Leaderman's data for the creep of silk and other fibres[8] (these results will be discussed further here). They showed that this model gave a good fit at a given level of stress by suitable choice of the four parameters E_1, E_2, A and α, over the four decades of time observed. An attempt to fit the data to the three-parameter standard linear solid was only successful over about one and a half decades.

The particular attraction of this model with the 'activated' dashpot arises from the following. Although creep curves are sigmoidal over a very long time interval when plotted on a logarithmic time scale, over the middle time region they are to a good approximation a straight line. Now it so happens that this model gives just this algebraic form for the creep (i.e. $e = a' + b' \log t$).

Since Eyring undertook this work, the limitation of the model representations of *linear* viscoelasticity have been better appreciated, and the ideas of a relaxation-time spectrum fully accepted. This approach can therefore be regarded as a particular type of curve-fitting procedure, which does not necessarily make a real contribution to an understanding of the non-linear behaviour.

9.4 THE RHEOLOGIST'S APPROACH TO NON-LINEAR VISCOELASTICITY

9.4.1 Large-strain Behaviour of Elastomers

A semi-empirical extension of linear viscoelastic theory has been used with considerable success by T. L. Smith[9] to explain the large-strain behaviour of elastomers.[10]

Consider a Maxwell element, where the stress σ and the strain e are related by the equation

$$\frac{de}{dt} = \frac{\sigma}{\eta} + \frac{1}{E}\frac{d\sigma}{dt}$$

For a constant rate of increase of strain $de/dt = R$, it may be readily shown (see Section 5.2.5 above) that

$$\sigma = \eta R(1 - e^{-t/\tau}) \quad \text{where} \quad \tau = \eta/E$$

i.e.

$$\sigma = R\tau E(1 - e^{-t/\tau})$$

It would thus follow that for a continuous distribution of relaxation times $H(\tau)$ this generalizes to

$$\sigma = R \int_{-\infty}^{\infty} \tau H(\tau)(1 - e^{-t/\tau}) \, d \ln \tau$$

For the constant strain rate R, write $R = e/t$. Then

$$\frac{\sigma}{e} = \frac{1}{t} \int_{-\infty}^{\infty} \tau H(\tau)(1 - e^{-t/\tau}) \, d \ln \tau + E_e$$

where the term E_e is added to denote the equilibrium modulus.

The quantity $$\frac{\sigma}{e} = \frac{\sigma(e, t)}{e},$$

which is a function of time only, is called the 'constant strain-rate modulus' $F(t)$. Smith assumes that for large strain $F(t)$ can be written as

$$F(t) = \frac{g(e)\sigma(e, t)}{e}$$

i.e.

$$\log F(t) = \log \left\{ \frac{g(e)}{e} \right\} + \log \sigma(e, t) \tag{9.2}$$

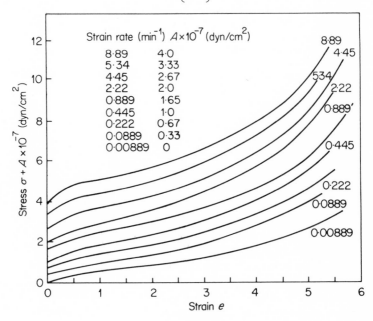

Figure 9.8. (a).

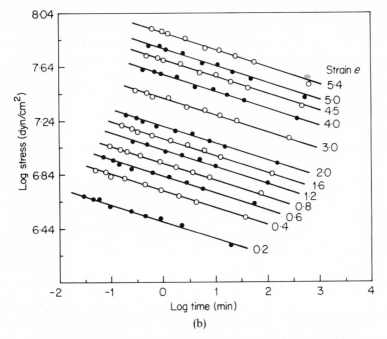

Figure 9.8. (a) Tensile stress–strain curves of SBR vulcanized rubber at $-34.4°C$ and at strain rates between 8.89×10^{-3} and 8.89 min^{-1}. The stress ordinates are displaced by an amount A to enable distinction between curves for different strain rates; (b) Variation of log stress with log time at different strain values for SBR vulcanized rubber. The data were obtained from analysis of the curves shown in Figure 9.8a and the strain values are indicated for each case (after Smith).

where $g(e)$ is some function of strain which approaches unity as the strain approaches zero.

This approach is very successful empirically, as shown in Figures 9.8a and 9.8b. The stress–strain curves at each strain rate are analysed by selecting the results for each value of strain and constructing plots of log σ, (σ = nominal stress i.e. load) versus log t. The resulting parallel plots (Figure 9.8b) show that the results can be represented by Equation (9.2), the displacements being the factor log $g(e)/e$. This implies that log $g(e)/e$ is independent of time. From what has been discussed of time–temperature equivalence it might well be anticipated that $g(e)/e$ will also be independent of temperature. Smith showed that this was true over a wide temperature range, only breaking down at the lowest temperatures.

Up to extensions of 100% the function $g(e)/e$ has a simple form. It was found that

$$\lambda\sigma = Ee$$

where λ is the extension ratio (ratio of extended length to initial length). To the approximation that elastomers are incompressible, $\lambda\sigma$ is the true stress. The result that true stress is proportional to strain is physically plausible.

Above 100% extension Smith used an empirical formula

$$\sigma = E\frac{e}{\lambda^2}\exp A\left(\lambda - \frac{1}{\lambda}\right)$$

proposed by Martin, Roth and Stiehler[11].

These results are interpreted as implying that the non-linearity can be attributed to the large strains, and that the time and strain dependence are separable. It is not clear how applicable the treatment is to the general viscoelastic behaviour, i.e. to recovery or superposition. However, similar treatments have been used with some success for polyisobutylene and other elastomers by Guth and his colleagues[12] and by Tobolsky[13].

9.4.2 Creep and Recovery of Plasticized Polyvinyl Chloride

Leaderman[14] carried the type of analysis used by Smith one stage further in analysing the creep and recovery of a sample of plasticized polyvinyl chloride. The apparently remarkable result was obtained here that the initial rate of recovery from a given load was larger than the initial creep under that load. (See also Section 9.5 below.) The situation is illustrated in Figure 9.9.

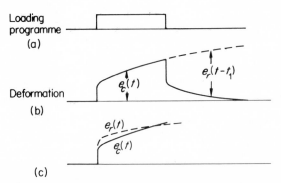

Figure 9.9. (a) Loading programme; (b) deformation; and (c) direct comparison of creep $e_c(t)$ and recovery $e_r(t)$ for a non-linear viscoelastic solid.

Leaderman showed that if $\frac{1}{3}(\lambda - 1/\lambda^2)$ is used as a measure of the deformation, both creep and recovery, and creep curves at different load levels can be described by a single time-dependent function. This is shown in Figures 9.10a and 9.10b. The quantity $\frac{1}{3}(\lambda - 1/\lambda^2)$ is the equivalent quantity to the Lagrangian strain measure in the theory of finite elasticity.

Let us consider why using $\frac{1}{3}(\lambda - 1/\lambda^2)$ as a measure of the deformation brings the creep and recovery curves into coincidence. As Leaderman defines recovery, (this is *not* how we have defined recovery previously in this text book) recovery measures the quantity

$$\frac{1}{3}\left(\lambda_1 - \frac{1}{\lambda_1{}^2}\right) - \frac{1}{3}\left(\lambda_2 - \frac{1}{\lambda_2{}^2}\right)$$

where λ_1 is the extension at the time of unloading, and λ_2 is the extension at a chosen time after unloading. If e_1 is the conventional strain at the time of unloading and e_2 is the conventional strain at a chosen time after unloading, $\lambda_1 = 1 + e_1$, $\lambda_2 = 1 + e_2$ and the recovery as defined by Leaderman in terms of conventional strain is $e_1 - e_2$.

Now a given change in the quantity $\frac{1}{3}(\lambda - 1/\lambda^2)$ at large λ (e.g. in the recovery situation where we change from λ_1 to λ_2), will involve a greater change in conventional strain $e_1 - e_2$ than it will at small λ (e.g. from $\lambda = 0$ in the creep situation). Thus recovery curves which coincide with creep curves using $\frac{1}{3}(\lambda - 1/\lambda^2)$ as a measure of the deformation will be larger than creep curves in the conventional strain representation.

9.4.3 Empirical Extension of the Boltzmann Superposition Principle

The rheological approaches to non-linear viscoelasticity discussed so far are only applicable to behaviour at large strains. Leaderman, in his extensive studies of the creep and recovery behaviour of textile fibres[8], was one of the first to appreciate that non-linear behaviour can occur at small strains.

Leaderman built the whole of his interpretation of viscoelastic behaviour on the Boltzmann Superposition Principle. As discussed in Chapter 5, this representation has three simple consequences which can be tested experimentally.

(1) Creep curves for different levels of one-step loading give a unique creep-compliance curve.

(2) Creep and recovery curves are identical for a given stress level, i.e. the creep under a particular constant load is identical to the recovery from creep under this load.

(3) In a two-step loading programme where a second load is added

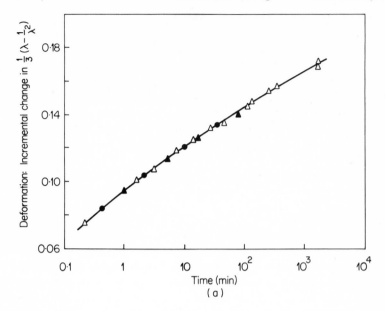

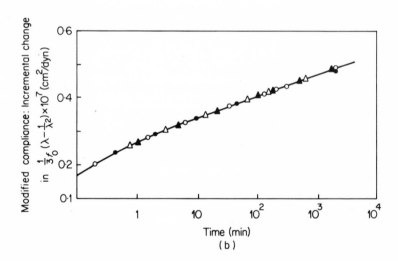

Figure 9.10. (a) Comparison of creep $\triangle$ and recovery $\blacktriangle$ for plasticized polyvinyl chloride under a constant nominal stress of $3\cdot554 \times 10^6$ dyn/cm^2; (b) Creep of plasticized polyvinyl chloride under constant nominal stress. $\bigcirc f_0 = 4\cdot443 \times 10^6$ dyn cm^{-2}; $\bullet f_0 = 3\cdot554 \times 10^6$ dyn cm^{-2}; $\triangle f_0 = 2\cdot666 \times 10^6$ dyn cm^{-2}; $\blacktriangle f_0 = 1\cdot777 \times 10^6$ dyn cm^{-2} (after Leaderman).

after an initial period of creep, the 'additional' creep due to the second load (i.e. total creep less creep under the initial load) is identical to the creep in a one-step loading programme under the second load only.

Leaderman found that for nylon and cellulosic fibres the creep-compliance curves did not coincide for different levels of stress. On the other hand, the creep and recovery curves for a given level of stress did coincide. It was also found that the compliance at the shortest time of measurement was identical for all levels of stress. The situation is summarized in Figure 9.11.

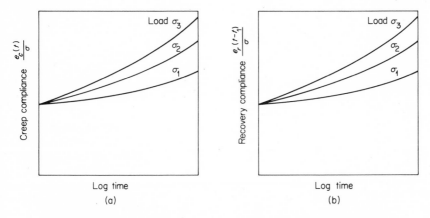

Figure 9.11. Comparison of creep compliance (a) and recovery compliance (b) at three load levels σ_1, σ_2, σ_3 for a non-linear viscoelastic material obeying Leaderman's modified Boltzmann superposition principle. Note that the creep and recovery curves for a given load level are identical.

The following explanation of these results was proposed:

(1) There is an instantaneous elastic deformation which is always proportional to the stress.

(2) The delayed creep and recovery for any level of load are a unique function of the stress.

This leads to a modified superposition principle,

$$e(t) = \frac{\sigma}{E} + \int_{-\infty}^{t} \frac{df(\sigma)}{d\tau}(t-\tau)\,d\tau$$

where $f(\sigma)$ is an empirical function of stress.

For different fibres apparently arbitrary functions of stress were obtained.

Leaderman's extension of the Boltzmann Superposition Principle provided a satisfactory representation of the behaviour in creep and recovery for the fibres which he examined. We will shortly see that it does not apply to the creep and recovery of all textile fibres. It will also appear that it does not describe the behaviour in more complicated loading programmes than creep and recovery.

9.5 FORMAL EXTENSION OF THE BOLTZMANN SUPER-POSITION PRINCIPLE

It has been proposed[15-22] that a more general formulation for non-linear viscoelasticity is required than Leaderman's empirical modification of the Boltzmann Superposition Principle.

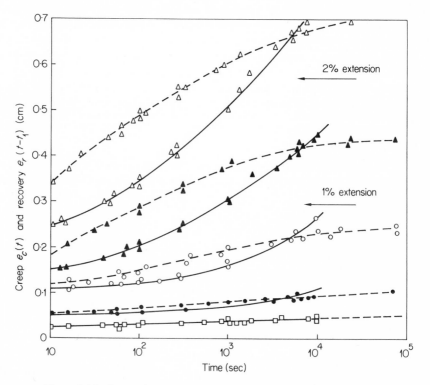

Figure 9.12. Successive creep ——— and recovery ————for an oriented monofilament of polypropylene of total length 30·2 cm. The load levels are △ 587 g; ▲ 401·8 g; ○ 281 g; ● 129·6 g; □ 67·6 g (after Ward and Onat).

The case of polypropylene fibres will be considered in detail, as the experimental results obtained illustrate the formal inadequacies of previous representations.

Figure 9.12 shows the creep and recovery curves against time for a polypropylene fibre at a number of different stress levels; this illustrates two features of the behaviour very clearly.

(1) Creep and recovery only coincide at the lowest stress level. This suggests that there is a linear viscoelastic region at low stresses, but that the behaviour is strikingly non-linear at higher stresses.

(2) The 'instantaneous' or short-time recovery is always greater than the instantaneous or short-time creep.

The second feature eliminates the possibility of using Leaderman's original formulation. He considered that the creep and recovery curves could be separated into two parts; an instantaneous or elastic response and a delayed deformation. Clearly for this polypropylene fibre there is no definitive measure of the elastic response.

The following illustrations (Figures 9.13a and b) show the creep and recovery compliances for different levels of stress. The creep-compliance curves are approximately independent of stress at lowest stress levels, as would be anticipated from Figure 9.12, but the recovery curves never

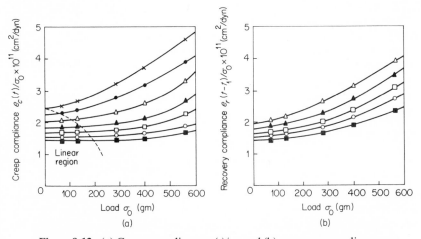

Figure 9.13. (a) Creep compliance $e_c(t)/\sigma_0$ and (b) recovery compliance $e_r(t-t_1)/\sigma_0$ as a function of applied load σ_0 for an oriented polypropylene monofilament (after Ward and Onat). The time of loading for the recovery test was $9 \cdot 3 \times 10^3$ secs. The times in seconds are given for both creep and recovery by the following key: × 9300; ● 3000; △ 1000; ▲ 300; □ 100; ○ 40; ■ 15.

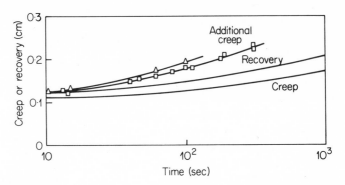

Figure 9.14. Comparison of creep and recovery curves for a load of 281 gm with additional creep due to addition of a further 281 gm after 3000 sec □ and 1000 sec △ respectively (after Ward and Onat).

coincide. Figure 9.14 shows the comparison of creep and recovery with additional creep. The additional creep curves are always in excess of the initial creep and show the same increase in 'instantaneous' elasticity as the recovery curves.

It can be seen that the simple tests of the Boltzmann Superposition Principle are not fulfilled for any case. Furthermore Leaderman's modification is not applicable for two reasons:

(1) Creep and recovery curves are not identical.

(2) There is not a linear 'instantaneous response'.

A more drastic modification of the Boltzmann Principle has therefore been proposed[16,23].

Consider the loading programme of Figure 9.15, in which incremental stresses $\Delta\sigma_1(\tau_1)$, $\Delta\sigma_2(\tau_2)$, $\Delta\sigma_3(\tau_3)$ ... etc., are added at the times τ_1, τ_2, τ_3, etc. For a linear system the deformation $e(t)$ at time t is given by:

$$e(t) = \Delta\sigma_1 J_1(t-\tau_1) + \Delta\sigma_2 J_1(t-\tau_2) \ldots$$

where $J_1(t)$ is the creep-compliance function. Let us now admit terms which arise from the *joint* contributions of the loading steps $\Delta\sigma_1(\tau_1)$, $\Delta\sigma_2(\tau_2)$, etc., to the final deformation.

These terms are taken to be of the form:

$$+\Delta\sigma_1\Delta\sigma_2 J_2(t-\tau_1, t-\tau_2) + \Delta\sigma_1\Delta\sigma_2\Delta\sigma_3 J_3(t-\tau_1, t-\tau_2, t-\tau_3) + \text{etc.},$$

where the 'memory functions' J_2, J_3 etc., are functions of *the differences*

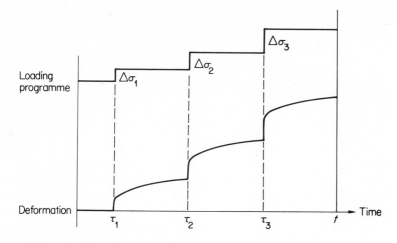

Figure 9.15. A multiple-step loading programme.

in time $t - \tau_1$, $t - \tau_2$, $t - \tau_3$, etc., between the instant in time at which the deformation is measured and the instant at which a given increment of stress $\Delta\sigma$ is applied. This condition was also used in setting up the Boltzmann principle.

For a continuous load history, the deformation $e(t)$ can then be written as an integral,

$$
\begin{aligned}
e(t) = {} & \int_{-\infty}^{t} J_1(t-\tau)\frac{d\sigma(\tau)}{d\tau}\,d\tau \\
& + \int_{-\infty}^{t}\int_{-\infty}^{t} J_2(t-\tau_1, t-\tau_2)\frac{d\sigma(\tau_1)}{d\tau_1}\frac{d\sigma(\tau_2)}{d\tau_2}\,d\tau_1\,d\tau_2 \\
& + \ldots + \int_{-\infty}^{t}\ldots\int_{-\infty}^{t} J_N(t-\tau_1\ldots t-\tau_N)\frac{d\sigma(\tau_1)}{d\tau_1}\ldots\frac{d\sigma(\tau_N)}{d\tau_N} \\
& \hspace{8cm} d\tau_1\ldots d\tau_N
\end{aligned}
$$

The first term is the Boltzmann Superposition Principle term, i.e. the linear term.

This representation is at first sight extremely complicated and the discussion will now be directed towards two features:

(1) The mathematical rigour of the representation.
(2) The practical application of the representation.

9.5.1 Mathematical Rigour

There are two theoretical aspects concerning this representation.

(1) It is being assumed that the elongation of the specimen at time t depends on all the previous values of the rate of loading to which the specimen has been subjected, i.e. the elongation is assumed to be a function of the history of the rate of loading

$$e(t) = F\left[\frac{d\sigma(\tau)}{d\tau}\right]_{\tau = -\infty}^{t}$$

In mathematical language F is a functional. There is then a theorem by Fréchèt quoted by Volterra and Pérès[24] which states that where F is continuous and non-linear the functional F can be represented to any degree of accuracy in the following manner:

$$e(t) = \int_{-\infty}^{t} J_1(t-\tau)\frac{d\sigma(\tau)}{d\tau}\,d\tau +$$

$$\int_{-\infty}^{t}\int_{-\infty}^{t} J_2(t-\tau_1, t-\tau_2)\frac{d\sigma(\tau_1)}{d\tau_1}\frac{d\sigma(\tau_2)}{d\tau_2}\,d\tau_1\,d\tau_2$$

$$+\ldots+ \int_{-\infty}^{t}\ldots\int_{-\infty}^{t} J_N(t-\tau_1,\ldots t-\tau_N)\frac{d\sigma(\tau_1)}{d\tau_1}\ldots\frac{d\sigma(\tau_N)}{d\tau_N}\,d\tau_1\ldots d\tau_N$$

This theorem gives formal justification for the multiple integral representation.

(2) The representation as discussed so far is for a one-dimensional situation. Green and Rivlin[23] have given a more complete development which does not suffer from this limitation.

Green and Rivlin[23] consider stress relaxation rather than creep. It is assumed that the stress at time t depends on the displacement gradients at time t and at N previous instants of time in the interval 0 to t. After considering the restrictions imposed by invariance under a rigid rotation, Green and Rivlin allow N to approach infinity, and obtain a multiple integral representation for general non-linear viscoelastic behaviour. Their representation describes stress relaxation. For the most general type of deformation it is not possible simply to invert the relationship to describe creep. This is because the stress functional involves the displacement gradients. Thus, during the deformation the stress components, in a fixed coordinate system, will change depending on the rotation of the body.

These objections do not apply to a one-dimensional situation, to which the remainder of this chapter is devoted.

9.5.2   The Practical Application of the Multiple-Integral Representation

The immediate objection to the multiple integral representation is that it is in principle so general that the data *must* be fitted, because there is an infinite number of curve-fitting constants.

This representation is therefore only useful if the following apply:

(1) The data can be fitted to a small number of multiple-integral terms.

(2) The behaviour under complex loading programmes can be predicted from a few simple loading programmes.

(3) The representation should at least assist an empirical understanding of the relationships between viscoelastic behaviour and the structure of the polymer.

With these qualifications in mind, the results for a polypropylene fibre[16,25,26] will be examined in detail.

For creep at a constant level of stress σ_0 the multiple-integral representation gives:

$$e_c(t) = J_1(t)\sigma_0 + J_2(t, t)\sigma_0^2 \dots J_N(t, \dots t)\sigma_0^N \dots \quad (9.3)$$

This immediately suggests that it will be of greater interest to examine the creep data in a different fashion, by plotting compliance $e(t)/\sigma_0$ against stress σ_0 for various fixed values of t, i.e. isochronous compliance curves.

The results are shown in Figure 9.13a. For a linear viscoelastic material, there would be a series of straight lines parallel to the load axis. In fact a small region approximates to this and has been so designated. In general the curves are parabolic which suggests that:

$$\frac{e_c(t)}{\sigma_0} = A + B\sigma_0^2,$$

where $A = J_1(t)$ and $B = J_3(t, t, t)$ are functions of time only.

It appears therefore that the non-linear behaviour of this particular specimen may be represented by retaining the first and third-order terms only. If this is correct the recovery response (Figure 9.16) in a creep and recovery test where $\sigma = 0, \tau < 0; \sigma = \sigma_0, 0 < \tau < t_1; \sigma = 0, \tau > t_1$, is given by

$$e_r(t - t_1) = J_1(t - t_1)\sigma_0 + J_3(t - t_1, t - t_1, t - t_1)\sigma_0^3$$
$$+ 3[J_3(t, t, t - t_1) - J_3(t, t - t_1, t - t_1)]\sigma_0^3 \dots \quad (9.4)$$

This shows that if our representation is adequate, recovery will take the form

$$\frac{e_r(t - t_1)}{\sigma_0} = A' + B'\sigma_0^2, \qquad \text{where } A' = J_1(t - t_1)$$

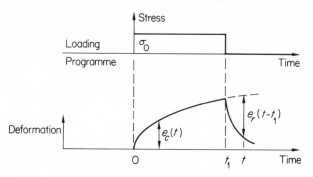

Figure 9.16. Creep and recovery programme.

and

$$B' = J_3(t-t_1, t-t_1, t-t_1) + 3[J_3(t, t, t-t_1) - J_3(t, t-t_1, t-t_1)]$$

are functions of time only.

The results are shown in Figure 9.13b, and demonstrate that this is correct. Equation (9.4) also shows that the recovery $e_r(t-t_1)$ in this test would be expected to be greater than the creep $e_c(t-t_1)$ in a creep test under constant stress σ_0 because we would expect $J_3(t, t, t-t_1) > J_3(t, t-t_1, t-t_1)$. The latter inequality is self-evident for linear creep behaviour, where it only states that $J(t) > J(t-t_1)$. This follows from the experimental fact that $J(t)$ is an increasing function of time t. The difference between 15-sec recovery and 15-sec creep was measured, and, as would be expected by comparing Equations (9.3) and (9.4), this (additional recovery compliance), $\dfrac{(e_r - e_c)}{\sigma}$ is also a parabolic function of the stress.

The third type of test employed is a two-step loading test with the loading programme

$$\sigma = 0, \quad \tau < 0; \qquad \sigma = \sigma_0, \quad 0 < \tau < t_1; \qquad \sigma = 2\sigma_0, \quad \tau > t_1$$

The additional creep response is defined as

$$e'_c(t-t_1) = e'(t, \sigma_0, t_1, \sigma_0) - e_c(t, \sigma_0), t > t_1$$

where $e'(t, \sigma_0, t_1, \sigma_0)$ is the elongation measured after the application of the second step of loading and $e_c(t, \sigma_0) = J(t)\sigma_0$. Evaluation of the multiple integral representation retaining the first and third-order term only gives

$$e'_c(t-t_1) = J_1(t-t_1)\sigma_0 + J_3(t-t_1, t-t_1, t-t_1)\sigma_0{}^3$$
$$+ 3[J_3(t, t, t-t_1) + J_3(t_1, t-t_1, t-t_1)]\sigma_0{}^3 \dots \quad (9.5)$$

The results shown in Figure 9.17 confirm that only first and third-order

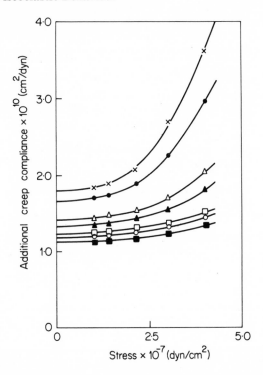

Figure 9.17. Additional creep compliance as a function of applied stress for an oriented polypropylene monofilament (after Ward and Wolfe).

terms are required for this specimen. Comparison of Equations (9.4) and (9.5) implies that the additional creep should be greater than the recovery for a given stress level, and this is confirmed by the experimental results shown in Figure 9.14.

This analysis affords a consistent if somewhat more complicated interpretation of the complexity of the viscoelastic behaviour, and offers some insight into the failure of simple superposition laws as discussed by Turner and reviewed in Section 9.2.1 above.

Attempts have also been made[26] to predict the behaviour in complex loading programmes from creep, recovery and superposition of identical loads. The predictions for these same polypropylene fibres for loading programmes involving superposition of a constant stress σ_0 and stresses of $2\sigma_0$ and $3\sigma_0$ respectively, after a fixed interval of time are shown in Table 9.1. The results are somewhat surprising. Although the creep and

Table 9.1. Comparisons of measured creep in complex loading programmes with the creep response predicted on the basis of the multiple-integral representation (predicted creep; second column) and with the creep response calculated by simple addition of individual creep responses (fourth column). The last column gives the correction factor due to the multiple integral term.

Loading programme	Predicted creep (cm)	Measured creep (cm)	Simple addition (cm)	Correction factor (cm)
1 $\sigma_0 + 2\sigma_0$	0.84_6	0.87_2	0.81_1	0.035
2 $2\sigma_0 + \sigma_0$	1.06_7	1.06_8	1.00_5	0.062
3 $\sigma_0 + 3\sigma_0$	0.83_4	0.86_8	0.79_8	0.036
4 $3\sigma_0 + \sigma_0$	1.15_9	1.15_6	1.10_6	0.053

Table 9.2. Additional creep e'_c, recovery e_r and creep e_c for the loading programmes detailed in Table 9.1.

Loading programme (from Table 9.1)	Individual terms e'_c, e_r, and e_c(cm) for $t - t_1 = 100$ s			Correction factor (cm)
	$e'_c(t - t_1)$	$e_r(t - t_1)$	$e_c(t - t_1)$	
1	0.236	0.227	0.223	0.035
2	0.610	0.509	0.463	0.062
3	0.173	0.168	0.166	0.036
4	0.737	0.608	0.537	0.053

recovery are markedly non-linear and very significantly different (Table 9.2) the corrections obtained by this sophisticated representation are small. It is only just convincing that this representation is better than a simple addition superposition law (compare second and fourth columns in Table 9.1). These results show, as the multiple integral representation emphasizes, that creep and superposition measurements can be very misleading with regard to recovery behaviour. In fact, a modified superposition rule proposed by Pipkin and Rogers[27] can bring the predicted and measured creep under complex loading programmes to very close agreement in this case. Their proposal is to add the *incremental* difference between the creep under stress σ_1 to that under stress σ_2 (see Figure 9.18).

This gives for a loading programme σ_1 at zero time followed by σ_2 at time $t = t_1$

$$e(t) = e_c(t, \sigma_1) + e_c(t - t_1, \sigma_1 + \sigma_2) - e_c(t - t_1, \sigma_1)$$

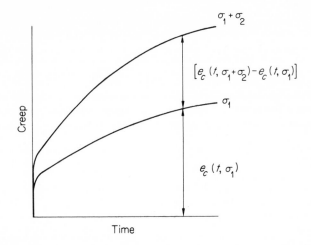

Figure 9.18. Comparison of creep under applied load σ_1 and $(\sigma_1 + \sigma_2)$ for a non-linear viscoelastic solid.

Such simplifications are extremely interesting, but they cannot be applied to an unknown polymer without first establishing the relative magnitude of the different multiple integral terms.

Finally, it is worth noting that there have been few attempts to make systematic examinations of polymers using the multiple-integral representation. Some preliminary work along these lines by Hadley and Ward[25] showed that the nature of the non-linear behaviour in polypropylene fibres was very much affected by the degree of molecular orientation and possibly by the morphological structure. Results for different levels of molecular weight were similar, although the absolute magnitude of the compliance level was significantly reduced with increasing molecular weight.

REFERENCES

1. S. Turner, *Polymer Engineering and Science*, **6**, 306 (1966).
2. W. N. Findley and G. Khosla, *J. Appl. Phys.*, **26**, 821 (1955).
3. P. G. Nutting, *J. Franklin Inst.*, **235**, 513 (1943).
4. K. Van Holde, *J. Polymer Sci.*, **24**, 417 (1957).
5. E. N. da C. Andrade, *Proc. Roy. Soc.*, **A84**, 1 (1910).
6. D. J. Plazek, *J. Colloid Sci.*, **15**, 50 (1960).
7. G. Halsey, H. J. White and H. Eyring, *Text. Res. J.*, **15**, 295 (1945).
8. H. Leaderman, *Elastic and Creep Properties of Filamentous Materials and Other High Polymers*, Textile Foundation, Washington, D.C., 1943.
9. T. L. Smith, *Trans. Soc. Rheol.*, **6**, 61 (1962).

10. I. H. Hall, *J. Polymer Sci.*, **54**, 505 (1961).
11. G. M. Martin, F. L. Roth and R. D. Stiehler, *Trans. Inst. Rubber Ind.*, **32**, 189 (1956).
12. E. Guth, P. E. Wack and R. L. Anthony, *J. Appl. Phys.*, **17**, 347 (1946).
13. A. V. Tobolsky and R. D. Andrews, *J. Chem. Phys.*, **13**, 3 (1945).
14. H. Leaderman, *Trans. Soc. Rheol.*, **6**, 361 (1962).
15. H. Leaderman, F. McCracken and O. Nakada, *Trans. Soc. Rheol.*, **7**, 111 (1963).
16. I. M. Ward and E. T. Onat, *J. Mech. Phys. Solids*, **11**, 217 (1963).
17. K. Onaran and W. N. Findley, *Trans. Soc. Rheol.*, **9**, 299 (1965).
18. W. N. Findley and J. S. Y. Lai, *Trans. Soc. Rheol.*, **11**, 361 (1967).
19. J. S. Y. Lai and W. N. Findley, *Trans. Soc. Rheol.*, **12**, 243 (1968).
20. J. S. Y. Lai and W. N. Findley, *Trans. Soc. Rheol.*, **12**, 259 (1968).
21. F. J. Lockett, *Intern. J. Eng. Sci.*, **3**, 59 (1965).
22. J. M. Lifshitz and H. Kolsky, *Technical Rept. No. A.M.22*, Brown University, Division of Applied Mathematics, 1966.
23. A. E. Green and R. S. Rivlin, *Arch. Rat. Mech.*, **1**, 1 (1957).
24. V. Volterra and J. Pérès, *Theorie Generale des Fonctionelles p-61*, Gauthier–Villars, Paris, 1936.
25. D. W. Hadley and I. M. Ward, *J. Mech. Phys. Solids*, **13**, 397 (1965).
26. I. M. Ward and J. M. Wolfe, *J. Mech. Phys. Solids*, **14**, 131 (1966).
27. A. C. Pipkin and T. G. Rogers, *J. Mech. Phys. Solids*, **16**, 59 (1968).

10

Anisotropic Mechanical Behaviour

10.1 THE DESCRIPTION OF ANISOTROPIC MECHANICAL BEHAVIOUR

An oriented polymer is in the strictest terms an anisotropic non-linearly viscoelastic material. A comprehensive understanding of anisotropic mechanical behaviour is therefore a very considerable task. In this chapter we will restrict the discussion to cases where the strains are small.

The mechanical properties of an anisotropic elastic solid for small strains are defined by the generalized Hooke's Law.

$$\varepsilon_{ij} = s_{ijkl}\sigma_{kl}$$

$$\sigma_{ij} = c_{ijkl}\varepsilon_{kl}$$

where the s_{ijkl} are the compliance constants and the c_{ijkl} are the stiffness constants. This has been discussed in Chapter 2. The use of this representation does not necessarily restrict the discussion to time-independent behaviour. The compliance and stiffness constants could be time-dependent, defining creep compliances and relaxation stiffnesses in step-function loading experiments, or complex compliances and complex stiffnesses in dynamic mechanical measurements. For simplicity, the methods of measurement are usually carefully standardized, e.g. by measuring each creep compliance after the same loading programme and the same time interval. It will be assumed that for such measurements there is an exact equivalence between elastic and linear viscoelastic behaviour, as proposed by Biot[1].

In an elastic material, the presence of symmetry elements leads to a reduction in the number of independent elastic constants and corresponding reductions will be assumed for anisotropic linear viscoelastic behaviour, although there is not enough experimental evidence to confirm that exactly the same rules hold in every case[2].

There are two important points to emphasize regarding the description of data:

(1) It is usually more convenient to work in terms of compliance constants than stiffness constants. This is because in the experimental procedures it is easier to apply a simple stress of a given type, e.g. a tensile stress or a shear stress, and measure the corresponding strains, e.g.

$$\varepsilon_{xx} = s_{1111}\sigma_{xx} + s_{1122}\sigma_{yy} + s_{1133}\sigma_{zz} + s_{1113}\sigma_{xz} + s_{1123}\sigma_{yz} + s_{1112}\sigma_{xy}$$

The compliance constants $s_{1111}, s_{1122}, s_{1133}$ etc., can be found by applying stresses $\sigma_{xx}, \sigma_{yy}, \sigma_{zz}$ etc., and measuring ε_{xx} in each case. The procedure will become clearer as the various experimental methods are discussed.

(2) In practice an abbreviated notation is often used in which

$$e_p = s_{pq}\sigma_q$$

$$\sigma_p = c_{pq}e_q$$

As explained in Section 2.5, σ_p represents σ_{xx}, σ_{yy}, etc., and e_q represents e_{xx}, e_{yy}, etc., and in the compliance and stiffness matrices s_{pq} and c_{pq} p, q take the values 1, 2 ... 6.

The conversion rules from the s_{ijkl} and c_{ijkl} notation to the abbreviated notation are given in Section 2.5 above.

It is important to remember that the engineering strains e_p are not the components of a tensor. Similarly, the 6×6 compliance matrix s_{pq}, does not represent a tensor and therefore tensor-manipulation rules do not apply. As will be demonstrated, for working out problems involving transformation of coordinates from one system of axes to another it is always desirable to use the original tensor notation in terms of ε_{ij}, σ_{kl} and s_{ijkl} or c_{ijkl}.

10.2 MECHANICAL ANISOTROPY IN POLYMERS

10.2.1 The Elastic Constants for Specimens Possessing Fibre Symmetry

Studies of mechanical anisotropy in polymers have for the most part been restricted to drawn fibres and uniaxially drawn films, both of which show isotropy in a plane perpendicular to the direction of drawing. The number of independent elastic constants is reduced to five (Reference 3, p. 138). Choosing the z direction as the axis of symmetry the compliance tensor

s_{ijkl} reduces to

$$\begin{pmatrix} s_{1111} & s_{1122} & s_{1133} & 0 & 0 & 0 \\ s_{1122} & s_{1111} & s_{1133} & 0 & 0 & 0 \\ s_{1133} & s_{1133} & s_{3333} & 0 & 0 & 0 \\ 0 & 0 & 0 & s_{2323} & 0 & 0 \\ 0 & 0 & 0 & 0 & s_{2323} & 0 \\ 0 & 0 & 0 & 0 & 0 & \frac{1}{2}(s_{1111}-s_{1122}) \end{pmatrix}$$

Extensional Modulus and Poisson's ratio

$$S_{33} = \frac{1}{E_3}, \nu_{13} = \frac{-S_{13}}{S_{33}}$$

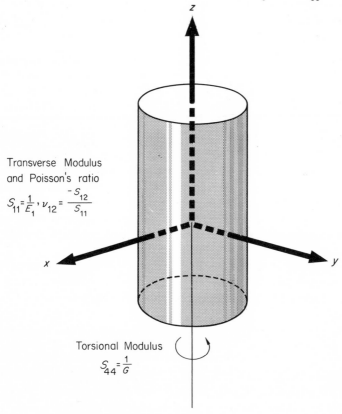

Transverse Modulus
and Poisson's ratio

$$S_{11} = \frac{1}{E_1}, \nu_{12} = \frac{-S_{12}}{S_{11}}$$

Torsional Modulus

$$S_{44} = \frac{1}{G}$$

Figure 10.1. The fibre compliance constants.

and the abbreviated matrix s_{pq} to

$$\begin{pmatrix} s_{11} & s_{12} & s_{13} & 0 & 0 & 0 \\ s_{12} & s_{11} & s_{13} & 0 & 0 & 0 \\ s_{13} & s_{13} & s_{33} & 0 & 0 & 0 \\ 0 & 0 & 0 & s_{44} & 0 & 0 \\ 0 & 0 & 0 & 0 & s_{44} & 0 \\ 0 & 0 & 0 & 0 & 0 & 2(s_{11}-s_{12}) \end{pmatrix}$$

The various compliance constants are illustrated diagrammatically in Figure 10.1. The situation is most easily appreciated for a fibre specimen, but a uniaxially oriented sheet possesses identical symmetry.

The relationships of these compliance constants to the better known Young's moduli and Poisson's ratios are as follows:

(1) Consider application of a stress along the z direction, i.e. along the fibre axis, or the draw direction for the polymer film.

Then $e_{zz} = s_{33}\sigma_{zz}$

$$\text{Young's modulus } E_3 = \frac{\sigma_{zz}}{e_{zz}} = \frac{1}{s_{33}} \qquad \text{giving } s_{33} = \frac{1}{E_3}$$

(2) Similarly the strain in the plane transverse to the fibre axis for a stress σ_{zz} along the fibre axis is given by

$$e_{xx} = e_{yy} = s_{13}\sigma_{zz}$$

$$\text{Poisson's ratio } v_{13} = -\frac{e_{xx}}{e_{zz}} = -\frac{s_{13}}{s_{33}}$$

(The negative sign ensures that Poisson's ratio is the conventionally positive quantity since e_{xx} is negative, i.e. a contraction.)

(3) In a similar manner, s_{11}, s_{12} and s_{13} are related to the modulus E_1 (the transverse modulus) and the corresponding Poisson's ratios $v_{21} = v_{12}$ and $v_{31} = v_{13}$ for application of a stress in a plane perpendicular to the fibre axis, i.e.

$$s_{11} = \frac{1}{E_1} \quad \text{and} \quad v_{21} = -\frac{s_{21}}{s_{11}} = -\frac{s_{12}}{s_{11}}, \qquad v_{31} = -\frac{s_{31}}{s_{11}} = -\frac{s_{13}}{s_{11}}$$

(4) The shear compliance is the reciprocal of the shear or torsional modulus G. There are two equivalent shear compliances $s_{44} = s_{55} = 1/G$. These relate to torsion about the symmetry axis z, i.e. shear in the yz or xz planes.

The shear compliance s_{66} relates to shear in the xy plane and is related to the compliance constants s_{11} and s_{12}, such that $s_{66} = 2(s_{11} - s_{12})$. This relationship expresses the fact that these specimens are isotropic in a plane perpendicular to the symmetry axis, i.e. that the elastic behaviour in this plane is specified by only two elastic constants as for an isotropic material. It will be seen that this property is very important in determining the elastic constants for fibres.

10.2.2 The Elastic Constants for Specimens Possessing Orthorhombic Symmetry

Oriented polymer films which are prepared by either rolling, rolling and annealing, or some commercial one-way draw processes, may possess orthorhombic rather than transversely isotropic symmetry. For such films the elastic behaviour is specified by nine independent elastic constants. Choose the initial drawing or rolling direction as the z axis for a system of rectangular Cartesian coordinates; the x axis to lie in the plane of the film and the y axis normal to the plane of the film (Figure 10.2). The compliance matrix is

$$
\begin{pmatrix}
s_{11} & s_{12} & s_{13} & 0 & 0 & 0 \\
s_{12} & s_{22} & s_{23} & 0 & 0 & 0 \\
s_{13} & s_{23} & s_{33} & 0 & 0 & 0 \\
0 & 0 & 0 & s_{44} & 0 & 0 \\
0 & 0 & 0 & 0 & s_{55} & 0 \\
0 & 0 & 0 & 0 & 0 & s_{66}
\end{pmatrix}
$$

There are three Young's moduli

$$
E_1 = \frac{1}{s_{11}}, \quad E_2 = \frac{1}{s_{22}} \quad \text{and} \quad E_3 = \frac{1}{s_{33}}
$$

and six Poisson's ratios

$$
\nu_{21} = -\frac{s_{21}}{s_{11}}, \quad \nu_{31} = -\frac{s_{31}}{s_{11}}, \quad \nu_{32} = -\frac{s_{32}}{s_{22}}, \quad \nu_{12} = -\frac{s_{12}}{s_{22}},
$$

$$
\nu_{13} = -\frac{s_{13}}{s_{33}}, \quad \nu_{23} = -\frac{s_{23}}{s_{33}},
$$

corresponding to situations where a tensile stress is applied along the x, y and z directions.

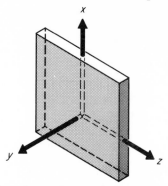

Figure 10.2. Choice of axes for a polymer sheet possessing orthorhombic symmetry.

There are three independent shear moduli $G_1 = 1/s_{44}$, $G_2 = 1/s_{55}$ and
$G_3 = 1/s_{66}$ corresponding to shear in the yz, xz, and xy planes respectively.
For a sheet of general dimensions, torsion experiments where the sheet
is twisted about the x, y, or z axis will involve a combination of shear
compliances. This will be discussed in greater detail later, when methods
of obtaining the elastic constants are described.

10.3 MEASUREMENT OF ELASTIC CONSTANTS

The measurement of elastic constants is a very different undertaking for
the two situations of a sheet and a fibre. The experimental methods
employed for these two cases will therefore be discussed separately.

10.3.1 Measurements on Films or Sheets

Extensional Moduli

The simplest measurement on a polymer film is to determine the
Young's modulus in various directions in the film by cutting long thin
strips in the selected directions.

We will consider a film of orthorhombic symmetry, the x and z axes
lying in the plane of the film and the y axis normal to the film as in
Section 10.2.2 above.

Consider a long strip cut in a direction making an angle θ with the z
direction (Figure 10.3a).

The Young's modulus for this strip $E_\theta = 1/s_\theta$ where s_θ is the compliance
in a direction making an angle θ with the z direction.

To calculate s_θ in terms of the compliance constants we will use the
full tensor notation.

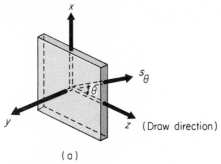

(a)

Figure 10.3. (a) The compliance s_θ is at an angle θ to the initial draw direction in the plane of the sheet.

The compliance constants s_{ijkl} referred to one system of Cartesian axes are related to those s'_{pqmn} referred to a second system of Cartesian axes by the tensor-transformation rule.

$$s'_{pqmn} = a_{pi}a_{qj}a_{mk}a_{nl}s_{ijkl}$$

where $a_{pi}, a_{qj}, \ldots$ define the cosines of the angles between the p axis in the second system and the i axis in the first, the q axis in the second system and the j axis in the first, ... and p, q, m, n take the values 1, 2, 3 in the second system of axes and i, j, k, l take the values $1'2'3'$ in the first system of axes.

We will take the direction of the strip to be the z' direction in a second system of Cartesian axes.

Then $s_\theta = s_{3'3'3'3'}$ is given by

$$s_{3'3'3'3'} = a_{3'1}a_{3'1}a_{3'1}a_{3'1}s_{1111} + a_{3'3}a_{3'3}a_{3'3}a_{3'3}s_{3333}$$
$$+ a_{3'1}a_{3'1}a_{3'3}a_{3'3}s_{1133} + a_{3'3}a_{3'3}a_{3'1}a_{3'1}s_{3311}$$
$$+ a_{3'1}a_{3'3}a_{3'3}a_{3'1}s_{1331} + a_{3'3}a_{3'1}a_{3'1}a_{3'3}s_{3113}$$
$$+ a_{3'3}a_{3'1}a_{3'3}a_{3'1}s_{3131} + a_{3'1}a_{3'3}a_{3'1}a_{3'3}s_{1313}$$

Note that all compliance terms containing the suffix 2 will vanish, because $a_{3'2} = 0$.

The change in coordinate systems corresponds to a rotation of the coordinate axes through an angle θ about the y direction as axis.

We therefore put $a_{3'1} = \sin \theta$ and $a_{3'3} = \cos \theta$.

and $\quad s_{3'3'3'3'} = \sin^4 \theta s_{1111} + \cos^4 \theta s_{3333} + 2 \sin^2 \theta \cos^2 \theta s_{1133}$
$$+ 4 \sin^2 \theta \cos^2 \theta s_{1313}$$

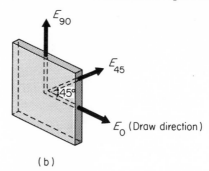

(b)

Figure 10.3. (b) The E_0, E_{45} and E_{90} moduli.

In the abbreviated notation

$$s_\theta = s_{3'3'} = \sin^4 \theta s_{11} + \cos^4 \theta s_{33} + \sin^2 \theta \cos^2 \theta(2s_{13} + s_{55}) \quad (10.1)$$

(Note factor 4 in converting from s_{ijkl} to s_{pq} when p and $q = 4, 5, 6$, i.e. 23, 13, 12).

It is thus possible to undertake three independent measurements on these sheets. For convenience choose these to be the Young's modulus on strips at 0, 45, and 90° to the initial draw direction and denote these by E_0, E_{45} and E_{90}, respectively (see Figure 10.3b). From Equation (10.1)

$$E_0 = \frac{1}{s_{33}}, \qquad E_{90} = \frac{1}{s_{11}} \quad \text{and} \quad \frac{1}{E_{45}} = \frac{1}{4}[s_{11} + s_{33} + (2s_{13} + s_{55})] \quad (10.2)$$

Such measurements therefore yield two of the nine independent elastic constants, s_{11} and s_{33}, directly, give a combination of s_{13} and s_{55}, but do not involve s_{12}.

For a transversely isotropic sheet where z is the symmetry axis, there are only five independent elastic constants, and $s_{55} = s_{44}$.

Torsion of Oriented Polymer Sheets

Torsion of oriented polymer sheets was undertaken by Raumann[4] to determine the shear compliances s_{44} and s_{66} for uniaxially oriented (transversely isotropic) low-density polyethylene. Torsion of oriented sheets can also be used to determine the shear compliances s_{44}, s_{55} and s_{66} for sheets possessing orthorhombic symmetry. As this situation is more general than that of transverse isotropy it will be considered first.

For the orthorhombic sheets, a solution can only be found to the elastic torsion problem when the sheets are cut as rectangular prisms with their

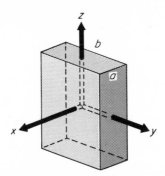

Figure 10.4. The orthorhombic sheet.

surfaces normal to the three axes of orthorhombic symmetry, and where the torsion axis coincides with one of these three axes.

A typical situation is illustrated in Figure 10.4. Torsion about the z axis involves the shear compliances in the yz and xz planes which are s_{44} and s_{55}, respectively.

The St. Venant Theory (see Reference 5, p. 201) gives the torque Q_z required to produce the twist T in a specimen of length l, thickness a and width b.

$$Q_z = \frac{ab^3 T}{s_{55} l} \beta(c_z) = \frac{ba^3 T}{s_{44} l} \overset{+}{\beta}(c_z) \qquad \text{i.e. } \frac{Q_z}{T} = \frac{\text{torsional rigidity of the}}{\text{specimen}}$$

where

$$c_z = \frac{1}{\overset{+}{c_z}} = \frac{a}{b} \left(\frac{s_{55}}{s_{44}} \right)^{1/2}$$

and $\beta(c_z)$ is a rapidly converging function of c_z which for $c_z > 3$ can be approximated to:

$$\beta(c_z) = \frac{1}{3} \left\{ 1 - \frac{0 \cdot 630}{c_z} \right\}$$

For a transversely isotropic sheet (with z direction as axis of symmetry) a similar expression describes torsion about an axis perpendicular to the symmetry axis. In this case $s_{44} = s_{55}$ and the torque Q_z is given by

$$Q_z = \frac{bc^3 T}{s_{66} l} \overset{*}{\beta}(c) = \frac{cb^3 T}{s_{44} l} \overset{+}{\beta}(c)$$

where

$$\overset{+}{c} = \frac{1}{\overset{*}{c}} = \frac{c}{b}\left(\frac{s_{44}}{s_{66}}\right)^{1/2},$$

with b the thickness, c the width and l the length of the specimen.

These formulae show that the relative contribution of the various shear compliances to the torque depend on their relative magnitude and the aspect ratios a/b or b/c. In principle the compliance can therefore be obtained from measurements on sheets of different aspect ratios. It has, however, been found that for very anisotropic polymer sheets of low-density polyethylene these St. Venant formulae do not hold[6].

For transversely isotropic sheets a much simpler formula applies for torsion around the symmetry axis z where the torque Q_z is given by

$$Q_z = \frac{ab^3 T}{s_{44} l}\beta(c) \qquad \text{where } c = a/b$$

$\beta(c)$ is now the same function of $c = a/b$ only.

10.3.2 Measurements on Fibres and Monofilaments

Extensional Modulus $E_3 = 1/s_{33}$

Dynamic mechanical measurements have been used to study the influence of molecular orientation on the extensional moduli of fibres drawn to different draw ratios* and also to compare the extensional moduli for a wide range of textile fibres produced by conventional manufacturing processes. The most extensive studies of this type are those of Wakelin and coworkers[7] and Meredith[8].

Detailed measurements of the extensional modulus of monofilaments have been made by longitudinal wave-propagation methods, where the relationship of the extensional modulus to molecular orientation and crystallinity has been examined. Early investigations using this technique were made by Kolsky and Hillier[9], Ballou and Smith[10], Nolle[11] and Hamburger[12]. The experimental method of Hillier and Kolsky, which was very similar to that of Ballou and Smith, is described in detail in Section 6.5.

More recently the measurement of the extensional modulus has been reexamined as a possible method for the measurement of molecular

---

* The draw ratio is the ratio of the length of a line parallel to the draw direction in the drawn material to its length before drawing. For synthetic fibres it is often determined by measuring the ratio of the initial diameter D_i to final diameter D_f, assuming that volume is conserved, i.e. draw ratio = D_i/D_f.

orientation in textile yarns by Charch and Moseley[13,14] and by Morgan[15]. Morgan has developed Hamburger's pulse-propagation method.

The Torsional Modulus $G = 1/s_{44}$

A convenient dynamic method for measuring the torsional modulus of synthetic fibre filaments was developed by Wakelin, Voong, Montgomery and Dusenbury[7].

A simpler method is that adopted by Meredith[8], where the fibre undertakes free torsional vibrations supporting known inertia bars at its free end.

The Extensional Poisson's Ratio $v_{13} = -\dfrac{s_{13}}{s_{33}}$

Measurements of the extensional Poisson's ratio v_{13} have been attempted using optical diffraction and mercury-displacement techniques by Davis[16] and Frank and Ruoff[17], respectively. Satisfactory data were only obtained for nylon because this fibre is a particularly favourable case, showing no permanent deformation up to 5% extension.

More recently, measurements have been made by observing in a microscope the radial contraction, together with the corresponding lateral extension of a fibre monofilament[18]. The monofilament was extended between two moveable grips which were part of a specially constructed microscope stage. Two ink marks were placed on the monofilament to act as reference points for the measurement of length and changes in length. An immersion liquid was used to reduce diffraction effects at the edges of the monofilament. The method was of limited accuracy, and errors of at least 10% were reported for 95% confidence limits on the mean value.

The Transverse Modulus $E_1 = 1/s_{11}$

The two remaining elastic constants for fibres, the compliances s_{11} and s_{12} have to be determined by more sophisticated methods. Both can be obtained from the compression of fibre monofilaments between parallel plates under conditions of plane strain. The transverse modulus is involved in the contact width $2b$[18,19] (Figure 10.5a).

The monofilament is a transversely isotropic solid; thus it is isotropic in a plane perpendicular to the fibre axis. This implies that under compressive loading normal to the fibre axis the stresses in the transverse plane will be identical in form to those for the compression of an isotropic cylinder. As the length of monofilament under compression is comparatively long, friction ensures that the compression occurs under

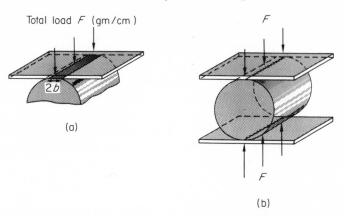

Total load F (gm/cm)

$2b$

(a)

F

F

(b)

Figure 10.5. The contact zone in the compression of a fibre monofilament (a) and for consideration of deformation in the central zone of the compressed monofilament it is sufficient to assume line contacts (b).

plane-strain conditions. There is, therefore, no change in dimension along the fibre axis ($e_{zz} = 0$) and only a normal stress acts along the fibre axis σ_{zz} which can be found in terms of the normal stresses σ_{xx} and σ_{yy} in the plane perpendicular to the fibre axis. We have

$$\sigma_{zz} = -\frac{s_{13}}{s_{33}}(\sigma_{xx} + \sigma_{yy})$$

All the stresses can therefore be obtained from the solution to the problem of compression of an isotropic cylinder. The corresponding strains can then be obtained using the constitutive equations $e_p = s_{pq}\sigma_q$.

The contact zone is arranged to be small compared with the radius of the monofilament. It is therefore adequate to assume that we are dealing with the contact between two semi-infinite solids and follow Hertz's classic solution for the compression of an isotropic cylinder[20]. In this solution the displacement of the cylinder within the contact zone is assumed to be parabolic and the boundary conditions are satisfied along the boundary plane only. For purely algebraic reasons it is most convenient to use the complex-variable method of McEwen[20] to obtain an analytical solution for b. It is found that

$$b^2 = \frac{4FR}{\pi}\left(s_{11} - \frac{s_{13}^2}{s_{33}}\right)$$

where F is the load per unit length of monofilament in gms/cm, and R

is the radius of the monofilament. This expression may be written as

$$b^2 = \frac{4FR}{\pi}(s_{11} - v_{13}^2 s_{33})$$

Highly oriented polymers are usually much stiffer along their axis than transverse to it. The quantity s_{33} is therefore usually very small compared with s_{11}. Since the Poisson's ratio v_{13} is typically near to 0·5, it follows that the term $v_{13}^2 s_{33}$ is only a small correction factor and that the contact width depends primarily on s_{11}. Thus the contact problem provides a good method in principle for determining s_{11}.

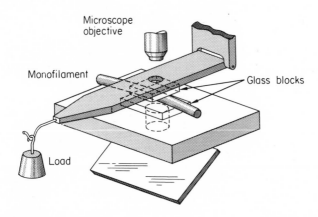

Figure 10.6. Schematic diagram of the compression experiment.

The apparatus is shown schematically in Figure 10.6. The monofilament is compressed between two parallel glass plates on a microscope stage. This is arranged as follows: A light but rigid metal bar is attached at one end to a pivot capable of slight vertical adjustments. A small vertical hole is bored through this lever arm, and immediately below the hole is cemented an optically flat block of glass, of thickness sufficient for only negligible distortion during loading. By hanging weights from the free end of the lever arm the monofilament is compressed between this block and a lower transparent glass flat placed over a small hole in a rigid base plate.

The loading device is fastened to the stage of a microscope equipped with a vertical illuminator, and the 45° plane mirror gives an enlarged image of the contact zone on a screen. At low loads, asperities and irregularities of the surface are very evident, and interference fringes are

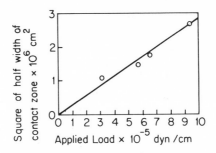

Figure 10.7. Compression of polyethylene terephthalate monofilament (diameter 2.82×10^{-2} cm): measurement of contact zone width as a function of applied load.

observed on each side of the contact zone. These fringes can be used to extrapolate to the true contact width, but this is a small correction which is well within the experimental error. A typical result for a polyethylene terephthalate monofilament is shown in Figure 10.7, showing the anticipated proportionality between the applied load and the square of the contact width.

The Transverse Poisson's Ratio $v_{12} = -s_{12}/s_{11}$

The transverse Poisson's ratio can be determined by measuring u_1, the change in diameter parallel to the plane of contact in the compression of the monofilament under conditions of plane strain as described in *the Transverse Modulus* section above.

A simple analysis of this problem follows from the condition that the contact zone can be arranged to be small compared with the radius of the monofilament. To calculate the deformations in the diametral plane it is then adequate to consider the problem as the compression of a cylinder under concentrated loads (Figure 10.5b). For an isotropic cylinder this is a well-known problem to be found in text books on elasticity (see Reference 22, p. 107). It is necessary to satisfy the boundary conditions on the surface of the cylinder, and this is done by addition of an isotropic tension in the plane perpendicular to the fibre axis.

The stresses for the transversely isotropic monofilament correspond exactly to those for the isotropic case. It is therefore very straightforward to calculate the strains and hence evaluate the diametrical expansion u_1. It is found that

$$u_1 = F\left\{\left(\frac{4}{\pi} - 1\right)\left(s_{11} - \frac{s_{13}^2}{s_{33}}\right) - \left(s_{12} - \frac{s_{13}^2}{s_{33}}\right)\right\}$$

For most oriented monofilaments, s_{13}^2/s_{33} is small compared with s_{11}, as discussed previously. Hence u_1 will depend primarily on s_{12}, with a substantial term in s_{11}, which is about $\frac{1}{4}s_{11}$. Thus the measurement of the diametral expansion provides a method for determining s_{12}, provided that s_{11} is determined from a measurement of the contact width b, as described in (4) above.

For the measurement of u_1, the apparatus shown in Figure 10.6 is again used, but the monofilament is surrounded by an immersion liquid, and the diameter is measured directly with a calibrated eyepiece. The immersion liquid is chosen to have a refractive index approximately equal to that of the monofilament, hence reducing diffraction effects without making the monofilament invisible. Very careful focusing of the microscope is necessary in these experiments. Inaccuracy in focusing can cause errors in the diameter measurements of the order of u_1 itself.

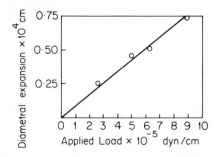

Figure 10.8. Compression of polyethylene terephthalate monofilament (diameter $2 \cdot 82 \times 10^{-2}$ cm): measurement of diametral expansion as a function of applied load.

A typical set of results for polyethylene terephthalate is shown in Figure 10.8. It can be seen that the change in diameter is proportional to the applied load, as predicted theoretically.

10.4 EXPERIMENTAL STUDIES OF MECHANICAL ANISOTROPY IN POLYMERS

The first attempts to determine a set of elastic constants for oriented polymers were Raumann and Saunders' measurements on low-density polyethylene sheets[4,23] and the combination of measurements on fibres and films of polyethylene terephthalate undertaken by Pinnock and Ward[24,25]. Later work included measurements on polyethylene terephthalate sheets by Raumann[26] and extensive measurements on various fibre monofilaments by Ward and coworkers[18,19,27]. In addition to these more complete surveys there are a few measurements of the extensional and torsional moduli of filaments (e.g. extensional and torsional modulus measurements on nylon and polyethylene terephthalate by Wakelin and coworkers[7], the sonic modulus measurements of Kolsky and Hillier[9], Ballou and Smith[10], and Morgan[15], and torsional measurements by Meredith[18]. All these measurements were confined to room temperature. Recently the low-density polyethylene data have been extended to cover a range of temperatures[28]. In addition there are a number of measurements on the viscoelastic behaviour of oriented polymers by Takayanagi and coworkers[29] and by Ward and coworkers[25,30].

Most measurements of mechanical anisotropy in polymers are for transversely isotropic systems, but a few measurements have been undertaken on rolled and annealed low-density polyethylene sheets possessing orthorhombic symmetry[31].

These experimental studies will now be discussed in some detail.

10.4.1 Low-density polyethylene sheets

Raumann and Saunders[23] prepared a series of oriented low-density polyethylene sheets by uniaxial stretching of isotropic sheets to varying final extensions. They measured the tensile modulus in directions making various angles with the initial draw direction, and presented their results in two types of diagram. Figure 10.9 shows the plot of Young's modulus $(1/s_\theta)$ for deformation in the plane of the sheet for a highly oriented

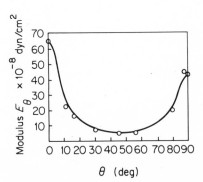

Figure 10.9. Comparison of the observed variation of modulus E_θ with angle θ to draw direction and the theoretical relation (i.e. full curve) calculated from E_0, E_{45} and E_{90} for low-density polyethylene sheet drawn to a draw ratio of 4·65 (after Raumann and Saunders).

sample. The unusual feature is that the sheet shows the lowest stiffness in a direction making an angle of about 45° to the initial draw direction, whereas intuitively one might expect the lowest stiffness direction to be at right angles to the draw direction, the latter being the direction of overall molecular orientation. Recalling the compliance equation (10.1)

$$s_\theta = s_{11} \sin^4 \theta + s_{33} \cos^4 \theta + (2s_{13} + s_{44}) \sin^2 \theta \cos^2 \theta$$

this experimental result implies that $(2s_{13} + s_{44})$ is much greater than either s_{11} or s_{33}, since when $\theta = 45°$ these terms will be equally weighted; see Equation (10.2) for E_{45}.

The second type of diagram is a plot of E_0, E_{45} and E_{90} as a function of draw ratio (Figure 10.10). Again the results are somewhat unexpected in that E_0 first falls with increasing draw ratio and $E_{90} > E_0$ at low draw

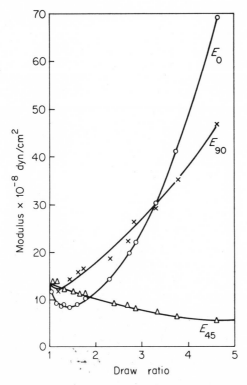

Figure 10.10. The variation of E_0, E_{45} and E_{90} with draw ratio in cold-drawn sheets of low-density polyethylene. Modulus measurements taken at room temperature (after Raumann and Saunders).

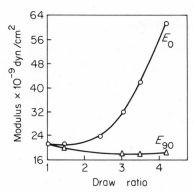

Figure 10.11. The variation of E_0 and E_{90} with draw ratio in cold-drawn sheets of low-density polyethylene. Modulus measurements taken at $-125°C$.

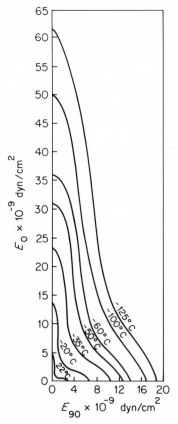

Figure 10.12. Polar representation of the mechanical anisotropy in a highly oriented low-density polyethylene sheet at different temperatures.

ratios, E_{90} and E_0 crossing at a draw ratio of about 3, as E_0 rises to a larger final value then E_{90} at the highest draw ratio.

Recent work by Gupta and Ward[28] has shown that this unexpected behaviour is specific to the room-temperature measurements on this polymer. Reducing temperature eventually produced a less complicated pattern for E_0, E_{45} and E_{90} as a function of draw ratio (Figure 10.11). We will shortly show that this is the more conventional situation. At the same time the polar diagram of the modulus in a highly oriented sheet changed markedly with decreasing temperature. (Figure 10.12.)

10.4.2 Nylon and Polyethylene Terephthalate Monofilaments

One of the earliest attempts to examine the mechanical anisotropy of fibres are results obtained by Wakelin and coworkers for the extensional and torsional moduli of nylon and polyethylene terephthalate filaments[7]. The results are shown in Figures 10.13 and 10.14. It can be seen that the extensional moduli increase steadily with increasing draw ratio in both cases, whereas the torsional moduli are comparatively unaffected.

10.4.3 Polyethylene Terephthalate, Nylon, Polyethylene and Polypropylene Monofilaments

In a comprehensive study of several fibre monofilaments, Hadley, Pinnock and Ward[18,19,27] have attempted to determine the five independent elastic constants for oriented polyethylene terephthalate, nylon, low and high-density polyethylene and polypropylene. All measurements were confined to room temperature, and the elastic constants were obtained as a function of molecular orientation as determined by the draw ratio and birefringence.

The results are summarized in Table 10.1 and Figures 10.15–10.19. (The calculated curves will be discussed in Section 10.6.)

The precise development of mechanical anisotropy in these fibres does depend on details of their chemical composition and the exact nature of the drawing process. However, the following general features can be distinguished.

The principal effect of increasing draw ratio (i.e. increasing molecular orientation) is to increase the Young's modulus E_3 measured in the direction of the fibre axis. Thus s_{33} for the highly oriented fibres is much less than the isotropic extensional compliance $(s_{33} = s_{11})$ for unoriented polymers. In nylon and polyethylene terephthalate there is a corresponding but small decrease in the transverse modulus E_1 with increasing draw ratio, for polypropylene and high-density polyethylene E_1 remains

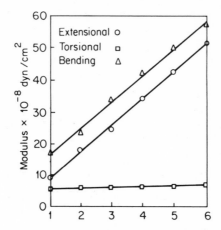

Figure 10.13. Elastic moduli of Nylon 66 of various draw ratios (after Wakelin et al.).

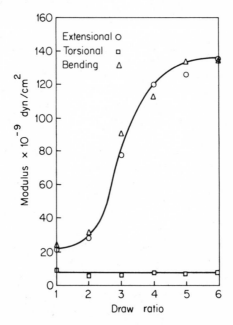

Figure 10.14. Elastic moduli of polyethylene terephthalate of various draw ratios (after Wakelin et al.).

Table 10.1. Elastic compliances of oriented fibres (units of compliance are cm^2 dyn^{-1} $\times$ 10^{11}; errors quoted are 95% confidence limits (see Reference 27).

Material	Birefringence (Δn)	s_{11}	s_{12}	s_{33}	s_{13}	s_{44}	$v_{13} = -\dfrac{s_{13}}{s_{33}}$	$v_{12} = -\dfrac{s_{12}}{s_{11}}$
Low-density polyethylene film[4]	—	22	-15	14	-7	680	0.50	0.68
Low-density polyethylene 1	0.0361	40± 4	$-25\pm$ 4	20± 2	$-11\pm$ 2	878± 56	0.55± 0.08	0.61± 0.20
Low-density polyethylene 2	0.0438	30± 3	$-22\pm$ 3	12± 1	$-7\pm$ 1	917± 150	0.58± 0.08	0.73± 0.20
High-density polyethylene 1	0.0464	24± 2	$-12\pm$ 1	11± 1	$-5.1\pm$ 0.7	34± 1	0.46± 0.15	0.52± 0.08
High-density polyethylene 2	0.0594	15± 1	$-16\pm$ 2	2.3± 0.3	$-0.77\pm$ 0.3	17± 2	0.33± 0.12	1.1± 0.14
Polypropylene 1	0.0220	19± 1	$-13\pm$ 2	6.7± 0.3	$-2.8\pm$ 1.0	18± 1.5	0.42± 0.16	0.68± 0.18
Polypropylene 2	0.0352	12± 2	$-17\pm$ 2	1.6± 0.04	$-0.73\pm$ 0.3	10± 2	0.47± 0.17	1.5± 0.3
Polyethylene terephthalate 1	0.153	8.9± 0.8	$-3.9\pm$ 0.7	1.1± 0.1	$-0.47\pm$ 0.05	14± 0.5	0.43± 0.06	0.44± 0.09
Polyethylene terephthalate 2	0.187	16± 2	$-5.8\pm$ 0.7	0.71± 0.04	$-0.31\pm$ 0.03	14± 0.2	0.44± 0.07	0.37± 0.06
Nylon	0.057	7.3± 0.7	$-1.9\pm$ 0.4	2.4± 0.3	$-1.1\pm$ 0.15	15± 1	0.48± 0.05	0.26± 0.08

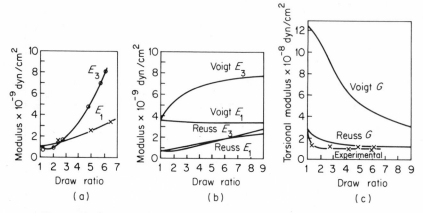

Figure 10.15. Low-density polyethylene filaments: extensional (E_3), transverse (E_1) and torsional moduli (G); comparison between experimental results and simple aggregate theory for E_3 and E_1 (a) and (b) and for G (c).

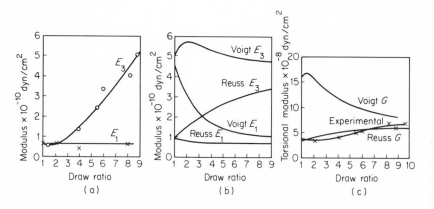

Figure 10.16. High-density polyethylene filaments: extensional (E_3), transverse (E_1) and torsional moduli (G); comparison between experimental results and simple aggregate theory for E_3 and E_1 (a) and (b) and for G (c).

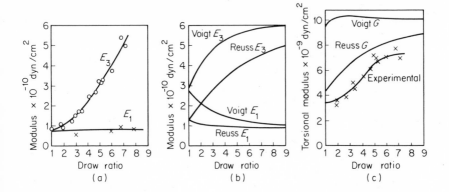

Figure 10.17. Polypropylene filaments: extensional (E_3), transverse (E_1) and torsional moduli (G); comparison between experimental results and simple aggregate theory for E_3 and E_1 (a) and (b) and for G (c).

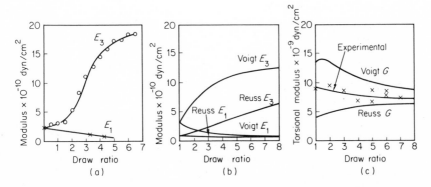

Figure 10.18. Polyethylene terephthalate filaments: extensional E_3, transverse E_1 and torsional moduli (G); comparison between experimental results and simple aggregate theory for E_3 and E_1 (a) and (b) and for G (c).

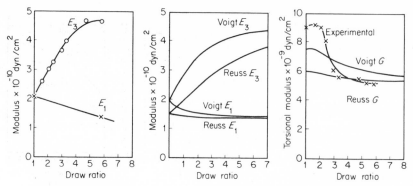

Figure 10.19. Nylon filaments: extensional (E_3), transverse (E_1) and torsional moduli (G); comparison between experimental results and simple aggregate theory for E_3 and E_1 (a) and (b) and for G (c).

almost constant, and for low-density polyethylene E_1 increases significantly with increasing draw ratio. The overall effect is that the extensional modulus E_3 for highly oriented filaments is greater than the transverse modulus E_1 (see Table 10.2). Polyethylene terephthalate is the most anisotropic fibre in this respect, with

$$\frac{E_3}{E_1} = \frac{s_{11}}{s_{33}} \sim 27.$$

Note the anomalous behaviour of low-density polyethylene at draw ratios less than 2, which is in agreement with the film data of Raumann and Saunders.

A further striking difference between low-density polyethylene and other fibres is shown by the change of the shear modulus G with draw ratio, a decrease greater than three times occurring over the range of molecular orientations examined, compared with only small changes for

Table 10.2. Comparison of elastic compliances of highly oriented and unoriented fibres (units of compliance are $cm^2 \, dyn^{-1} \times 10^{-11}$).

	Highly oriented			Unoriented	
	s_{11}	s_{33}	s_{44}	$s_{11} = s_{33}$	s_{44}
Low-density polyethylene	30	12	917	81	238
High-density polyethylene	15	2·3	17	17	26
Polypropylene	12	1·6	10	14	27
Polyethylene terephthalate	16	0·71	14	4·4	11
Nylon	7·3	2·4	15	4·8	12

the other materials. For the other fibres s_{44} lay close in value to s_{11}: in polyethylene terephthalate, high-density polyethylene and polypropylene $s_{44}/s_{11} \sim 1$; in nylon $s_{44}/s_{11} \sim 2$. Low-density polyethylene appears, as far as the present room-temperature measurements are concerned, to be an exceptional polymer with the extensional compliance s_{33} having the same order of magnitude as the transverse compliance s_{11}, and the shear compliance s_{44} being more than an order of magnitude greater than either s_{33} or s_{11}. This exceptional behaviour has been exemplified by a detailed analysis of the anisotropy in this polymer (see Section 8.4.4).

It is interesting to note the similarity between the elastic constants of high-density polyethylene and polypropylene: the relative values of the elastic compliances show very similar trends (see Table 10.1) and the major differences can be expressed as a marginally greater stiffness for polypropylene as might be expected from its higher melting point.

The compliance s_{13} is in all cases low, and appears to decrease rapidly with increasing draw ratio, in a similar manner to s_{33}. Thus the extensional Poisson's ratio $v_{13} = -s_{13}/s_{33}$ is rather insensitive to draw ratio and, with the exception of high-density polyethylene, does not differ significantly from 0.5. The assumption that the fibres are incompressible is thus generally a valid approximation. (Note that for anisotropic bodies, v_{13} is not confined to values less than 0.5, but is limited solely by the inequalities necessary for a positive strain energy:

$$s_{12}^2 < s_{11}^2; \; s_{13}^2 < \tfrac{1}{2}s_{33}(s_{11} + s_{12})$$

(see for example Reference 3).

The transverse Poisson's ratio $v_{12} = -s_{12}/s_{11}$ was subject to large experimental errors but even so the values can be seen to range widely, with those for polypropylene and high-density polyethylene being considerably higher than for the other materials.

Two small unusual features of the mechanical anisotropy can be noted. There is a small minimum in the extensional modulus E_3 of high-density polyethylene at low draw ratios, and a very small maximum in the torsional modulus of nylon at a draw ratio of 1.5.

10.4.4 Rolled and Annealed Polyethylene Sheets of Orthorhombic Symmetry[31]

Rolling and annealing processes recently established by Hay and Keller[32] (see also Point[33]) have enabled the production of oriented low-density polyethylene sheets which show orthorhombic symmetry in wide-angle X-ray diffraction measurements. Two types of sheet can be produced.

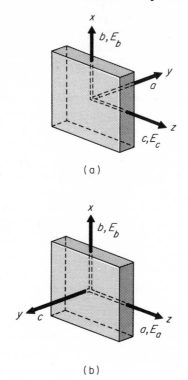

Figure 10.20. (a) The orthorhombic $b-c$ sheet; (b) the orthorhombic $a-b$ sheet.

In the first type (Figure 10.20a) the c-axes of the crystallites lie along the initial draw direction, b-axes lie in the plane of the sheet and a-axes normal to the plane of the sheet. In the second type (Figure 10.20b) the a-axes lie along the draw direction, b-axes again in the plane of the sheet and c-axes normal to the plane of the sheet.

Some of the salient results are shown in Tables 10.3 and 10.4, and Figures 10.21 and 10.22, which also include data for drawn and annealed sheets of fibre symmetry. Several features are to be noted:

(1) The $a-b$ and $b-c$ sheets are approximately complementary; E_b values for both sheets are very similar. Furthermore, the appropriate average of E_a and E_b in the ab plane, E_{ab}, is close to the observed E_{90} value in the fibre symmetry sheet.

Table 10.3. Value of tensile modulus $\times 10^{-9}$ dyn/cm² of some low-density polyethylene sheets in different directions over a range of temperatures[31].

Temp. °C	$a-b$ sheet (annealing temperature 105°C) orthorhombic symmetry				Drawn and annealed low-density sheet (annealing temperature 105°C) fibre symmetry			$b-c$ sheet (annealing temperature 95°C) orthorhombic symmetry		
	E_a	E_{45}	E_b	Average E_{ab} calculated	E_{90}	E_0	E_{45}	E_c	E_b	E_{45}
50	0.51	0.74	1.00	0.71	0.69	0.44	0.30	0.82	1.08	0.27
22	1.14	1.63	2.11	1.65	1.57	1.11	0.82	2.05	2.10	0.67
−125	9.92	15.41	10.06	12.20	12.00	15.30	9.87	22.32	10.45	11.32

Table 10.4. Shear compliances $\times 10^{11}$ cm^2/dyn of low-density polyethylene sheets based on E_{45} measurements at room temperature.[32]

	Shear compliances		
Structure	s_{44}	s_{55}	s_{66}
Cold-drawn sheet (fibre symmetry)	760	760	—
Drawn and annealed sheet (fibre symmetry)	482	482	—
$a-b$ sheet (orthorhombic symmetry)	—	—	155
$b-c$ sheet (orthorhombic symmetry)	510	—	—

(2) There is a cross-over in modulus with temperature for the annealed sheet and for the orthorhombic sheets in that $E_0 > E_{90}$, $E_c > E_a \sim E_b$ at low temperatures, whereas at high temperatures $E_0 < E_{90}$ and $E_c < E_a$ or E_b. Possible interpretations of this behaviour will be discussed in Section 10.9 below.

(3) The shear compliances s_{44} and s_{55} are appreciably higher than s_{66}, which is consistent with the behaviour of cold-drawn sheets, showing some correlation of properties between annealed and unannealed sheets.

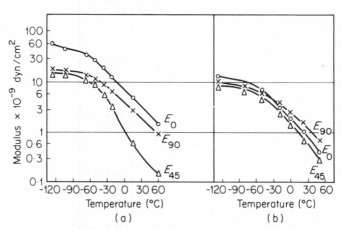

Figure 10.21. Mechanical anisotropy as a function of temperature for low-density polyethylene sheet: (a) cold drawn; (b) cold drawn and annealed.

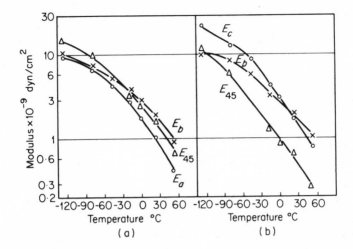

Figure 10.22. Mechanical anisotropy as a function of temperature for low-density polyethylene: (a) $a-b$ sheet; (b) $b-c$ sheet.

10.5 CORRELATION OF THE ELASTIC CONSTANTS OF AN ORIENTED POLYMER WITH THOSE OF AN ISOTROPIC POLYMER: THE AGGREGATE MODEL

It is to be expected that the mechanical properties of polymers will depend on the exact details of the molecular arrangements, i.e. both the crystalline morphology and the molecular orientation, these being intimately related so that any attempt to separate their influence must be an artificial one to a greater or lesser degree. In the case of polyethylene terephthalate it was found that the degree of molecular orientation (as measured from birefringence, for example) was the primary factor in determining the mechanical anisotropy. Table 10.5 shows some extensional and torsional modulus results for a number of polyethylene terephthalate fibres measured at room temperature. It can be seen that the influence of crystallinity on these moduli is small compared with the effect of molecular orientation on the extensional modulus. It has therefore been proposed that to a first approximation the unoriented fibre or polymer can be regarded as an aggregate of anisotropic elastic units whose elastic properties are those of the highly oriented fibre or polymer[34,35]. The average elastic constants for the aggregate can be obtained in two ways, either by assuming uniform stress throughout the aggregate (which will imply a summation of compliance constants) or uniform strain (which will imply

Table 10.5. Physical properties of polyethylene terephthalate fibres at room temperature[25].

Birefringence	X-ray crystallinity	Extensional modulus $\times 10^{-10}$ dyn/cm^2	Torsional modulus $\times 10^{-10}$ dyn/cm^2
0	0	2·0	0·77
0	33	2·2	0·89
0·142	31	9·8	0·81
0·159	30	11·4	0·62
0·190	29	15·7	0·79

a summation of stiffness constants). Because in general the principal axes of stress and strain do not coincide for an anisotropic solid these two approaches both involve an approximation. With the first assumption of uniform stress, the strains throughout the aggregate are not uniform; with the alternative assumption of uniform strain, non-uniformity of stress occurs. It was shown by Bishop and Hill[36] that for a random aggregate the correct value lies between the two extreme values predicted by these alternative schemes.

Consider the case of uniform stress. This can be imagined as a system of N elemental cubes arranged end-to-end forming a 'series' model (Figure 10.23a). Assume that each elemental cube is a transversely isotropic elastic solid, the direction of elastic symmetry being defined by the angle θ which its axis makes with the direction of applied external stress σ. The strain in each cube e_1 is then given by the compliance formula.

$$e_1 = [s_{11} \sin^4 \theta + s_{33} \cos^4 \theta + (2s_{13} + s_{44}) \sin^2 \theta \cos^2 \theta] \sigma$$

where s_{11}, s_{33}, etc., are the compliance constants of the cube. We ignore the fact that the cubes in general distort under the applied stress and do not satisfy compatibility of strain throughout the aggregate. Then the average strain e is:

$$e = \frac{\Sigma e_1}{N} = [s_{11} \overline{\sin^4 \theta} + s_{33} \overline{\cos^4 \theta} + (2s_{13} + s_{44}) \overline{\sin^2 \theta \cos^2 \theta}] \sigma$$

where $\overline{\sin^4 \theta}$, etc., now define the average values of $\sin^4 \theta$, etc., for the aggregate of units. For a random aggregate it is found that

e/σ = average extensional compliance = $\overline{s'}_{33}$

$$= \tfrac{8}{15} s_{11} + \tfrac{1}{5} s_{33} + \tfrac{2}{15} (2s_{13} + s_{44}) \qquad (10.3)$$

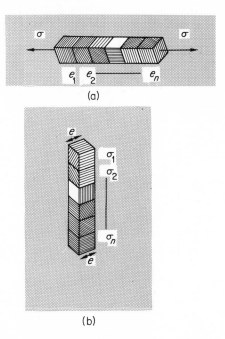

Figure 10.23. The aggregate model (a) for uniform stress; (b) for uniform strain.

In a similar manner the case of uniform strain can be imagined as a system of N elemental cubes stacked in a 'parallel' model (Figure 10.23b). For this case the stress in each cube σ_1 is given by the stiffness formula.

$$\sigma_1 = [c_{11} \sin^4 \theta + c_{33} \cos^4 \theta + 2(c_{13} + 2c_{44}) \sin^2 \theta \cos^2 \theta]e$$

where c_{11}, c_{33}, etc., are the stiffness constants of the cube. The average stress σ is then

$$\sigma = \frac{\Sigma \sigma_1}{N} = [c_{11} \overline{\sin^4 \theta} + c_{33} \overline{\cos^4 \theta} + 2(c_{13} + 2c_{44}) \overline{\sin^2 \theta \cos^2 \theta}]e$$

where $\overline{\sin^4 \theta}$ etc., are the average values of $\sin^4 \theta$. For a random aggregate

$$\frac{\sigma}{e} = \overline{c'_{33}} = \frac{8}{15}c_{11} + \frac{1}{5}c_{33} + \frac{4}{15}(c_{13} + 2c_{44}) \qquad (10.4)$$

Equations (10.3) and (10.4) define one compliance constant and one stiffness constant for the isotropic polymer. For an isotropic polymer there are two independent elastic constants, and these two schemes

Table 10.6. Comparison of calculated and measured extensional and torsional compliances ($\times 10^{11} cm^2/dyn$) for unoriented fibres.

| | Extensional compliance ($\overline{s'_{11}} = \overline{s'_{33}}$) | | | Torsional compliance ($\overline{s'_{44}}$) | | |
| | Calculated | | Measured | Calculated | | Measured |
	Reuss average	Voigt average		Reuss average	Voigt average	
Low-density polyethylene	139	26	81	416	80	238
High-density polyethylene	10	2·1	17	30	6	26
Polypropylene	7·7	3·8	14	23	11	2·7
Polyethylene terephthalate	10·4	3·0	4·4	25	7·6	11
Nylon	6·6	5·2	4·8	17	13	12

predict a value for the isotropic shear compliance $\overline{s'_{44}}$ and the isotropic shear stiffness $\overline{c'_{44}}$ respectively. These are:

$$\overline{s'_{44}} = \tfrac{14}{15}s_{11} - \tfrac{2}{3}s_{12} - \tfrac{8}{15}s_{13} + \tfrac{4}{15}s_{33} + \tfrac{2}{5}s_{44} \qquad (10.5)$$

$$\overline{c'_{44}} = \tfrac{7}{30}c_{11} - \tfrac{1}{6}c_{12} - \tfrac{2}{15}c_{13} + \tfrac{1}{15}c_{33} + \tfrac{2}{5}c_{44} \qquad (10.6)$$

Averaging the compliance constants defines the elastic properties of the isotropic aggregate in terms of $\overline{s'_{33}}$ and $\overline{s'_{44}}$. This is called the 'Reuss average'[37]. Averaging the stiffness constants defines the elastic properties of the aggregate in terms of $\overline{c'_{33}}$ and $\overline{c'_{44}}$. This is called the 'Voigt average'[38]. In the latter case it is desirable to invert the matrix and obtain the $\overline{s'_{33}}$ and $\overline{s'_{44}}$ corresponding to these values of $\overline{c'_{33}}$ and $\overline{c'_{44}}$ in order to compare directly the values obtained by the two averaging procedures.

The results of such a comparison are summarized in Table 10.6 for five polymers. For polyethylene terephthalate and low-density polyethylene the measured isotropic compliances lie between the calculated bounds, suggesting that in these polymers the molecular orientation is indeed the primary factor determining the mechanical anisotropy. In nylon the measured compliances lie just outside the bounds, suggesting that although molecular orientation is important in determining the mechanical anisotropy, other structural factors also play an important part. Finally, in high-density polyethylene and polypropylene the measured values for the isotropic compliances $\overline{s'_{11}} = \overline{s'_{33}}$ lie well outside the calculated bounds, suggesting that factors other than orientation play a major role in the mechanical anisotropy. In polypropylene Pinnock and Ward[39] suggested that simultaneous changes occur in morphology and molecular mobility, both of which affect the mechanical properties.

10.6 MECHANICAL ANISOTROPY OF FIBRES AND FILMS OF INTERMEDIATE MOLECULAR ORIENTATION

Fibres and films of intermediate molecular orientation are often produced by a two-stage process in which the first stage consists of making an approximately isotropic specimen which is then uniaxially stretched or drawn. The aggregate model can be extended to determine the mechanical anisotropy as a function of the draw ratio.

The starting point for such a theory was the observation that in general terms the birefringence–draw-ratio curves for several crystalline polymers take a similar form, as noted previously by several workers (Crawford and Kolsky for low-density polyethylene[40] and Cannon and Chappel

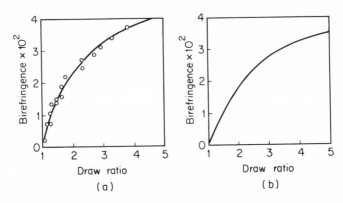

Figure 10.24. (a) Experimental and (b) theoretical curves for the birefringence of low-density polyethylene as a function of draw ratio.

for nylon[41]), with the birefringence increasing rapidly at low draw ratios, but approaching the maximum value asymptotically at draw ratios greater than about five. Results for low-density polyethylene are shown in Figure 10.24a.

Crawford and Kolsky concluded that the birefringence was directly related to the permanent strain, and they proposed a model of rod-like units rotating towards the draw direction on drawing. The essential mathematical step in the theory is illustrated in Figure 10.25. Each unit is considered to be transversely isotropic. The orientation of a single unit is therefore defined by the angle θ between its symmetry axis and the draw direction, and the angle ϕ which is the angle between the projection of the symmetry axis on a plane perpendicular to the draw direction and any direction in this plane. It is assumed that the symmetry axes of the anisotropic units rotate in the same manner as lines joining pairs of points in the macroscopic body, which deforms uniaxially at constant volume. This assumption is similar to the 'affine' deformation scheme of Kuhn and Grün for the optical anisotropy of rubbers[42], (see Section 4.1.2 above for a definition of 'affine'), but it ignores the required change in length of the units on deformation. We will therefore call it the 'pseudo-affine' deformation scheme. Kuhn and Grün did in fact

Initial draw
direction

Figure 10.25. The pseudo-affine deformation.

consider this scheme and reject it in their discussion of rubber-like behaviour. The angle θ in Figure 10.25 thus changes to θ', $\phi = \phi'$, and it
can be shown that

$$\tan \theta' = \frac{\tan \theta}{\lambda^{3/2}} \qquad \text{where } \lambda \text{ is the draw ratio}$$

This relationship can be used to calculate the orientation distribution
function for the units in terms of the draw ratio.

On this model the birefringence Δn of a uniaxially oriented polymer is
given by

$$\Delta n = \Delta n_{\max}(1 - \tfrac{3}{2}\overline{\sin^2 \theta})$$

where $\overline{\sin^2 \theta}$ is the average value of $\sin^2 \theta$ for the aggregate of units and
$\Delta n_{\max}$ is the maximum birefringence observed for full orientation.

This pseudo-affine deformation scheme gives a reasonable first order
fit to the birefringence data for low density polyethylene[40], nylon[41],
polyethylene terephthalate[43] and polypropylene[39]. Figure 10.24b shows
the first case. It is to be noted that this formulation of the birefringence
equation ignores the distinction between different structural elements
in the polymer (e.g. crystalline regions and disordered regions). With this
reservation in mind, the aggregate model is now extended to predict the
mechanical anisotropy in the manner outlined in Section 10.5 above.
This gives the following equations for the compliance constants s'_{11},
$s'_{12}, s'_{13}, s'_{33}$ and s'_{44} and the stiffness constants $c'_{11}, c'_{12}, c'_{13}, c'_{33}$ and
c'_{44} of the partially oriented polymer.

$$s'_{11} = \tfrac{1}{8}(3I_2 + 2I_5 + 3)s_{11} + \tfrac{1}{4}(3I_3 + I_4)s_{13} + \tfrac{3}{8}I_1 s_{33} + \tfrac{1}{8}(3I_3 + I_4)s_{44}$$

$$c'_{11} = \tfrac{1}{8}(3I_2 + 2I_5 + 3)c_{11} + \tfrac{1}{4}(3I_3 + I_4)c_{13} + \tfrac{3}{8}I_1 c_{33} + \tfrac{1}{2}(3I_3 + I_4)c_{44}$$

$$s'_{12} = \tfrac{1}{8}(I_2 - 2I_5 + 1)s_{11} + I_5 s_{12} + \tfrac{1}{4}(I_3 + 3I_4)s_{13} + \tfrac{1}{8}I_1 s_{33} + \tfrac{1}{8}(I_3 - I_4)s_{44}$$

$$c'_{12} = \tfrac{1}{8}(I_2 - 2I_5 + 1)c_{11} + I_5 c_{12} + \tfrac{1}{4}(I_3 + 3I_4)c_{13} + \tfrac{1}{8}I_1 c_{33} + \tfrac{1}{2}(I_3 - I_4)c_{44}$$

$$s'_{13} = \tfrac{1}{2}I_3 s_{11} + \tfrac{1}{2}I_4 s_{12} + \tfrac{1}{2}(I_1 + I_2 + I_5)s_{13} + \tfrac{1}{2}I_3 s_{33} - \tfrac{1}{2}I_3 s_{44}$$

$$c'_{13} = \tfrac{1}{2}I_3 c_{11} + \tfrac{1}{2}I_4 c_{12} + \tfrac{1}{2}(I_1 + I_2 + I_5)c_{13} + \tfrac{1}{2}I_3 c_{33} - 2I_3 c_{44}$$

$$s'_{33} = I_1 s_{11} + I_2 s_{33} + I_3(2s_{13} + s_{44})$$

$$c'_{33} = I_1 c_{11} + I_2 c_{33} + 2I_3(c_{13} + 2c_{44})$$

$$s'_{44} = (2I_3 + I_4)s_{11} - I_4 s_{12} - 4I_3 s_{13} + 2I_3 s_{33} + \tfrac{1}{2}(I_1 + I_2 - 2I_3 + I_5)s_{44}$$

$$c'_{44} = \tfrac{1}{4}(2I_3 + I_4)c_{11} - \tfrac{1}{4}I_4 c_{12} - I_3 c_{13} + \tfrac{1}{2}I_3 c_{33} + \tfrac{1}{2}(I_1 + I_2 - 2I_3 + I_5)c_{44}$$

$$(10.7)$$

In these equations s_{11}, s_{12}, etc., are the compliance constants and c_{11}, c_{12}, etc., are the stiffness constants for the anisotropic elastic unit, which in practice means those of the most highly oriented specimen obtained. The terms I_1, I_2, I_3, I_4, I_5 are the orientation functions, defining the average values of $\sin^4 \theta(I_1), \cos^4 \theta(I_2), \cos^2 \theta \sin^2 \theta(I_3), \sin^2 \theta(I_4)$ and $\cos^2 \theta$ (I_5) for the aggregate. Note that only two of these orientation functions are independent parameters (e.g. $I_4 = I_1 + I_3$, $I_5 = I_2 + I_3$, $I_4 + I_5 = 1$).

The orientation functions can be calculated on the pseudo-affine deformation scheme and Figures 10.15–19 show that the aggregate model then predicts the general form of the mechanical anisotropy. It is particularly interesting that the predicted Reuss average curves for low-density polyethylene show the correct overall pattern, including the minimum in the extensional modulus. This arises as follows. On the pseudo-affine deformation scheme $\overline{\sin^4 \theta}$ and $\overline{\cos^4 \theta}$ decrease and increase monotonically respectively with increasing draw ratio whereas $\overline{\sin^2 \theta \cos^2 \theta}$ shows a maximum value at a draw ratio of about 1.2. Thus s'_{33} can pass through a maximum with increasing draw ratio (giving a minimum in the Young's modulus E_0) provided that $(2s_{13} + s_{44})$ is sufficiently large compared with s_{11} and s_{33} which should be approximately equal. The theory assumes elastic constants for the units which are identical with those measured for the highly oriented polymer. In low-density polyethylene s_{44} is much larger than s_{11} and s_{33}, which are fairly close in value; hence these conditions are fulfilled and the anomalous mechanical anisotropy is predicted.

At low temperatures, as discussed above (see Figure 10.11), a more conventional pattern of mechanical anisotropy is observed for low-density polyethylene. At the same time the polar diagram of the modulus changes (Figure 10.12) and s_{44} is no longer very much greater than the other elastic constants. These results are thus consistent with the aggregate model.

The theoretical curves of Figures 10.15–19 differ from those obtained experimentally in two ways. First there are features of detail (a small minimum in the transverse modulus of low-density polyethylene; a small minimum in the extensional modulus of high-density polyethylene) which are not predicted at all. It has been shown elsewhere[44] that such effects may be associated with mechanical twinning. Secondly, the predicted development of mechanical anisotropy with increasing draw ratio is much less rapid than is observed in practice. Deficiences in the pseudo-affine deformation scheme are not unexpected due to the simplifying nature of the assumptions made. Recently attempts have been made to determine the quantities $\overline{\sin^4 \theta}, \overline{\cos^4 \theta}$, and $\overline{\sin^2 \theta \cos^2 \theta}$ directly from wide-angle X-ray diffraction and nuclear magnetic resonance[45,46,47]. In low-density polyethylene a considerably improved fit was obtained in this manner (Figure 10.26). The conclusion from these results is that the mechanical anisotropy of low-density polyethylene relates to the orientation of the crystalline regions and that it is predicted to a very good degree of approximation by the Reuss averaging scheme.

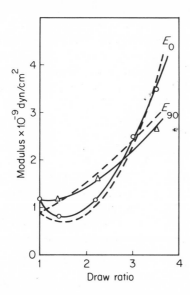

Figure 10.26. Comparison of experimental (full line) variation of E_0 and E_{90} for cold drawn low-density polyethylene with those predicted by the aggregate model using orientation functions from nuclear magnetic resonance.

The aggregate model predicts only that the elastic constants should lie between the Reuss and Voigt average values. In polyethylene terephthalate it is clear that the experimental compliances lie approximately midway between the two bounds. For cold-drawn fibres it has been shown that this median condition applies almost exactly.[47]

For low-density polyethylene, the Voigt averaging scheme does not predict the anomalous behaviour. However, the Reuss average does, and therefore appears to describe the physical situation more closely. A similar conclusion was reached by Odajima and Maeda[49] who compared theoretical estimates of the Reuss and Voigt averages of single crystals of polyethylene with experimental values.

In nylon the Voigt average is closest to the experimentally observed data. It is interesting to note that both averaging schemes predict a maximum in the torsional modulus as a function of draw ratio.

The aggregate model would not appear to be generally applicable to high-density polyethylene and polypropylene. It appears that for polypropylene the aggregate model is applicable only at low draw ratios[39]. As discussed above, there are simultaneous changes in morphology and molecular mobility at higher draw ratios.

It is interesting that in different polymers the Reuss or Voigt averages or a mean of these is closest to the measured values. It is likely that these conclusions will relate to the detailed nature of the stress and strain distributions at a molecular level in the polymers and should in turn be related to the structure.

10.7   THE SONIC VELOCITY

It has been suggested by Morgan[15] and others[13] that the sonic modulus (i.e. the extensional modulus measured at high frequencies by a wave-propagation technique) can be used to obtain a direct measure of molecular orientation in a manner analogous to the derivation of the so-called optical orientation function $f_c = (1 - \frac{3}{2}\overline{\sin^2 \theta})$ from the birefringence.

Consider the equations for the extensional modulus of the aggregate

$$s'_{33} = \overline{\sin^4 \theta} s_{11} + \overline{\cos^4 \theta} s_{33} + \overline{\sin^2 \theta \cos^2 \theta}(2s_{13} + s_{44})$$

Table 10.2 summarizes the measured values of s_{11}, s_{33} and s_{44} for a number of polymers, as obtained from the monofilament data of Hadley, Pinnock and Ward[27]. It can be seen that in all cases except that of low-density polyethylene s_{11} and s_{44} are of approximately the same value, and that s_{33} is comparatively small. Remembering that Poisson's ratio is usually close to 0.5, this implies that s_{13} will also be comparatively small.

This suggests that in certain cases for low degrees of molecular orientations

$$\text{(where } \overline{\cos^4 \theta} \ll \overline{\sin^4 \theta} \quad \text{and} \quad \overline{\sin^2 \theta \cos^2 \theta})$$

we may approximate and write:

$$s'_{33} = \overline{\sin^4 \theta} s_{11} + \overline{\sin^2 \theta \cos^2 \theta} s_{44} \tag{10.8}$$

$$= \overline{(\sin^4 \theta + \sin^2 \theta \cos^2 \theta)} s_{11}$$

$$= \overline{\sin^2 \theta} s_{11} \tag{10.9}$$

Remembering that the birefringence is given by $\Delta n = \Delta n_{\text{max}}(1 - \frac{3}{2}\overline{\sin^2 \theta})$

it can be seen that the extensional compliance, the reciprocal of the extensional modulus, should be directly related to the birefringence through $\overline{\sin^2 \theta}$ independent of the mechanism of molecular orientation[50]. To this degree of approximation it then follows that

$$s'_{33} = \tfrac{2}{3}s_{11}(\Delta n_{max} - \Delta n) \qquad (10.10)$$

We would therefore predict a linear relationship between the extensional compliance s'_{33} and the birefringence Δn, which extrapolates to zero extensional compliance at the maximum birefringence value.

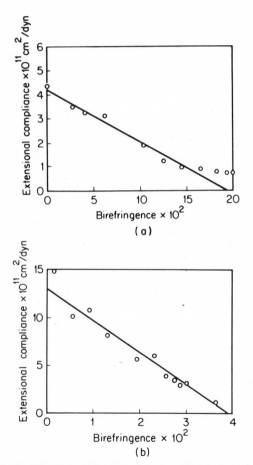

Figure 10.27. Experimental curves showing the relationship between the extensional compliance and birefringence for fibre of (a) polyethylene terephthalate; (b) polypropylene.

Figures 10.27a and b show results for polyethylene terephthalate and polypropylene which suggest that this is a reasonable approximation. But the values of s_{11} obtained from these plots do not agree with that measured experimentally for the most highly oriented fibre monofilament, suggesting that this approximate treatment is not very soundly based.

10.8   OTHER THEORIES OF MECHANICAL ANISOTROPY

It has been proposed by Hennig[51] that the isotropic extensional modulus E_u can be expressed as $3/E_u = s_{33} + s_{11}$ where s_{33} and s_{11} are the compliance constants of a uniaxially oriented polymer. This is not a special case of the aggregate theory but follows from simple considerations, neglecting the shear-compliance components. It has been shown to hold for certain glassy polymers at low degrees of molecular orientation.

Another interpretation of the development of mechanical anisotropy uses a model medium composed of oriented linear spring elements[52]. The variation of the mechanical properties are expressed in terms of the extensional modulus of the spring only, which is itself determined from the moduli of the unoriented material. Not surprisingly, the theory predicts the same relative distributions of moduli regardless of the polymer, i.e. implying interdependence of the five elastic constants. The data discussed above show that this is not the case.

The elastic modulus along the molecular chain axis has been calculated for a number of polymers from the force constants of the chemical bonds by Lyons[53] and Treloar[54]. The calculated moduli are an order of magnitude greater than the normal measured moduli of the materials, but are in reasonably good agreement with those determined from the change in the X-ray diffraction pattern upon stressing the sample[55,56].

10.9 ANISOTROPIC VISCOELASTIC BEHAVIOUR

As discussed in the introduction to this chapter, a formal extension to the use of the generalized Hooke's Law for anisotropic elastic solids can be made for anisotropic linear viscoelastic solids. This is an area which has been comparatively little studied. The earliest investigations were the comparisons of stress relaxation and dynamic mechanical behaviour for nylon in extension and torsion by Hammerle and Montgomery[57], and the studies of the dynamic mechanical behaviour of polyethylene terephthalate in extension and torsion by Pinnock and Ward[25].

These studies were very much of a phenomenological nature, and did not lead to a greater understanding of the underlying relaxation mechanisms. Recently, more extensive investigations by Takayanagi and his colleagues[29] and by Stachurski and Ward[30] have proved more rewarding in this respect.

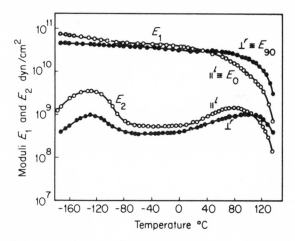

Figure 10.28. Temperature dependence of E_1 and E_2, the components of the dynamic modulus, in directions parallel ($\parallel^l$) and perpendicular ($\perp^r$) to the initial draw direction for annealed samples of high density polyethylene (after Takayanagi et al.).

Takayanagi prepared a series of drawn sheets and drawn and annealed sheets of several polymers including high-density polyethylene and polypropylene. Uniaxial orientation was assumed to be obtained in all cases. The dynamic extensional modulus was measured along the draw direction ($\parallel^l$ direction in Figure 10.28) and perpendicular to the draw direction ($\perp^r$ in Figure 10.28). For the drawn and annealed samples it is extremely interesting to observe that the parallel modulus (E_0 in our previous nomenclature) crosses the perpendicular modulus (E_{90} previously) at high temperatures. Thus, although $E_0 > E_{90}$ at low temperatures as expected intuitively from the molecular orientation, $E_0 < E_{90}$ at high temperatures.

Takayanagi proposed a simple model to explain this behaviour, which is shown in Figure 10.29. The basic feature of this model is illustrated by graphs of the variation of modulus with temperature (which Takayanagi terms 'dispersion curves'). Application of stress in the parallel direction

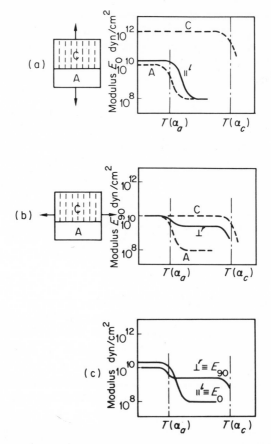

Figure 10.29. Schematic representations of change in modulus E with temperature on the Takayanagi model for (a) the $\|^l$ and (b) the $\perp^r$ situations corresponding to E_0 and E_{90} respectively. Calculations assume amorphous relaxation at temperature $T(\alpha_a)$ and crystalline relaxation at temperature $T(a_c)$ and (c) shows combined results. C crystalline phase, A amorphous phase (after Takayanagi et al.).

(Figure 10.29a) gives a large fall in modulus with increasing temperature, the stiffness at high temperatures being primarily determined by the amorphous regions which are very compliant above the relaxation transition. In the perpendicular direction the fall in modulus is not so great, as the crystalline region still supports the applied stress and a comparatively high stiffness is maintained (Figure 10.29b). This model results in the cross-over behaviour shown in Figure 10.29c. It can be criticized on the grounds that it does not analyse the mechanics of the true situation even on this model, which is a laminated structure. In such a structure the amorphous material is constrained by the crystalline material and may not become compliant in the manner proposed by Takayanagi.

Gupta and Ward[31] observed similar cross-over points in drawn and annealed sheets of low-density polyethylene, and also in the orthorhombic drawn rolled and annealed sheets of this polymer (see Section 10.4.4 above). They attributed the fall in modulus in the parallel direction, i.e. in E_0, or in the a and c directions in the $a-b$ and $b-c$ sheets respectively to an interlamellar shear process. Structural studies[32,58] imply that these annealed sheets possess a lamellar texture, low-angle X-ray diffraction measurements showing that the planes of the lamellae make angles of approximately 40° with the initial draw direction (or with the c-direction in the rolled and annealed sheets). Thus when the tensile stress is applied along the initial draw direction (E_0 measurement) or along either the a or c direction in the orthorhombic sheets (E_a and E_c measurements) the direction of maximum shear stress is approximately parallel to the lamellar planes. At temperatures above the relaxation transition (which involves mobility of material between the lamellae) interlamellar shear can occur, with a corresponding fall in the modulus. A tensile stress in the b direction, on the other hand, will not favour interlamellar shear, as the lamellar planes are approximately parallel to the b axis. This gives $E_b > E_a \sim E_c$ above the relaxation transitions, which is the observed behaviour. The results for the fibre symmetry sheets are explained by considering that the lamellar planes make angles of 35°–40° to the initial draw direction (the fibre axis) and follow a conical distribution around this axis. Application of a tensile stress parallel to the initial draw direction results in a maximum shear stress nearly parallel to all the lamellar planes. Application of the tensile stress in the 90° direction gives maximum shear stress parallel to only some lamellar planes. This is consistent with the observation that $E_0 < E_{90}$ above the relaxation transition.

A comparative study of the viscoelastic behaviour of these sheets by Stachurski and Ward[30] confirmed these conclusions and the anisotropy of the loss factor tan δ gave a striking characterization for the assignment of the relaxations to different molecular processes. This has been discussed in Chapter 8. Figures 8.20a and b show that there are two relaxation processes in this region with very different anisotropy, and as discussed previously these are attributed to the interlamellar shear relaxation and the c-axis shear relaxation respectively.

REFERENCES

1. M. A. Biot, *International Union of Theoretical and Applied Mechanics Colloquium*, (Madrid), Springer–Verlag, Berlin, 1955, p. 251.
2. T. G. Rogers and A. C. Pipkin, *J. Appl. Maths and Physics*, **14**, 334 (1963).
3. J. F. Nye, *Physical Properties of Crystals*, Clarendon Press, Oxford, 1957.
4. G. Raumann, *Proc. Phys. Soc.*, **79**, 1221 (1962).
5. S. G. Lekhnitskii, *Theory of Elasticity of an Anisotropic Elastic Body*, Holden Day, San Francisco, 1964.
6. N. H. Ladizesky and I. M. Ward, *J. Macromol. Sci. (Phys.)*, **B5(4)**, 759 (1971).
7. J. H. Wakelin, E. T. L. Voong, D. J. Montgomery and J. H. Dusenbury, *J. Appl. Phys.*, **26**, 786 (1955).
8. R. Meredith, *J. Text. Inst.*, **45**, 489 (1954).
9. H. Kolsky and K. W. Hillier, *Proc. Phys. Soc.*, **B62**, 111 (1949).
10. J. W. Ballou and J. C. Smith, *J. Appl. Phys.*, **20**, 493 (1949).
11. A. W. Nolle, *J. Polymer Sci.*, **5**, 1 (1949).
12. W. J. Hamburger, *Text. Res. J.*, **18**, 705 (1948).
13. W. H. Charch and W. W. Moseley, *Text Res. J.*, **29**, 525 (1959).
14. W. W. Moseley, *J. Appl. Polymer Sci.*, **3**, 266 (1960).
15. H. M. Morgan, *Text. Res. J.*, **32**, 866 (1962).
16. V. Davis, *J. Text. Inst.*, **50**, 1688 (1960).
17. F. I. Frank and A. L. Ruoff, *Text Res. J.*, **28**, 213 (1958).
18. D. W. Hadley, I. M. Ward and J. Ward, *Proc. Roy. Soc.*, **A285**, 275 (1965).
19. P. R. Pinnock, I. M. Ward and J. M. Wolfe, *Proc. Roy. Soc.*, **A291**, 267 (1966).
20. H. Hertz, *Miscellaneous papers*, Macmillan, London: 1896, p. 146.
21. E. McEwen, *Phil. Mag.*, **40**, 454 (1949).
22. S. Timoshenko and J. N. Goodier, *Theory of Elasticity*, McGraw–Hill, New York, 1951.
23. G. Raumann and D. W. Saunders, *Proc. Phys. Soc.*, **77**, 1028 (1961).
24. I. M. Ward, *Text. Res. J.*, **31**, 650 (1961).
25. P. R. Pinnock and I. M. Ward, *Proc. Phys. Soc.*, **81**, 260 (1963).
26. G. Raumann, *Brit. J. Appl. Phys.*, **14**, 795 (1963).
27. D. W. Hadley, P. R. Pinnock and I. M. Ward, *J. Mater. Sci.*, **4**, 152 (1969).
28. V. B. Gupta and I. M. Ward, *J. Macromol. Sci. (Phys.)* **B(1)2**, 373 (1967).
29. M. Takayanagi, K. Imada and T. Kajiyama, *J. Polymer Sci.*, **C-15**, 263 (1966).
30. Z. H. Stachurski and I. M. Ward, *J. Polymer Sci.*, **A-2, 6**, 1083 (1969); *J. Polymer Sci.*, **A-2, 6**, 1817 (1968); *J. Macromol. Sci.*, **B3(3)**, 427 (1969) and *J. Macromol. Sci.*, **B3(3)**, 445 (1969).
31. V. B. Gupta and I. M. Ward, *J. Macromol. Sci. (Phys.)*, **B2**, 89 (1968).
32. I. L. Hay and A. Keller, *J. Materials Sci.*, **1**, 41 (1966).
33. J. J. Point, *J. Chim. Phys.*, **50**, 76 (1953).
34. I. M. Ward, *Proc. Phys. Soc.*, **80**, 1176 (1962).
35. H. H. Kausch-Blecken Von Schmeling, *Kolloidzeitschrift* **237**, 251 (1970).
36. J. Bishop and R. Hill, *Phil. Mag.*, **42**, 414, 1298 (1951).
37. A. Reuss, *Z. Angew. Math. Mech.*, **9**, 49 (1929).
38. W. Voigt, *Lehrbuch der Kistallphysik*, Teubner, Leipsig, 1928, p. 410.
39. P. R. Pinnock and I. M. Ward, *Brit. J. Appl. Phys.*, **17**, 575 (1966).
40. S. M. Crawford and H. Kolsky, *Proc. Phys. Soc.*, **B64**, 119 (1951).
41. C. G. Cannon and F. P. Chappel, *Brit. J. Appl. Phys.*, **10**, 68 (1959).

42. W. Kuhn and F. Grün, *Kolloid zeitschrift*, **101**. 248 (1942).
43. P. R. Pinnock and I. M. Ward, *Brit. J. Appl. Phys.*, **15**, 1559 (1964).
44. F. C. Frank, V. B. Gupta and I. M. Ward, *Phil. Mag.*, **21**, 1127 (1970).
45. V. B. Gupta, A. Keller and I. M. Ward, *J. Macromol. Sci. (Phys.)*, **B2(1)**, 139 (1968).
46. V. B. Gupta and I. M. Ward, *J. Macromol. Sci. (Phys.)*, **B4(2)**, 453 (1970).
47. V. J. McBrierty and I. M. Ward, *Brit. J. Appl. Phys.*, **21**, 1529 (1968).
48. S. W. Allison and I. M. Ward, *Brit. J. Appl. Phys.*, **18**, 1151 (1967).
49. A. Odajima and I. Maeda, *Rep. Prog. Polymer Phys. (Japan)*, **9**, 169 (1966).
50. I. M. Ward, *Text. Res. J.*, **34**, 806 (1964).
51. J. Hennig, *Kolloidzeitschrift*, **200**, 46 (1964).
52. S. R. Kao and C. C. Hsiao, *J. Appl. Phys.*, **35**, 3127 (1964).
53. W. J. Lyons, *J. Appl. Phys.*, **29**, 1429 (1958).
54. L. R. G. Treloar, *Polymer* **1**, 95 (1960); *Polymer* **1**, 279 (1960), and *Polymer* **1**, 290 (1960).
55. W. J. Dulmage and L. E. Contois, *J. Polymer Sci.*, **28**, 275 (1958).
56. I. Sakurada, Y. Nukushina and T. Ito, *J. Polymer Sci.*, **57**, 651 (1962).
57. W. G. Hammerle and D. J. Montgomery, *Text. Res. J.*, **23**, 595 (1953).
58. T. Seto and T. Hara, *Rept. Prog. Polymer Phys. (Japan)*, **7**, 63 (1963).

11

The Yield Behaviour of Polymers

Until comparatively recently, the yield behaviour of polymers has not received much attention. This is mainly because it was not thought profitable to treat it as a distinct mode of mechanical behaviour, different in kind from either the viscous flow processes which occur at high temperatures or the large extensions observed in the temperature range above the glass transition. The yield process in a polymer was often considered to be a softening due to a local rise in temperature and was referred to as a localized 'melting'.

A number of different factors have contributed to the appreciably greater interest in yield behaviour of polymers since about 1960. In the first instance it has been recognized that the classical concepts of plasticity are relevant to forming, rolling and drawing processes in polymers. Secondly, there have been a number of striking experimental studies of 'slip bands' and 'kink bands' in polymers which suggest that deformation processes in polymers might be similar to those in crystalline materials such as metals and ceramics. Finally, it is now evident that distinct yield points are observed and there is much interest in understanding these in the context of other ideas in polymer science.

Our first task in this chapter is to discuss the relevance of classical ideas of plasticity to the yielding of polymers. Although the yield behaviour is temperature and strain-rate dependent it will be shown that provided the test conditions are chosen suitably, yield stresses can be measured which satisfy conventional yield criteria.

This part of the discussion is at a purely phenomenological level. Two aspects of the yield behaviour which provide information at a molecular level will also be considered. These are the temperature and strain-rate sensitivity, and the molecular reorientation associated with plastic deformation.

The temperature and time dependence often obscure some generalities of the yield behaviour. For example, it might be concluded that some polymers show necking and cold-drawing whereas others are brittle and

270

fail catastrophically. Yet another type of polymer (a rubber) extends homogeneously to rupture. A salient point to recognize is that polymers in general show all these types of behaviour depending on the exact conditions of test (Figure 1.1). This is quite irrespective of their chemical nature and physical structure. Thus explanations of yield behaviour which involve, for example, cleavage of crystallites or lamellar slip or amorphous mobility are only relevant to specific cases. As in the case of linear viscoelastic behaviour or rubber elasticity what we must first seek is an understanding of the relevant phenomenological features, decide on suitable measurable quantities and then provide a molecular interpretation of the subsequent constitutive relations.

11.1 DISCUSSION OF LOAD–ELONGATION CURVE

The most dramatic manifestation of yield is seen in tensile tests when a neck or deformation band occurs, as in Figure 11.1 (see Plate VIA). In these cases the plastic deformation is concentrated either entirely or primarily in a small region of the specimen. The precise nature of the plastic deformation depends both on the geometry of the specimen and on the nature of the applied stresses. This will be discussed more fully later.

The characteristic necking and cold-drawing behaviour is as follows. On the initial elongation of the specimen, homogeneous deformation occurs and the conventional load-extension curve shows a steady increase in load with increasing elongation (AB in Figure 11.2). At the point B the specimen thins to a smaller cross-section at some point, i.e. a neck is formed. Further elongation brings a fall in load. Continuing extension

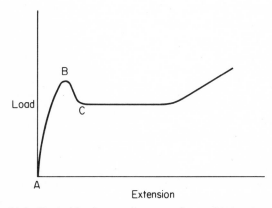

Figure 11.2. Typical load–extension curve for a cold-drawing polymer.

is achieved by causing the shoulders of the neck to travel along the specimen as it thins from the initial cross-section to the drawn cross-section. The existence of a finite or natural draw ratio is an important aspect of polymer deformation and is discussed in Section 11.8.2 below.

Ductile behaviour in polymers does not always give a stabilized neck. Before analysing the requirements for necking and cold-drawing the distinction between true and conventional stress–strain curves will be considered.

11.1.1   Necking and the Ultimate Stress

Necking and cold-drawing are accompanied by a non-uniform distribution of stress and strain along the length of a test specimen. Let us consider these phenomena in terms of the true stress–strain curve of a material, rather than the conventional stress–strain curve which relates the applied tension to the overall extension. The cross-section of the sample is decreasing with increasing extension, so that the true stress may be increasing when the apparent or conventional stress or load may be remaining constant or even decreasing. This has been very well discussed by Nadai[1] and Orowan[2] and their argument will be followed here.

Consider the conventional stress–strain curve or the load–elongation curve for a ductile material (Figure 11.3). The ordinate is equal to the stress σ_a obtained by dividing the load P by the original cross-sectional area A_0

$$\sigma_a = \frac{P}{A_0}$$

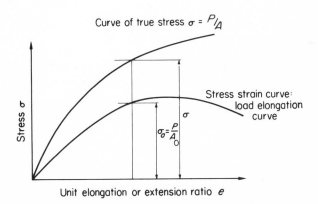

Figure 11.3. Comparison of load–elongation curve and true stress elongation curve.

This gives a stress–strain curve of the form shown. The load reaches its maximum value at the instant the uniform extension of the sample stops. At this elongation the specimen begins to neck and consequently the load falls as shown by the last part of the stress–strain curve. Finally the sample fractures at the narrowest point of the neck.

It is more instructive to plot the true tensile stress at any elongation rather than the apparent stress σ_a.

The true stress $\sigma = P/A$ where A is the actual cross-section at any time.

We now assume that the deformation takes place at constant volume, this assumption being usual for plastic deformation.

Then $Al = A_0 l_0$, and if we put $\dfrac{l}{l_0} = 1 + e$ where $e =$ elongation per unit length

$$A = \frac{A_0 l_0}{l} = \frac{A_0}{1+e}$$

The true stress

$$\sigma = \frac{P}{A} = \frac{(1+e)P}{A_0} = (1+e)\sigma_a$$

This gives the load

$$P = \frac{A_0 \sigma}{1+e}$$

Thus if we know σ, the true stress, as a function of e, i.e. the true stress–strain curve, P can be computed for any elongation. In particular P_{max}, the maximum load, is related directly to the apparent stress by the relationship:

$$\sigma_a = \frac{P}{A_0}$$

P_{max} is defined by the condition $dP/de = 0$, i.e.

$$\frac{dP}{de} = \frac{A_0}{(1+e)^2}\left[(1+e)\frac{d\sigma}{de} - \sigma\right] = 0$$

or

$$\frac{d\sigma}{de} = \frac{\sigma}{1+e}$$

The measured ultimate stress can be obtained from the true stress–strain curve by the simple construction shown in Figure 11.4. The ultimate

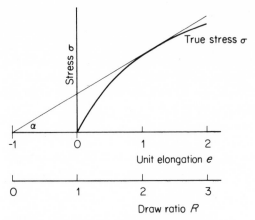

Figure 11.4. The Considère construction.

stress is obtained when the tangent to the true stress–strain curve $d\sigma/de$ is given by the line from the point -1 on the elongation axis. The angle α in Figure 11.4 is defined by:

$$\tan \alpha = \frac{d\sigma}{de} = \frac{\sigma}{1+e} = \frac{\sigma}{R}$$

where R is the draw ratio. This construction is called the 'Considère construction' and is useful in discussing whether a polymer will neck and cold-draw.

The significance of the argument at this stage relates to the failure of plastics in the ductile state. Orowan[2] first pointed out that for ductile materials the ultimate stress is entirely determined by the stress–strain curve, that is by the plastic behaviour of the material, without any reference to its strength properties, provided that fracture does not occur before the load maximum corresponding to $d\sigma/de = \sigma/1+e$ is reached. This explains why the yield stress is such an important property in many plastics. In fact it defines the practical limit of behaviour much more than the ultimate fracture, unless the plastic fails by brittle fracture.

11.1.2 Necking and Cold-Drawing: A Phenomenological Discussion

Figure 11.5 shows that there are three distinct regions on the true stress–strain curve of a typical cold-drawing polymer.

(1) Initially the stress rises in an approximately linear manner as the applied strain increases.

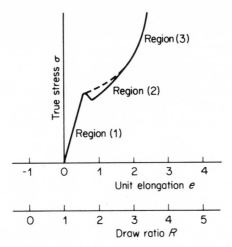

Figure 11.5. The true stress–strain curve for a cold-drawing polymer.

(2) At the yield point there is a fall in true stress and a region where the stress rises less steeply with the strain. In some earlier discussions[3] it was not recognized that a fall in true stress can occur, and it was proposed only that the true stress rises less steeply with increasing strain (dotted line in Figure 11.5). This region was attributed to 'strain softening'.

(3) Finally at large extensions the slope of the true stress–strain curve increases again, i.e. a 'strain-hardening' effect occurs.

There are two ways in which a neck may be initiated. First, if, for a given applied load one element is subjected to a higher true stress, because its effective cross-sectional area is smaller, that element will reach the yield point at a lower tension than any other point in the sample. Secondly, a fluctuation in material properties may cause a localized reduction of the yield stress in a given element so that this element reaches the yield point at a lower applied tension. When a particular element has reached its yield point it is easier to continue deformation entirely within this element because it has a lower effective stiffness than the surrounding material. Hence further deformation of the sample is accomplished by straining in only one region and a 'neck' is formed.

This localized deformation will continue until strain-hardening increases the effective stiffness of the element, i.e. it reaches the third part (3) of the true stress–strain curve in Figure 11.5. At this point the deformation will stabilize in the highly strained element and in order to accommodate further extension of the sample as a whole, new elements will be brought to their yield point. In this way a neck propagates along the length of a sample as successive elements are brought to a degree of strain-hardening which is greater than the yield stress of the undeformed material.

11.1.3   Use of the Considère Construction

In Figure 11.6 two tangent lines have been drawn to the true stress–strain curve from the point $e = -1$ or $R = 0$. In terms of the extension ratio or draw ratio R, Considère's construction gives the conventional stress

$$\frac{P}{A_0} = \frac{\sigma}{1+e} = \frac{\sigma}{R}$$

Thus a line from the point $R = 0$ to a point on the true stress–strain curve has slope σ/R and gives us the conventional stress, i.e. the applied tension at that point.

The first tangent line has been drawn from O to D, i.e. to the yield point. At this point σ/R and hence the conventional stress is a maximum. As further deformation takes place the slope of σ/R decreases continuously as the true stress–strain curve is traced, until the point E is reached. This is where strain-hardening occurs. The final tension settles down to a value represented by the slope of the line OE, the second tangent line, and drawing takes place by deforming successive elements by an amount corresponding to this extension ratio. After the entire sample has been drawn, further deformation may take place along the steeper part of the stress–strain curve until fracture occurs.

It should be noted that these arguments do not satisfactorily explain how the sample draws at constant tension throughout its length (along the line BE in the diagram) and at the same time apparently follows the

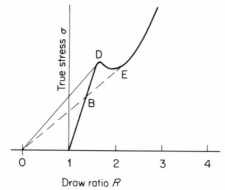

Figure 11.6. A true stress–strain curve for a cold-drawing polymer and the Considère construction.

true stress–strain curve along BDE. This anomaly has been attributed by Vincent[3] to our ignorance regarding the actual stress conditions in the neck, where from the corresponding work in metals[4] it can be implied that there is a complex combination of tensile and hydrostatic stresses.

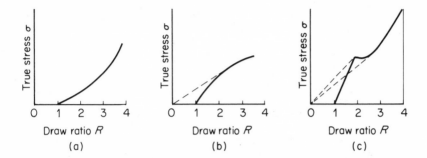

Figure 11.7. The three types of true stress–strain curves for polymers. (a) Case 1: $d\sigma/dR > \sigma/R$; (b) Case 2: $d\sigma/dR = \sigma/R$ at one point; and (c) Case 3: $d\sigma/dR = \sigma/R$ at two points (after Vincent).

The Considère construction can be used as a criterion to decide whether a polymer will neck, or will neck and cold-draw. There are three possible situations:

Case 1. $d\sigma/dR$ is always greater than σ/R

This is shown in Figure 11.7a, and it can be seen that there is no tangent line which can be drawn to the true stress–strain curve from the point $R = 0$. The polymer therefore extends uniformly with increasing load, and no neck is formed.

Case 2. $d\sigma/dR = \sigma/R$ at one point

In this case (Figure 11.7b) the polymer extends uniformly up to the point where $d\sigma/dR = \sigma/R$ and then necks. The neck gets steadily thinner and then the measured load decreases until fracture occurs. This case has been discussed previously, where the treatment of Nadai and Orowan[1,2] was presented (Section 11.1.1).

Case 3. $d\sigma/dR = \sigma/R$ at two points

This is the case where necking and cold-drawing occurs (Figure 11.7c), and it is only necessary to state that the requirement of two tangents becomes a criterion for necking *and* cold-drawing.

11.1.4   Definition of Yield Stress

The Considère construction shows that the existence of a yield drop can be defined by the requirement that a tangent line can be drawn to the true stress–strain curve from the point $R = 0$. We will define the yield stress as the true stress at the maximum observed load. Because this stress is achieved at a comparatively low elongation of the sample it is often adequate to use the engineering definition of yield stress as maximum observed load divided by *initial* cross-sectional area.

In some cases there is no observed load drop (e.g. shear tests in Figure 11.28), and a second definition of yield stress is used. This is the stress where the two tangents to the initial and final parts of the load–elongation curve intersect (Figure 11.28).

11.2   IDEAL PLASTIC BEHAVIOUR

The simplest theories of plasticity do not contain time as a variable and ignore any features of the behaviour which take place below the yield point. In other words we assume a rigid-plastic material whose stress–strain relationship in tension is shown in Figure 11.8. For stresses below the yield stress σ_y there is no deformation. For stresses above the yield stress the deformation is determined by the movement of the applied loads.

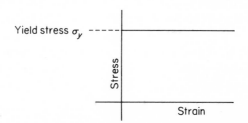

Figure 11.8.   Stress–strain relationship for an ideal rigid-plastic material.

The constitutive relations of elasticity and viscoelasticity relate the magnitudes of the stresses in the material to the magnitudes of the strains. In plasticity the situation is somewhat different in that, although the relationships between the different components of the plastic strain increment relate to the stresses in the material, the absolute magnitudes of the plastic strain increments are determined by the movement of the external applied loads.

There are two aspects to classical plasticity, or the behaviour of an ideal rigid-plastic material. First, we wish to define the stress situations in which plastic flow can occur. For a simple tensile test this merely involves a definition of the yield stress as in Section 11.1.4. Secondly we will seek to define the appropriate plastic strain-increment relationships in terms of the stress situation at yield.

11.2.1 The Yield Criterion: General Considerations

The first question which may be asked regarding the yielding of materials concerns the relationship between the measured yield point for different types of testing, e.g. tension and simple shear. The aim is to find a function of all the components of stress which reaches a critical value for all tests, i.e. for different combinations of stress. This function is called the 'yield criterion', and in its most general form can be written:

$$f(\sigma_{xx}, \sigma_{yy}, \sigma_{zz}, \sigma_{xy}, \sigma_{yz}, \sigma_{zx}) = \text{constant}.$$

The actual form of the function f can be restricted by several considerations. In the first instance it will be assumed that the material is isotropic. The term f must then be a function of the invariants of the stress tensor. If we refer our stresses to the principal axes of stress, the stress tensor

$$\begin{bmatrix} \sigma_{xx} & \sigma_{xy} & \sigma_{xz} \\ \sigma_{xy} & \sigma_{yy} & \sigma_{yz} \\ \sigma_{xz} & \sigma_{yz} & \sigma_{zz} \end{bmatrix}$$

becomes

$$\begin{bmatrix} \sigma_1 & 0 & 0 \\ 0 & \sigma_2 & 0 \\ 0 & 0 & \sigma_3 \end{bmatrix}$$

In terms of the principal components of stress $\sigma_1, \sigma_2, \sigma_3$, the three simplest stress invariants are

$$J_1 = \sigma_1 + \sigma_2 + \sigma_3$$
$$J_2 = \sigma_1\sigma_2 + \sigma_2\sigma_3 + \sigma_3\sigma_1$$
$$J_3 = \sigma_1\sigma_2\sigma_3$$

and it can be shown that all other invariants of stress can be expressed in terms of these three.

This gives the yield criterion as:

$$f(J_1, J_2, J_3) = \text{constant}.$$

The apparent number of variables in our yield criterion has been reduced (by eliminating the three shear-stress components) but at the expense of defining our reference axes (which requires three direction variables).

In metals it has been established that the yield behaviour is to a first approximation independent of the hydrostatic component of stress. This is not true for polymers, but in our preliminary discussions we will consider the simplification which such an approximation allows.

The yield behaviour will now depend only on the components of the deviatoric stress tensor σ'_{ij}, obtained by subtracting the hydrostatic components of stress from the total stress tensor. We have that $\sigma'_{ij} = \sigma_{ij} - p\delta_{ij}$ where the prime indicates the deviatoric stress tensor. In Cartesian notation σ'_{ij} is

$$\begin{bmatrix} \sigma_{xx} - p & \sigma_{xy} & \sigma_{xz} \\ \sigma_{xy} & \sigma_{yy} - p & \sigma_{yz} \\ \sigma_{xz} & \sigma_{yz} & \sigma_{zz} - p \end{bmatrix}$$

where $p = \frac{1}{3}(\sigma_{xx} + \sigma_{yy} + \sigma_{zz})$.

In terms of principal components of stress the yield criterion is a function of

$$\sigma'_1 = \sigma_1 - p$$
$$\sigma'_2 = \sigma_2 - p$$

and

$$\sigma'_3 = \sigma_3 - p$$

Since $\sigma'_1 + \sigma'_2 + \sigma'_3 = 0$ the yield criterion reduces to:

$$f(J'_2, J'_3) = 0$$

where

$$J'_2 = -(\sigma'_1\sigma'_2 + \sigma'_2\sigma'_3 + \sigma'_3\sigma'_1) = \tfrac{1}{2}(\sigma'^2_1 + \sigma'^2_2 + \sigma'^2_3) = \tfrac{1}{2}\sigma'_{ij}\sigma'_{ij}$$
$$J'_3 = \sigma'_1\sigma'_2\sigma'_3 = \tfrac{1}{3}(\sigma'^3_1 + \sigma'^3_2 + \sigma'^3_3) = \tfrac{1}{3}\sigma'_{ij}\sigma'_{jk}\sigma'_{ki}$$

A further simplification is obtained by assuming that if a stress σ_{ij} can induce yield, then so can a stress $-\sigma_{ij}$, e.g. it is implied that yield stresses in simple tension and compression are equal. Analytically this means either that f does not involve J'_3 or that it involves only even powers of J'_3.

This simplification is not generally true for polymers, where the difference between tensile and compressive yield (the Bauschinger effect[5]) plays an important part in the yield behaviour. However as with the hydrostatic component of stress we will develop our initial argument by considering that there is no Bauschinger effect in polymers.

11.2.2 The Tresca Yield Criterion

The earliest yield criterion to be suggested for metals was Tresca's proposal that yield occurs when the maximum shear stress reaches a critical value,[6] i.e.

$$\sigma_1 - \sigma_3 = \text{constant}$$

with

$$\sigma_1 > \sigma_2 > \sigma_3$$

This yield criterion is of a similar nature to the Schmid critical resolved shear-stress law for the yield of metal single crystals (see Section 11.2.7). We will see that the Tresca yield criterion is only of limited interest in polymers.

11.2.3 The von Mises Yield Criterion

Von Mises[7] proposed that the yield criterion did not involve J'_3, but was a function of J'_2 only, i.e. that yield occurs when J'_2 reaches a critical value, i.e.

$$J'_2 = \text{constant} = \mathbf{K}^2 \text{ say}$$

It is important to discuss three alternative ways of expressing the von Mises yield criterion.

(1) In terms of the principal components of the deviatoric stress tensor

$$J'_2 = -(\sigma'_1\sigma'_2 + \sigma'_2\sigma'_3 + \sigma'_3\sigma'_1) = \tfrac{1}{2}(\sigma_1'^2 + \sigma_2'^2 + \sigma_3'^2)$$
$$= \tfrac{1}{2}\{(\sigma_1 - p)^2 + (\sigma_2 - p)^2 + (\sigma_3 - p)^2\}$$

This gives the yield criterion in terms of the principal components of the total stress tensor

$$(\sigma_1 - p)^2 + (\sigma_2 - p)^2 + (\sigma_3 - p)^2 = 2\mathbf{K}^2 \tag{11.1}$$

where $p = \tfrac{1}{3}(\sigma_1 + \sigma_2 + \sigma_3)$ is the hydrostatic pressure.

(2) An equally well-known representation of this criterion is

$$(\sigma_1-\sigma_2)^2+(\sigma_2-\sigma_3)^2+(\sigma_3-\sigma_1)^2 = 6K^2 \qquad (11.2)$$

which is readily obtained by algebraic manipulation.

(3) Finally, if we refer our stresses to a Cartesian set of axes x, y, z which are other than principal axes of stress the stress tensor

$$\begin{bmatrix} \sigma_1 & 0 & 0 \\ 0 & \sigma_2 & 0 \\ 0 & 0 & \sigma_3 \end{bmatrix}$$

becomes

$$\begin{bmatrix} \sigma_{xx} & \sigma_{xy} & \sigma_{xz} \\ \sigma_{xy} & \sigma_{yy} & \sigma_{yz} \\ \sigma_{xz} & \sigma_{yz} & \sigma_{zz} \end{bmatrix}$$

Equation (11.2) can be written as

$$2(\sigma_1^2+\sigma_2^2+\sigma_3^2)-2(\sigma_1\sigma_2+\sigma_2\sigma_3+\sigma_3\sigma_1) = 6K^2$$

i.e.

$$3(\sigma_1^2+\sigma_2^2+\sigma_3^2)-(\sigma_1+\sigma_2+\sigma_3)^2 = 6K^2$$

Remembering that the invariants of the stress tensor are

$$\sigma_1+\sigma_2+\sigma_3 = \sigma_{xx}+\sigma_{yy}+\sigma_{zz}$$

and

$$\sigma_1^2+\sigma_2^2+\sigma_3^2 = \sigma_{xx}^2+\sigma_{yy}^2+\sigma_{zz}^2+2(\sigma_{xy}^2+\sigma_{yz}^2+\sigma_{xz}^2)$$

the von Mises yield criterion becomes

$$3[(\sigma_{xx}^2+\sigma_{yy}^2+\sigma_{zz}^2)+2(\sigma_{xy}^2+\sigma_{yz}^2+\sigma_{zx}^2)]-(\sigma_{xx}+\sigma_{yy}+\sigma_{zz})^2 = 6K^2$$

i.e.

$$(\sigma_{xx}-\sigma_{yy})^2+(\sigma_{yy}-\sigma_{zz})^2+(\sigma_{zz}-\sigma_{xx})^2+6(\sigma_{xy}^2+\sigma_{yz}^2+\sigma_{zx}^2) = 6K^2 \qquad (11.3)$$

We will see that the von Mises yield criterion is of considerable interest in the case of polymers, but that we cannot ignore the pressure dependence of the yield behaviour. The von Mises criterion can be modified in a number of ways. One simple way is to allow **K** to be an arbitrary function of the hydrostatic pressure.

11.2.4 The Coulomb Yield Criterion

On the Tresca yield criterion, the critical shear stress for yield is independent of the normal pressure on the plane in which yield is occurring. Coulomb had previously proposed a more general yield criterion for the failures of soils[8]. This states that the critical shear stress τ for yielding to occur in any plane increases linearly with the pressure applied normal to this plane, i.e.

$$\tau = \tau_c + \mu\sigma_N \qquad (11.4)$$

τ_c = 'cohesion' of the material

μ = coefficient of friction. (Sometimes μ is written tan ϕ, for reasons which will be made evident below.)

σ_N = compressive stress on the shear plane

We will see that this yield criterion is of considerable interest in the case of polymers.

11.2.5 Geometrical Representations of the Tresca, von Mises and Coulomb Yield Criteria

The Tresca and von Mises Yield Criteria

The Tresca and von Mises yield criteria take very simple analytical forms when expressed in terms of the principal stresses. One reason for this is the assumption of material isotropy which implies that σ_1, σ_2 and σ_3 are interchangeable. Thus the yield criteria form simple surfaces in principal stress space, i.e. space where the three rectangular Cartesian axes are parallel to the principal stress directions.

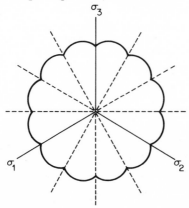

Figure 11.9. Cross-section of the yield surface normal to the [111] direction in principal stress space.

Because the yield criterion is independent of the hydrostatic component of stress we can replace σ_1, σ_2 and σ_3 by $\sigma_1 + p$, $\sigma_2 + p$ and $\sigma_3 + p$ respectively. Thus if the point $\sigma_1, \sigma_2, \sigma_3$ lies on the yield surface, so does the point $\sigma_1 + p$, $\sigma_2 + p$, $\sigma_3 + p$. This shows that the yield surface must be parallel to the [111] direction in principal stress space and can be represented geometrically by the cross-section normal to this direction (Figure 11.9). The material isotropy implies equivalence of σ_1, σ_2 and σ_3 and hence that the section has a three-fold symmetry about the [111] direction. The assumption that $f(\sigma_{ij}) = f(-\sigma_{ij})$ (i.e. no Bauschinger effect) implies equivalence of σ_1 and $-\sigma_1$, and hence we have finally six-fold symmetry about the [111] direction.

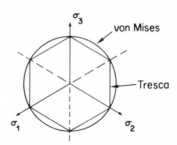

Figure 11.10. Cross-section of Tresca and von Mises yield surfaces normal to the [111] direction in principal stress space.

The cross-section normal to the [111] direction therefore consists of 12 equivalent parts (see Figure 11.9), and for the Tresca and von Mises yield criteria takes the particularly simple forms shown in Figure 11.10. The Tresca criterion gives a regular hexagon and the von Mises criterion a circle.

The Coulomb Yield Criteria

The Coulomb yield criterion is

$$\tau = \tau_c + \sigma_N \tan \phi \qquad (11.4)$$

For uniaxial compression under a compressive stress σ_1 where yield occurs on a plane whose normal makes an angle θ with the direction of σ_1 (Figure 11.11), the shear stress is

$$\tau = \sigma_1 \sin \theta \cos \theta$$

and the normal stress $\sigma_N = \sigma_1 \cos^2 \theta$

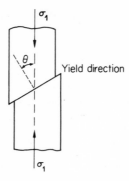

Figure 11.11. The yield direction for a material obeying the Coulomb criterion which is subjected to a compressive stress σ_1.

Yield occurs when

$$\sigma_1 \sin \theta \cos \theta = \tau_c + \sigma_1 \tan \phi \cos^2 \theta$$

i.e. when

$$\sigma_1 (\cos \theta \sin \theta - \tan \phi \cos^2 \theta) = \tau_c$$

In practice this is achieved by yield occurring in the plane which maximizes $(\cos \theta \sin \theta - \tan \phi \cos^2 \theta)$ so that yield occurs for the smallest value of σ_1. This gives

$$\tan \phi \tan 2\theta = -1 \quad \text{or} \quad \theta = \frac{\pi}{4} + \frac{\phi}{2} \qquad (11.5)$$

Thus the direction of yielding defines ϕ, where $\tan \phi$ is the 'coefficient of friction'.

We see that the Coulomb yield criterion therefore defines both the stress condition required for yielding to occur *and* the directions in which the material will deform. Where a deformation band forms the direction of the deformation band is the direction which is neither rotated nor distorted by the plastic deformation. This follows because the band direction marks the direction which establishes material continuity between the deformed material in the deformation band and the undistorted material in the rest of the specimen. If volume is conserved the band direction therefore denotes the direction of shear in a simple shear (by the definition of a shear strain). Thus for a Coulomb yield criterion the band direction is defined by Equation (11.5).

In the case of the von Mises yield criterion the prediction of the plastic deformation and hence the deformation-band direction requires further hypotheses other than those contained in the yield criterion. This will be discussed below.

11.2.6   Combined Stress States

For the analysis of combined stress states in the two-dimensional situation the Mohr circle diagram is of value. In Figure 11.12a two states of stress which produce yield with principal stresses σ_1 and σ_2 and σ_3 and σ_4 respectively are represented by two circles of identical radius, tangential to the yield surface. The yield criterion in this case is assumed to be that

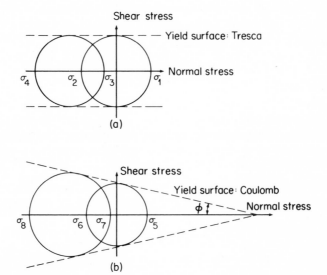

Figure 11.12. Mohr circle diagram for two states of stress which produce yield in a material satisfying (a) the Tresca yield criterion and (b) the Coulomb yield criterion.

of Tresca and the yield surface degenerates for the two-dimensional case to two lines parallel to the normal stress axis.

In Figure 11.12b two states of stress causing yield for a material which satisfies the Coulomb criterion are shown as σ_5 and σ_6 and σ_7 and σ_8, respectively. In this case the yield surface is two lines each making an angle ϕ with the normal stress axis.

11.2.7 Yield Criteria for Anisotropic Materials

A very simple yield criterion for anisotropic materials is the critical resolved shear-stress law of Schmid (see Reference 9, p. 4). This law states that yield occurs when the resolved shear stress in the slip direction in the slip plane reaches a critical value. For a tensile stress σ making angles α and β with the slip direction and the normal to the slip plane respectively this gives the critical resolved shear stress τ_c as

$$\tau_c = \sigma \cos \alpha \cos \beta \qquad (11.6)$$

We will see that although this law is extensively used in metal plasticity it is of restricted application in polymers.

For polymers, a generalization of the von Mises yield criterion by Hill[10] has proved more useful. Hill's yield criterion applies to anisotropic materials which possess three mutually orthogonal planes of symmetry at every point, i.e. the material possesses at least orthorhombic symmetry. It will also apply in a simplified form for solids possessing transverse isotropy.

The intersections of the three orthogonal planes define the principal axes of symmetry, and these directions are chosen as Cartesian axes of reference. The yield criterion then takes the form

$$F(\sigma_{yy}-\sigma_{zz})^2 + G(\sigma_{zz}-\sigma_{xx})^2 + H(\sigma_{xx}-\sigma_{yy})^2 + 2L\sigma_{yz}^2 + 2M\sigma_{zx}^2 + 2N\sigma_{xy}^2 = 1$$

$$(11.7)$$

where $F, G, H, L, M,$ and N are parameters which characterize the anisotropy of the yield behaviour.

This yield criterion satisfies several requirements:

(1) It reduces to the von Mises yield criterion for vanishingly small anisotropy.

(2) There is no Bauschinger effect, i.e. it contains no odd powers of stress components.

(3) The yield criterion is independent of the hydrostatic component of stress, i.e. the normal-stress terms appear as differences.

We may also note that if F, G, H, M and N are small the yield criterion reduces to

$$2L\sigma_{yz}^2 = 1$$

which is the Tresca yield criterion. In physical terms this means that a single yield process represented by the constant L dominates the yield behaviour.

11.2.8   The Stress–Strain Relations for Isotropic Materials

Once we achieve the combination of stresses required to produce yield in an idealized rigid-plastic material, deformation can proceed without altering the stresses, and is determined by the movements of the external constraints, e.g. the displacement of the jaws of the tensometer in a tensile test. This means that there is no unique relationship between the stresses and the *total* plastic deformation. Instead the relationships which do exist relate the stresses and the *incremental* plastic deformation. This was first recognized by St. Venant, who proposed that for an isotropic material the principal axes of the strain *increment* are parallel to the principal axes of stress.

If the material is assumed to remain isotropic after yield there is no dependence on the deformation or stress history. Furthermore, if we assume that the yield behaviour is independent of the hydrostatic component of stress then the principal axes of the strain increment are parallel to the principal axes of the stress deviator.

Lévy[11] and von Mises[7] independently proposed that the principal components of the strain-increment tensor

$$\begin{bmatrix} de_1 & 0 & 0 \\ 0 & de_2 & 0 \\ 0 & 0 & de_3 \end{bmatrix}$$

and the deviatoric stress tensor

$$\begin{bmatrix} \sigma'_1 & 0 & 0 \\ 0 & \sigma'_2 & 0 \\ 0 & 0 & \sigma'_3 \end{bmatrix}$$

are proportional, i.e.

$$\frac{de_1}{\sigma'_1} = \frac{de_2}{\sigma'_2} = \frac{de_3}{\sigma'_3} = d\lambda \qquad (11.8)$$

where $d\lambda$ is *not* a material constant but is determined by our choice of the extent of deformation of the material, e.g. by the displacement of the jaws of the tensometer.

Since $\sigma'_1 + \sigma'_2 + \sigma'_3 = 0$

and

$$d\lambda = \frac{de_1 + de_2 + de_3}{\sigma'_1 + \sigma'_2 + \sigma'_3}$$

it is implied that $de_1 + de_2 + de_3 = 0$, i.e. that the deformation takes place at constant volume, or that the material is incompressible.

If the stress–strain relations are referred to other than principal axes we have:

$$de_{ij} = \sigma'_{ij} d\lambda$$

i.e.

$$\frac{de_{xx}}{\sigma'_{xx}} = \frac{de_{yy}}{\sigma'_{yy}} = \frac{de_{zz}}{\sigma'_{zz}} = \frac{de_{yz}}{\sigma'_{yz}} = \frac{de_{zx}}{\sigma'_{zx}} = \frac{de_{xy}}{\sigma'_{zy}} \qquad (11.9)$$

These equations are called the Lévy–Mises equations.

11.2.9 The Plastic Potential

The Lévy–Mises equations can also be considered to arise from more sophisticated considerations based on the concept of the plastic potential and the 'normality' rule for ideal plastic behaviour.

These ideas follow from a general representation of plastic behaviour, extensively discussed by Hill[10], which assumes that the components of the plastic strain increment tensor are proportional to the partial derivatives of a function called the 'plastic potential', which is a scalar function of stress. This is analogous to the use of a strain-energy function to derive stresses in elastostatics, but it is important to emphasize that whereas the strain-energy function relates to the total strains in an elastic material, the plastic potential relates to the strain increments.

A particularly simple situation is obtained if the plastic potential is assumed to be a surface in stress space of the same shape as the yield surface. The St. Venant principle is now seen as the mathematical expression of the fact that the plastic strain increments occur in directions normal to the yield surface. This is sometimes called the 'normality' condition of ideal plasticity, and some workers (notably Drucker[12]) have attempted to justify this flow rule in terms of a maximum work criterion.

11.2.10 The Stress–Strain Relations for an Anisotropic Material of Orthorhombic Symmetry

By analogy with the Lévy–Mises equations for an isotropic plastic material, Hill has proposed the following plastic strain increment relations for an anisotropic material referred to the principal axes of anisotropy.

$$de_{xx} = d\lambda[H(\sigma_{xx} - \sigma_{yy}) + G(\sigma_{xx} - \sigma_{zz})], \qquad de_{yz} = d\lambda L\sigma_{yz}$$

$$de_{yy} = d\lambda[F(\sigma_{yy} - \sigma_{zz}) + H(\sigma_{yy} - \sigma_{xx})], \qquad de_{zx} = d\lambda M\sigma_{zx}$$

$$de_{zz} = d\lambda[G(\sigma_{zz} - \sigma_{xx}) + F(\sigma_{zz} - \sigma_{yy})], \qquad de_{xy} = d\lambda N\sigma_{xy} \qquad (11.10)$$

It is to be noted that in this case the principal axes of the plastic strain increment only coincide with the axes of anisotropy when the principal axes of stress coincide with the latter. It is also to be noted that the proportionality factor $d\lambda$ is dimensionally different from that in the Lévy–Mises equations for an isotropic plastic material.

11.3   THE YIELD PROCESS

We have seen that yield is often associated with a load drop on the load-extension curve, and always involves a change in slope on the true stress–strain curve. This load drop has sometimes been attributed either to adiabatic heating of the specimen or to the geometrical reduction in cross-sectional area on the formation of a neck. It is necessary to examine these explanations in detail, determine their shortcomings and establish the test conditions under which true yield behaviour can be observed. With this information it is then possible to consider the relevance of experimental data on the yield behaviour of polymers to our discussions of an ideal rigid-plastic material.

11.3.1   The Adiabatic Heating Explanation

It was soon recognized that under conventional conditions of cold-drawing, where the specimen is extended at strain rates of the order of 10^{-2} sec^{-1}, a considerable rise of temperature occurs in the region of the neck. Marshall and Thompson[13], following Müller[14], proposed that cold-drawing involves a local temperature rise and that necking occurs because of strain-softening produced by the consequent fall in stiffness with rising temperature. The stability of the drawing process was then attributed to the stability of an adiabatic process of heat transfer through the shoulders of the neck, with extension taking place at constant tension throughout the neck.

Let us consider Marshall and Thompson's argument as they presented it. The starting point was to assume that the isothermal load–extension curves for polyethylene terephthalate were of the form shown in Figure 11.13. It was then argued that under their conditions of drawing (between rollers at high speeds) the drawing process takes place under adiabatic conditions. The 'adiabatic' load-extension curve was then calculated by assuming that all the work done in stretching appears as

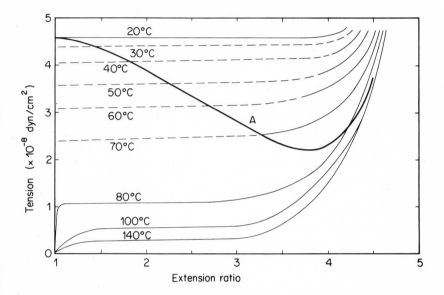

Figure 11.13. Tension–elongation–temperature properties of amorphous polyethylene terephthalate of various temperatures. Dashed curves are interpolated data, not obtainable in practice because necking occurred. *A* is estimated curve from 20°C (after Marshall and Thompson).

heat within the filament, i.e. no account was taken either of increases in elastic potential energy or of heat liberated due to crystallization. This calculation was performed by estimating the work done in successive 10% extensions, adjusting the temperature level at each 10% step, and finally checking the consistency of the integrated area under the resultant curve with the corresponding total temperature rise.

This produces a load–extension curve where the load falls with extension (curve A in Figure 11.13), which is an unstable situation. Marshall and Thompson suggested that instead of drawing each element unstably under variable tension with the temperature varying as for these adiabatic conditions, a condition of constant tension is obtained with heat being transferred through the neck to maintain the differences in temperature required to achieve this.

A heat-balance equation was proposed which predicted to a reasonable approximation the measured length of the shoulder of the neck in the cold-drawing of polyethylene terephthalate.

Hookway[15] later attempted to explain the cold-drawing of Nylon 66 on somewhat similar grounds, suggesting that there is actually a possibility of local melting in the neck due to a combination of hydrostatic tension and temperature. The idea of local melting raises an important point with regard to the structure of the cold-drawn polymer. Nylon is crystalline in the undrawn, unoriented state. If the crystals do not break down in the necking and drawing process, the morphological state of the undrawn polymer will be particularly important in determining the structure of the drawn polymer.

There is no doubt that an appreciable rise in temperature does occur at conventional drawing speeds, and the ideas of Marshall and Thompson are very relevant to an understanding of the complex situation. Calorimetric measurements by Brauer and Müller[16] have, however, shown that at slow rates of extension the increase in temperature is quite small ($\sim 10°C$) and not sufficient to give an explanation for necking and cold-drawing in terms of adiabatic heating. More recently it has been clearly demonstrated that necking can still take place under quasi-static conditions. This was noted by Lazurkin,[17] who observed cold-drawing in rubbers at very low speeds of drawing. (The drawing took place at temperatures below the glass-transition temperature.) Vincent confirmed this result by showing that cold-drawing occurs in polyethylene at very low extension rates at room temperature[3].

The adiabatic heating explanation arose at least in part because the initial yield process was not regarded as distinct from the drawing process. A further examination of the cold drawing of polyethylene terephthalate was undertaken by Allison and Ward[18]. Their results showed that although the *drawing* process is affected by adiabatic heat generation at high strain rates, the *yield* process is not affected. The principal evidence for this conclusion is a comparison of the yield stress and the drawing stress as a function of strain rate. The results, shown in Figure 11.14, demonstrate that the yield stress continues to rise with increasing strain rate, beyond the strain rate at which the drawing stress falls quite distinctly. It was argued that provided the drawing is carried out at a low strain rate, any heat which is generated will be conducted away from the neck sufficiently rapidly for no temperature rise to occur. As the strain rate is increased and the process becomes more nearly adiabatic, the effective temperature at which the drawing is taking place is increased. In particular, heat will be conducted into the unyielded portion of the sample. This will lower the yield stress of the undeformed material and reduce the force necessary to propagate the neck.

On the basis of the experimental results shown in Figure 11.14 it was possible to make an approximate calculation of the temperature rise in a sample caused by the drawing process. Two sets of information were required for this calculation: (1) the effect of increasing strain rate on the

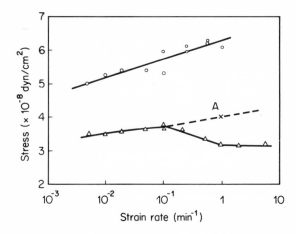

Figure 11.14. Comparison of yield stress ○ and drawing stress △ as a function of strain rate for polyethylene terephthalate.

yield and drawing stresses and (2) the effect of temperature on the yield stress. It was further assumed that to a first approximation both the yield stress and the drawing stresses are measures of the force necessary to initiate large-scale molecular motion and that similar mechanisms are operative in both initial yielding and the subsequent propagation of the neck.

The data shown in Figure 11.14 indicate that initially both the yield and drawing stresses increase by similar amounts when the strain rate is increased. It would be expected that, if the drawing process were isothermal, the drawing stress would continue to increase with increasing strain rate in a manner similar to the yield stress. The difference between the drawing stress predicted on the assumption of an isothermal process and that measured experimentally can therefore be attributed to an increase in the sample temperature.

Subsidiary experiments showed that over a wide range the yield stress is a linear function of temperature. The data are shown in Figure 11.15, and it was calculated that an increase of 10°c in temperature will cause

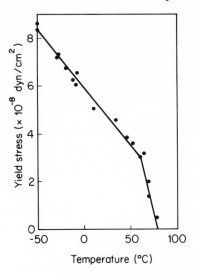

Figure 11.15. The yield stress of polyethylene terephthalate as a function of temperature.

the stress to decrease by about 0.48×10^8 dyn cm^{-2}. A typical temperature rise in the neck during drawing was then calculated as follows. From Figure 11.14 we obtain by extrapolation a drawing stress of 3.9×10^8 dyn cm^{-2} at a strain rate of 1 min^{-1} in the absence of any heating effects (point A in Figure 11.14). The measured drawing stress is 3.0×10^8 dyn cm^{-2} and the yield stress 6.0×10^8 dyn cm^{-2}. Assuming a similar dependence on temperature for the drawing stress and the yield stress, the fall in the drawing stress from the predicted value of 3.9 to 3.0×10^8 dyn cm^{-2} corresponds to a temperature rise of $0.9 \times 6/3.9 \times 10/0.48 = 29°$c. This gives the point B in Figure 11.16.

The calculation was performed for polyethylene terephthalate and nylon (the latter based on the experimental results of Hookway[15]). The results are shown in Figure 11.16, together with experimentally measured temperature rises reported by Vincent[3] for polyethylene and polyvinyl chloride. It is apparent that adiabatic heating effects become important as the strain rate is raised above about 10^{-1} min^{-1}.

This calculated temperature rise agrees approximately with that calculated from the work done in drawing, assuming that no heat is generated due to crystallization. In polyethylene terephthalate, X-ray diffraction diagrams of cold-drawn fibres show that very little crystallization has occurred.

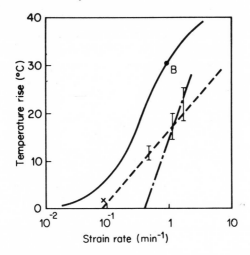

Figure 11.16. The temperature rise as a function of strain rate: —— calculated from Figures 11.14 and 11.15; —·— calculated from Hookway's (1958) data; ———— reported by Vincent (1960).

The work done per unit volume is then given by $W = \sigma_D(D_r - 1)$ where σ_D is the drawing stress and D_r the natural draw ratio. From the results obtained $\sigma_D = 2.3 \times 10^8$ dyn cm^{-2} when $D_r = 3.6$ giving $W = 19.8$ cal cm^{-3}. For polyethylene terephthalate the specific heat is 0.28 cal g^{-1} °C^{-1} and the density is 1.38 g cm^{-3}. This gives a calculated temperature rise of 57°C, compared with 42°C obtained from Figure 11.16.

11.3.2   The Isothermal Yield Process: the Nature of the Load Drop

There is no doubt that a temperature rise does occur in cold-drawing under many conditions of test. We have shown, however, that there is very good evidence to support the view that necking can still take place under quasi-static conditions where there is no appreciable temperature rise. Vincent[3] therefore proposed that the observed fall in load is a geometrical effect due to the fact that the fall in cross-sectional area during stretching is not compensated by an adequate degree of strain hardening. This effect was called strain softening and attributed to the reduction in the slope of the stress–strain curve with increasing strain.

Contrary to this latter explanation of the load drop in terms of geometric softening, results reported by Andrews and Whitney[19] showed a yield drop in compression for polystyrene and polymethylmethacrylate. This led Brown and Ward[20] to make a detailed investigation of yield drops in polyethylene terephthalate. They studied isotropic and oriented specimens, under a variety of test conditions (tension, shear and compression) and concluded that in most cases there is clear evidence for the existence of an intrinsic yield drop, i.e. that a fall in true stress can occur in polymers, as in metals.

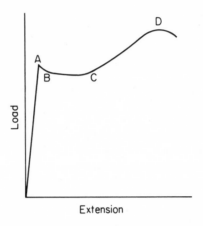

Figure 11.17. Load–extension curve in tension of mild steel.

There is, however, a significant difference between polymers and many metals with regard to yield behaviour. In a polymer, as shown in Figure 11.2, it is to be noted that only one maximum is observed on the load–extension curve. This contrasts with metals (illustrated by mild steel in Figure 11.17) where often two maxima are observed on a typical load–extension curve. The first maximum (point A in Figure 11.17) is the upper yield point. This represents a fall in true stress, an intrinsic load drop, and corresponds to a sudden increase in the amount of plastic strain which relaxes the stress. From B to C Lüders bands propagate throughout the specimen. Lüders bands have also been observed in polymers[21]. At C the specimen is homogeneously strained and the stress begins to rise as the material work hardens uniformly. A second maximum is observed at point D. This second type of maximum is always associated with the beginning of necking in the specimen. Necking occurs when the strain-hardening of the metal is exceeded by the geometrical softening due to the reduction in the cross-sectional area of the specimen as it is strained, i.e. the Orowan–Vincent explanation, discussed in Section 11.1.1 above.

The second maximum, as we have seen previously, is not observed if the true stress–strain curve is plotted instead of the load–extension curve.

The first maximum, on the other hand, would exist on the true stress–strain curve. It is called an intrinsic yield point, because it relates to the intrinsic behaviour of the material.

In polymers, as we have emphasized, only one maximum is observed in the load–extension curve. The investigations of Andrews and Whitney[19] and Brown and Ward[20], show that this maximum combines the effect of the geometrical changes and an intrinsic load drop, and cannot be attributed to the geometrical changes alone. In particular, the cold-drawing results are not accounted for by a decrease in the slope of the true stress–strain curve, as suggested in the explanation of Vincent. It is important to note that every element of the material does not follow the same true stress–strain curve, since the stress for initiation is greater than that for propagation of yielding. This confirms that it will not be possible to give a complete explanation of necking and cold-drawing in terms of the Considère construction on a true stress–strain curve as has already been remarked (Section 11.1.3).

11.4 EXPERIMENTAL EVIDENCE FOR YIELD CRITERIA IN POLYMERS

11.4.1   Isotropic Glassy Polymers

Polystyrene and Other Polymers

Whitney and Andrews[19] studied the behaviour of polystyrene, polymethylmethacrylate, polycarbonate and polyvinyl (formal). They were particularly concerned with the effect of the hydrostatic component of stress on the yield behaviour and the volume changes occurring before and during yielding. Their results for polystyrene are summarized in Figure 11.18 which shows the section of the yield surface normal to the principal stress σ_3. Tensile stress in the $\sigma_1\sigma_2$ plane is positive and compressive stress negative. The numbered circles represent the following tests: (1) uniaxial tension, (2) uniaxial compression, (3) torsional shear, (4) biaxial tension and (5) biaxial compression.

Solid cylinders were used for torsional shear. The biaxial tensile stress was produced by straining, in a single direction, a square film sample (length = width ≫ thickness) held in wide grips. The biaxial compression was produced by using a so-called 'double punch' test in which the sample is squeezed between two rectangular bars lying across the rectangular sample, under conditions of plane strain. It is not clear how point (5) is obtained in the $\sigma_1\sigma_2$ plane plot of Whitney and Andrews as the σ_3 stress is not zero in this case.

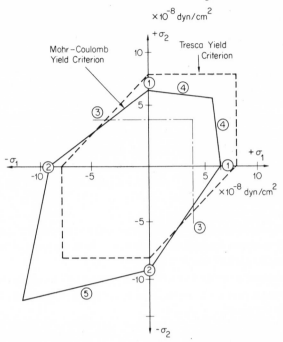

Figure 11.18. Yield and fracture envelopes for polystyrene in the $\sigma_1\sigma_2$ stress plane (after Whitney and Andrews, 1967).

With this reservation regarding point (5) it is still clear that the results do not fit either the Tresca or the von Mises yield criterion. There is a definite asymmetry between the uniaxial tensile and compressive yield stresses and this asymmetry is confirmed by the torsion and combined stress measurements. Whitney and Andrews propose that the yield behaviour is affected either by the mean normal stress or by the hydrostatic component of the stress (these are not, in fact, identical). They point out that the observed data will fit a Coulomb criterion where the yield stress is assumed to depend linearly on the mean normal stress as has been discussed in Section 11.2.4. above.

The Coulomb criterion also specifies the direction in which yielding will occur. It states that the material yields in shear in the plane where the shear stress reaches the critical value $\tau = \tau_c + \mu\sigma_N$, where σ_N is the normal stress on the plane. Whitney and Andrews suggested that the direction of yield, i.e. of the deformation bands, was consistent with their assumption of the Coulomb yield criterion.

Polymethylmethacrylate

In a review article, Thorkildsen[22] quotes a few measurements on the behaviour of thin-walled tubes of polymethylmethacrylate under combined tension and internal pressure. These suggest that the von Mises yield criterion is applicable. From the discussion it appears that the yield stress was considered as the stress at 0.2% strain, i.e. an engineering proof stress, but it is not clear that a consistent definition of the yield point was assumed for all experiments.

In a more detailed study, Bowden and Jukes[23] examined the yield behaviour of polymethylmethacrylate in a plane-strain compression test first developed by Ford[24] to study plastic deformation of metals. The experimental set-up is shown in Figure 11.19. A particular advantage of this technique is that yield behaviour can be observed in compression for materials which normally fracture in a tensile test. In this case PMMA was studied at room temperature, i.e. below its brittle–ductile transition temperature in tension.

The yield point in compression σ_1 was measured for various values of applied tensile stress σ_2, and the results (Figure 11.20) were analysed in terms of the Coulomb yield criterion.

The data give $\tau = 4.66 + 0.258\sigma_N$ (kg/mm^2) but it is to be noted that σ_1 is plotted as a true stress (applied compressive load divided by cross-sectional area of the dies), and σ_2 as a nominal stress (applied tensile load divided by initial cross-sectional area of sheet). A more consistent

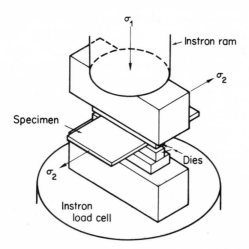

Figure 11.19. The plane-strain compression test (after Bowden and Jukes).

representation in terms of true stress for both σ_1 and σ_2 gives

$$\tau = 4\cdot74 + 0\cdot158\sigma_N$$

The raw data of Figure 11.20 give $\sigma_1 = -11\cdot1 + 1\cdot365\sigma_2$, when both σ_1 and σ_2 are expressed as true stresses. This illustrates directly the divergence from the Tresca criterion where

$$\sigma_1 - \sigma_2 = \text{constant at yield.}$$

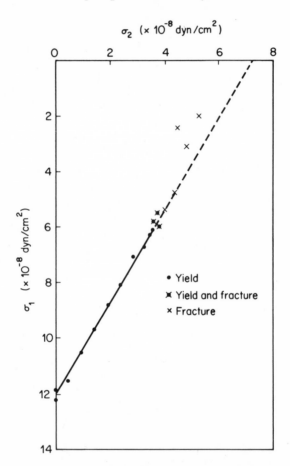

Figure 11.20. Measured values of the compressive yield stress σ_1 (true stress) plotted against applied tensile stress, σ_2 (nominal stress). The full circles denote ductile yield, the crosses, brittle fracture, and the combined points, tests where ductile yielding occurred, followed immediately by brittle fracture (after Bowden and Jukes).

For a von Mises yield criterion the third stress $\sigma_3 = v(\sigma_1 + \sigma_2)$ is involved (v = Poisson's ratio). It can also be shown that Bowden and Jukes' data do not fit a von Mises criterion. (As they remark, for $v = \frac{1}{2}$, the von Mises criterion for a plane strain test degenerates to $\sigma_1 - \sigma_2$ = constant.)

Attempts to fit the direction of the deformation band to the Coulomb criterion (see Section 11.2.5) in these tests were not very satisfactory. The discrepancies were attributed to the complications of establishing general strain-increment laws for materials where the yield criterion is not independent of the hydrostatic component of stress and where volume need not be conserved.

11.4.2   Anisotropic Polymers

If an oriented polymer is subjected to further extension in a direction not parallel to the initial draw direction, the deformation is sometimes concentrated into a narrow deformation band, as shown in Figure 11.21 (see Plate VIB). As previously mentioned, this is analogous to the formation of a neck in an unoriented polymer, and the development of the deformation band is associated with a yield drop, which cannot be accounted for by a geometrical softening[20].

The deformation bands which form are of two types. One type is nearly parallel to the initial draw direction and has the appearance of a slip band in metals; the other type is more diffuse, makes a larger angle with the draw direction, and has the general appearance of a kink band in metals.

It has been suggested by F. C. Frank[25] that the first type of band be called a 'slippy' band and the second a 'kinky' band to indicate their partial but incomplete kinship with these two modes of deformation observed in metal crystals.

A detailed examination of the plastic deformation in such specimens is especially rewarding for two major reasons. First, the yield criterion will be more complex than for unoriented polymers but clearly gives greater scope for an understanding of yield processes in molecular terms. Secondly, when we consider the molecular changes associated with the plastic deformation, we are now measuring changes in an already highly ordered structure, using techniques such as X-ray diffraction which are particularly suited for this purpose. Even macroscopic properties such as birefringence can be more precisely tested when there is a high degree of initial orientation.

In this chapter the yield behaviour of oriented polymers will be discussed first, and the molecular aspects of the plastic deformation deferred to a later section.

Nylon

Slip bands or deformation bands were first observed in oriented nylons by Zaukelies[26]. Singly or doubly oriented bristles of Nylon 66 and Nylon 610 were prepared by drawing, and drawing and rolling, respectively. The bristles were compressed in a tensometer and kink bands were observed. Examination of the deformed samples showed that there was very considerable reorientation of the material within the kink band.

Zaukelies explained his results in terms of crystal plasticity theory. In particular he suggested that the angle θ between the kink band plane and the slip plane was consistent with the Orowan equation.

$$\cot \theta = \frac{2md}{nc}$$

where

     c = crystal-lattice spacing along the slip plane
     d = crystal-lattice spacing between adjacent slip planes
     n = number of c-axis spacings per slip plane
     m = number of slip planes acting in unison

The kink-band angles at 25°c and 100°c were 43·5° and 35·5° respectively, which Zaukelies proposed was consistent with the Orowan equation for slip on 010 planes putting $m = 2$ and 3, respectively. He also proposed various dislocation motions as explanations for the slip processes.

Polyethylene

In a subsequent publication, Kurakawa and Ban[27] (and later Keller and Rider[28]) describe the observation of deformation bands in tensile tests on oriented high-density polyethylene sheets. Kurakawa and Ban concluded that in some cases, when there was only a small angle between the initial draw direction (the IDD) and the tensile axis in the redrawing, the band direction was in the c-axis direction (i.e. the (001) direction in the crystalline regions of the polymer) and coincided with the c-axis direction of the deformed material in the band. In other cases, the deformation band was noted to be a little inclined to the (001) direction. They therefore suggested that the basic deformation process was not simple slip in the (001) direction, but a combination of (001) slip and twinning. Mechanical twinning in polyethylene had been proposed as an explanation of crystal reorientation during rolling by Frank, Keller and O'Connor[29].

When the tensile axis made a large angle with the IDD, kink bands were observed. Wide-angle X-ray diffraction measurements showed that gross reorientation occurred in the kink bands, as in the case of nylon.

Keller and Rider, in a more detailed examination[28], reported the observation of similar deformation bands to those described by Kurakawa and Ban. A variety of drawn; drawn and rolled; drawn, rolled and annealed sheets were examined. In each case the alignment of the c-axis, as determined by wide-angle X-ray diffraction data, was concluded to be an important factor in determining the nature of the deformation bands.

It was noted that the band boundary was generally not parallel to the c-axis direction, and that the boundary of the kink bands did not bisect the angle between the c-axis directions on either side of the boundary. In spite of these anomalies it was concluded that the ductile deformation approximated to slip in the c-direction within the crystalline regions and the similarity to the plastic behaviour of hexagonal metal single crystals was emphasized.

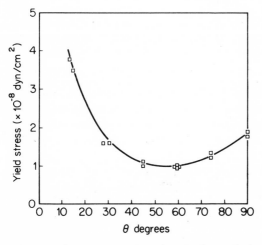

Figure 11.22. Yield stress plotted against θ, the angle between the tensile axis and the initial draw direction, for drawn high-density polyethylene sheets, measured at room temperature (after Keller and Rider).

Keller and Rider[28] measured the yield stress as a function of the angle θ between the tensile-testing direction and the initial draw direction. Their results are shown in Figure 11.22. They proposed that the data can be fitted to the Coulomb yield criterion.

In this case $\tau_c = \sigma(\sin\theta\cos\theta + \mathrm{k}\sin^2\theta)$ where θ is the angle between the tensile stress direction and the initial draw direction and k is a constant.

The Coulomb yield criterion also defines the plane in which yield will occur (Section 11.2.5 above). In this respect it did not satisfactorily describe the high-density polyethylene data, as the deformation band did not form in the direction predicted from the yield criterion.

Polyethylene Terephthalate

A further polymer to be studied in this area is polyethylene terephthalate, which is of much lower crystallinity than high-density poly-

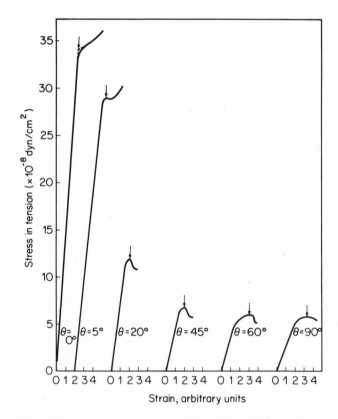

Figure 11.23. Stress–strain curves for various angles θ between the tensile axis and the initial draw direction for tensile tests on drawn polyethylene terephthalate sheets. The yield points are marked by arrows (after Brown, Duckett and Ward).

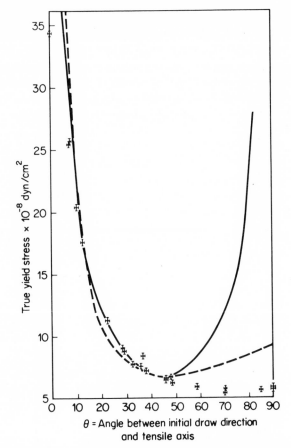

Figure 11.24. Tensile yield–stress data for oriented polyethylene terephthalate sheets showing the prediction of a critical resolved shear stress law. ——— resolving shear stress parallel to the initial draw direction, and – – – – resolving shear stress parallel to the deformation band direction (after Brown, Duckett and Ward).

ethylene ($\sim 30\%$ as against $\sim 85\%$)[20,30,31,32]. It was found that the direction of the deformation band differed significantly from the initial drawing direction for most directions of test.

A typical set of tensile tests for different angles θ between the test direction and the initial draw direction is shown in Figure 11.23. The corresponding variation in yield stress with angle θ between the tensile-testing direction and the initial draw direction of the film is plotted in Figure 11.24.

We have seen that the simplest yield criterion for an anisotropic solid would be the critical resolved shear-stress law of Schmid. For a tensile stress σ, $\tau_c = \sigma \sin \theta \cos \theta$. Figure 11.24 shows that it is not possible to fit the PET data to this equation. Attempts to fit the data to a Coulomb criterion were also unrewarding. Although the critical resolved shear-stress laws were moderately successful for angles θ between 10° and 60°, where the strain observed is basically shear parallel to the deformation band directions, agreement broke down at higher values of θ.

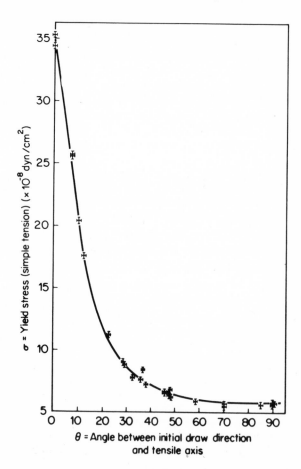

Figure 11.25. Tensile yield-stress data for oriented polyethylene terephthalate sheets showing the best fit obtained from the Hill theory for anisotropic sheets of orthorhombic symmetry (after Brown, Duckett and Ward).

The yield-stress data can, however, be shown to be consistent with the Hill anisotropic yield criterion, as shown in Figure 11.25.

For the plane-stress situation of the tensile tests the Hill yield criterion for a material of orthorhombic symmetry (Section 11.2.7) reduces to

$$\sigma^2\{(G+H)\cos^4\theta + (H+F)\sin^4\theta + 2(N-H)\sin^2\theta\cos^2\theta\} = 1 \qquad (11.11)$$

We have chosen rectangular Cartesian axes with x parallel to the initial draw direction, y in the plane of the sheet and z normal to the sheet. This gives the yield stress σ as

$$\sigma = \{(G+H)\cos^4\theta + (H+F)\sin^4\theta + 2(N-H)\sin^2\theta\cos^2\theta\}^{-1/2}$$

In Figure 11.25 we see a curve following this equation with constants chosen to give an exact fit to data for PET sheet at 0°, 45° and 90°. It can be seen that very good agreement with the experimental data is obtained.

11.4.3 The Band Angle in the Tensile Test

In Section 11.2.10 above we discussed the modified Lévy–Mises equations relating the plastic strain increments to the applied stresses for a material of orthorhombic symmetry.

In the simple tensile tests $\sigma_{zz} = 0$ and these equations reduce to

$$de_{xx} = d\lambda[(G+H)\sigma_{xx} - H\sigma_{yy}]$$
$$de_{yy} = d\lambda[(H+F)\sigma_{yy} - H\sigma_{xx}] \qquad de_{xy} = d\lambda N\sigma_{xy}$$
$$de_{zz} = -d\lambda[G\sigma_{xx} + F\sigma_{yy}] \qquad\qquad\qquad (11.12)$$

As discussed in Section 11.2.5, the deformation-band direction is the direction which is common to the deformed and the undeformed material, and must therefore define a direction which is neither rotated nor distorted by the plastic deformation. This means that the plastic strain increment must be zero in the band direction. There are two such directions in the material; one defines the direction of the 'slippy' band and the other the 'kinky' band.

We have chosen a rectangular set of axes x, y, z with x parallel to the initial draw direction, y in the plane of the sheet. Now consider a set of axes x', y', z', produced by a rotation about the z axis through an angle β. We refer the plastic strain increments to the new frame of reference by

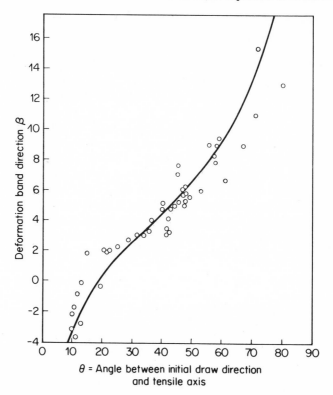

Figure 11.26. Graph of deformation band angle β against θ for tensile tests on drawn polyethylene terephthalate sheets. The curve shows the predictions of the Hill theory modified to include the Bauschinger effect using appropriate values of the constants, F, G, H, N and σ_i as described in the text (after Brown, Duckett and Ward).

using the following transformations:

$$de'_{xx} = de_{xx} \cos^2 \beta + de_{yy} \sin^2 \beta + 2\, de_{xy} \sin \beta \cos \beta$$

$$de'_{yy} = de_{xx} \sin^2 \beta + de_{yy} \cos 2\beta - 2\, de_{xy} \sin \beta \cos \beta$$

$$de'_{zz} = de_{zz}$$

$$de'_{xy} = -(de_{xx} - de_{yy}) \sin \beta \cos \beta + de_{xy}(\cos^2 \beta - \sin^2 \beta) \quad (11.13)$$

The condition for a band forming at an angle β to the initial draw direction is therefore that $de'_{xx} = 0$, i.e.

$$de_{xx} \cos^2 \beta + de_{yy} \sin^2 \beta + 2\, de_{xy} \sin \beta \cos \beta = 0$$

or

$$de_{yy} \tan^2 \beta + 2 \, de_{xy} \tan \beta + de_{xx} = 0$$

From Equations (11.12) above we have de_{xx}, de_{yy} and de_{xy}. Now put

$$\sigma_{xx} = \sigma \cos^2 \theta$$

$$\sigma_{yy} = \sigma \sin^2 \theta$$

$$\sigma_{xy} = \sigma \sin \theta \cos \theta$$

Substituting we have a quadratic in $\tan \beta$. Thus for each value of θ the theory predicts two possible band directions depending on θ, F, G, H and N.

Figure 11.26 shows the fit obtained to the 'slippy' band angle in polyethylene terephthalate[31] using the yield-stress data. Good agreement is obtained which confirms the applicability of both the modified von Mises yield criterion and the associated plastic strain-increment rules proposed by Hill.

11.5 SIMPLE SHEAR–STRESS YIELD TESTS

The yield stresses for oriented polyethylene terephthalate sheets were also measured in a series of simple shear tests[31,32], where the direction of shear displacement made varying angles ϕ with the initial draw direction (Figure 11.27). The stress–strain curves are shown in Figure 11.28, and the yield stress as a function of angle in Figure 11.29. It can be seen that the stress–strain curves vary markedly in form with angle ϕ. Note in particular that there is a clear yield drop when $\phi = 45°$ but no yield drop at 135°.

The applied shear stress can be resolved into a compressive stress and a tensile stress at right angles. When $\phi = 45°$, the initial draw direction is

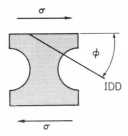

Figure 11.27. A typical specimen for the simple shear test with the direction of the applied shear stress σ making an angle ϕ with the initial draw direction.

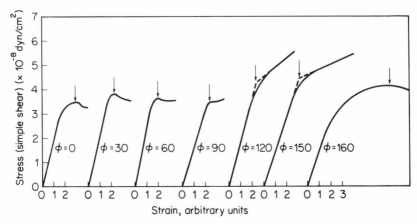

Figure 11.28. Stress–strain curves for various values of ϕ in simple shear tests on drawn polyethylene terephthalate sheets. Arrows mark the yield points (after Brown, Duckett and Ward).

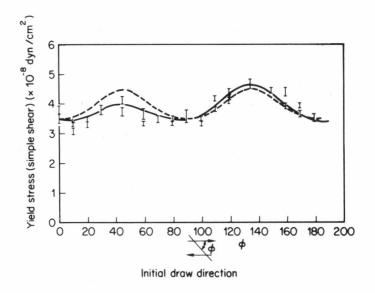

Figure 11.29. Graph of yield stress in simple shear against ϕ for drawn polyethylene terephthalate sheet. – – – – Best fit obtained from Hill theory for orthorhombic sheets. —— Best fit obtained from Hill theory modified to include the Bauschinger term (after Brown, Duckett and Ward).

parallel to the compressive component of stress. Thus it has been argued[31] that the lower yield stress in this direction occurs because the stress situation is favourable for compressing the chains, which is a comparatively easy process. When $\phi = 135°$, the mean direction of the chains is parallel to the tensile component of stress and the chains are under tension. This is a difficult process which may, for example, involve breaking chains, so that the yield stress is now much higher (see Figure 11.30). The different nature of the stress–strain curves in the 45° and 135° directions can be attributed to differences in strain hardening. At 45° the

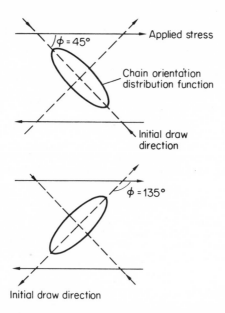

Figure 11.30. Line drawing showing the relative orientations of the initial draw direction and the applied stress for $\phi = 45°$ and $\phi = 135°$.

chains are disoriented by the plastic deformation, the polymer softens, and a clear yield drop is observed. At 135° the chains are further oriented by plastic deformation, the polymer stiffens and no clear yield drop can be observed.

In the simple shear tests $-\sigma_{xx} = +\sigma_{yy} = \sigma \sin 2\phi$

$$\sigma_{xy} = \sigma \cos 2\phi$$

and the Hill anisotropic yield criterion reduces to

$$\sigma^2\{(G+F+4H)\sin^2 2\phi + 2N \cos^2 2\phi\} = 1$$

Figure 11.29 shows the experimental value of σ versus ϕ and a curve obeying the above equation with appropriately chosen constants to give the best overall fit (dashed line). Whilst the overall nature of the curve appears to be correct, in that two maxima and two minima are correctly predicted at approximately the observed values of ϕ, the theory predicts that the two maxima be equal. It can be seen that this is not true.

11.5.1 The Bauschinger Effect

We have seen that the yield-stress values in the shear tests for $\phi = 45°$ and $\phi = 135°$ differ markedly. This was attributed to the difference between the situations where the compressive and tensile components of the applied stress are parallel to the initial draw direction respectively. A subsidiary experiment was performed in which specimens of oriented PET rods cut with their axes parallel to the draw direction were subjected to direct tension and compression[31]. It was found that the compressive yield stress parallel to the draw direction was appreciably less than the tensile yield stress in this direction. This gives direct evidence for a Bauschinger effect in oriented PET.

A simple method of introducing the Bauschinger effect into the yield criterion[31,32] is to include a term σ_i which represents the difference between the tensile and compressive yield points parallel to the initial drawing direction, i.e. the x direction in Equation (11.7). (A more complete treatment would add similar terms in the y and z directions.)

The Hill anisotropic yield criterion is modified to

$$F(\sigma_{yy} - \sigma_{zz})^2 + G(\sigma_{zz} - \sigma_{xx} + \sigma_i)^2 + H(\sigma_{xx} - \sigma_i - \sigma_{yy})^2$$
$$+ 2L\sigma_{yz}^2 + 2M\sigma_{zx}^2 + 2N\sigma_{xy}^2 = 1 \qquad (11.14)$$

The shear tests are then described by the equation

$$(G + H)\sigma_{xx}^2 + (H + F)\sigma_{yy}^2 - 2H\sigma_{xx}\sigma_{yy} + 2N\sigma_{xy}^2 = 1 \qquad (11.15)$$

where

$$\sigma_{xx} = -\sigma \sin 2\phi - \sigma_i, \qquad \sigma_{yy} = \sigma \sin 2\phi$$
$$\sigma_{xy} = \sigma \cos 2\phi$$

On substitution we have

$$\sigma^2 \{(G + F + 4H) \sin^2 2\phi + 2N \cos^2 2\phi\}$$
$$+ 2\sigma\sigma_i(G + 2H) \sin 2\phi = 1 + (G + H)\sigma_i^2 \qquad (11.16)$$

The full line in Figure 11.29 represents the predictions of this modified

Hill theory. It is immediately apparent that the introduction of the Bauschinger term gives a much better fit to the experimental data.

Somewhat similar shear measurements have been undertaken by Robertson and Joynson[33] on oriented polyethylene and polypropylene. In both cases the plots of shear stress versus angle ϕ between the direction of the shear displacement and the initial draw direction, showed two unequal maxima. This was attributed to the ease of chain kinking (which, it is suggested, predominates when $\phi = 45°$) compared with glide between fibrils (which, it is suggested, predominates when $\phi = 135°$). This explanation is similar to the introduction of the Bauschinger effect, which implies that it is more difficult to extend an already oriented structure than to contract it.

Finally it is necessary to note that the introduction of the Bauschinger term into the yield criterion will lead to a further modification of the Lévy–Mises equations for predicting the band angle in the tensile test. To do this on the simple scheme proposed above, we merely replace $\sigma_{xx} = \sigma \cos^2 \theta$ by $\sigma_{xx} = \sigma \cos^2 \theta - \sigma_i$. The quadratic in $\tan \beta$, then predicts for each value of β two possible band directions depending on θ, F, G, H, N and σ_i.

Figure 11.26 also shows the fit obtained to the 'slippy' band angle in PET[31] using the appropriate constants which are consistent with the tensile and shear yield-stress data. Good agreement is obtained, showing the validity of the modified von Mises equation. There is of course some latitude in the method of obtaining a fit when there are a large number of disposable constants. In particular, we have seen that the Bauschinger term σ_i, which is essential to describe the shear yield-stress data, can be omitted and a fit to the tensile yield-stress data and the band angle still be obtained, provided that appropriate changes are made in the other parameters F, G, H and N.

11.6 INFLUENCE OF HYDROSTATIC PRESSURE ON YIELD

We have already discussed evidence for the applicability of a Coulomb-type yield criterion in polymers (Section 11.4.1. above). Direct evidence for the influence of hydrostatic pressure on the yield behaviour of polymers has also been obtained. Ainbinder, Laka and Maiors[34] examined the tensile behaviour of polymethylmethacrylate, polystyrene, Kapron (Nylon 6), polyethylene and several other polymeric materials. In every case both the modulus and yield strength increased with increasing hydrostatic pressure, the effects being largest with the more amorphous polymers. Remarkable increases in ductility under hydrostatic pressure have also been observed in polypropylene[35].

In a recent investigation, Biglione, Baer and Radcliffe[36] determined the load–extension behaviour in tension for polystyrene as a function of hydrostatic pressure from atmospheric pressure to 6 kilobars (kbar)*. At atmospheric pressure this polymer shows brittle behaviour in tension. It was found that at pressures of 3 kbar and greater the polymer became ductile, and necked with the formation of deformation bands. In this region of ductility the yield stress increased linearly with increasing applied pressure.

The effect of hydrostatic pressure on the shear-yield behaviour of polymers has also been investigated. Rabinowitz, Ward and Parry[37] determined the torsional stress–strain behaviour of isotropic poly-methylmethacrylate, crystalline polyethylene terephthalate and poly-ethylene under hydrostatic pressures up to 7 kbar. There were complications in uncoated specimens of PMMA due to the onset of fracture, but apart from this, the shear yield stress for each polymer again increased linearly with increasing hydrostatic pressure. It was also found that the dependence of the yield stress on hydrostatic pressure was about twice as great for PMMA as for PET and polyethylene where it was approximately equal.

It may therefore seem inconsistent that the data for anisotropic polymers can be fitted to a modified von Mises criterion which assumes that yield is not affected by hydrostatic pressure. The yield surface should more correctly be regarded as a closed surface in stress space. It is however, not surprising that a small part of this surface can be represented by the modified von Mises equation when there are many disposable parameters.

To obtain results of greater significance it will be necessary to study the deformation under more complex states of stress and a point of particular interest will be to test the 'normality' rule for the plastic strain increments. The most profitable approach to these problems may well involve similar theories to those recently advanced in soil mechanics.

11.7 THE TEMPERATURE AND STRAIN-RATE DEPENDENCE OF YIELD AND DRAWING PROCESSES

It is very apparent that polymers are not ideal plastic materials. However, we have seen that provided temperature and strain rate are maintained, constant, and conditions chosen so that adiabatic heating does not

---

* 1 kbar $\equiv 10^9$ dyn cm^{-2}.

occur, the yield behaviour can be profitably discussed in terms of theories of ideal plasticity.

The regions where a yield drop may be observed are bounded by the brittle–ductile transition at low temperatures (Section 12.1) and by the glass transition at high temperatures. It is natural to ask how such situations fit into considerations of the temperature and strain-rate sensitivity of the yield process. Increasing strain rate increases the yield stress without much affecting the brittle stress and hence increases the temperature of the brittle–ductile transition which determines the low boundary of yield behaviour (Section 12.1). The upper boundary has been examined by Andrews and his collaborators[38] in research which will now be discussed.

The variation of the yield stress σ_Y and the drawing stress σ_D with temperature and strain-rate was studied for a number of amorphous polymers. Andrews prefers the nomenclature upper and lower yield stress for σ_Y and σ_D, respectively.

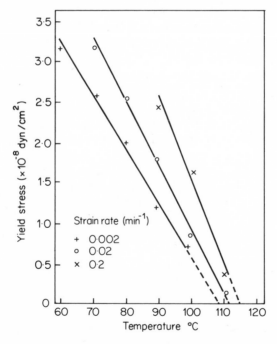

Figure 11.31. Variation in the yield stress with temperature at various strain rates for polymethylmethacrylate (after Langford, Whitney and Andrews).

Figure 11.31 shows the variation in yield stress with temperature and strain rate for polymethylmethacrylate. It is notable that approximately straight lines were obtained with higher dependence on temperature at higher strain rates. In addition the lines appear to converge and would extrapolate to zero yield stress at about 110°C, which is close to the glass-transition temperature for this polymer.

Figure 11.32 shows similar results for the drawing stress. This again shows a series of straight lines converging at a temperature close to the transition temperature.

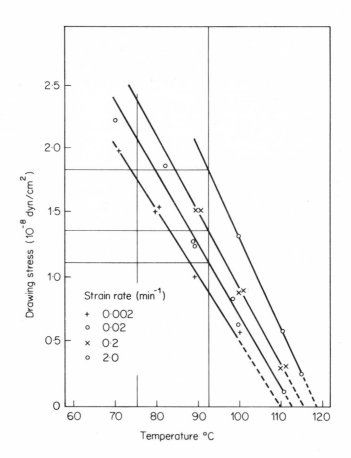

Figure 11.32. Variation in the drawing stress with temperature at various strain rates for polymethylmethacrylate (after Langford, Whitney and Andrews).

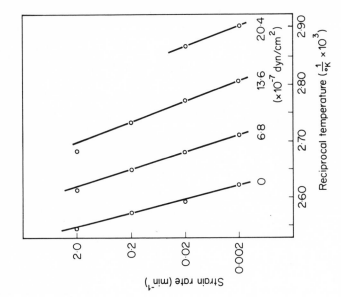

Figure 11.33. Strain rate versus reciprocal temperature at different values of yield stress for polymethyl-methacrylate (after Langford, Whitney and Andrews).

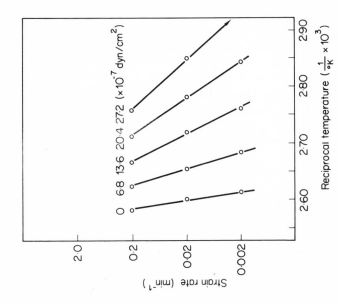

Figure 11.34. Strain rate versus reciprocal temperature at different values of the drawing stress for polymethyl-methacrylate (after Langford, Whitney and Andrews).

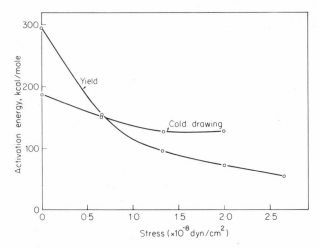

Figure 11.35. Activation energies derived for the yield and cold-drawing behaviour of polymethylmethacrylate using data of Figures 11.33 and 11.34 (after Langford, Whitney and Andrews).

From these results Andrews constructed plots of σ_Y and σ_D at various temperatures. As might be expected, σ_Y and σ_D converge to the same value at the glass-transition temperature where the polymer now stretches homogeneously.

Figures 11.33 and 11.34 show the strain rate as a function of reciprocal temperature for different yield and drawing stresses. These results are cross-plots from the previous graphs. On this basis, it was concluded that there is a simple activated process, with the activation energies dependent on the stress level. The actual values of the activation energies are shown in Figure 11.35. They are very high, which could be interpreted as due to the cooperative nature of the molecular processes involved.

11.8 THE MOLECULAR INTERPRETATIONS OF YIELD AND COLD-DRAWING

11.8.1 The Yield Process

Internal Viscosity Theories

Many workers[17,39–46] have considered that the applied stress induces molecular flow much along the lines of the Eyring viscosity theory where internal viscosity decreases with increasing stress. On this view the yield stress denotes the point at which the internal viscosity falls to a value such that the applied strain rate is identically the plastic strain rate $\dot{e}$ predicted by the Eyring equation. Then

$$\dot{e} = K \sinh \frac{v\tau}{2kT} \qquad (11.17)$$

where K is a constant, τ is the shear stress and v, the 'Eyring volume', represents the volume of the polymer segment which has to move as a whole in order for plastic deformation to occur.

For high values of τ we can approximate, and

$$\dot{e} = K \exp \left(\frac{v\tau}{2kT} \right) \qquad (11.18)$$

Haward and Thackray[45] have compared the flow volumes obtained in this way with the volume of the 'statistical random link'. The latter was obtained from solution studies, by assuming that the observed behaviour could be predicted on the basis of a polymer chain with freely jointed links of a particular length. It was found that for several polymers (e.g. polyvinylchloride, PMMA, polystyrene) the Eyring flow volume was between about two and 10 times that of the statistical links.

In an earlier paper, Lazurkin[17] rejected a previous proposal by Hookway[15] and by Horsley and Nancarrow[47] that the molecular flow occurs because the applied stress reduces the melting point of the crystals. He remarked that similar behaviour is observed for both crystalline and non-crystalline polymers, the dependence of the yield stress σ_y on strain rate $\dot{e}$ having the same form $\sigma_y = A + B \log \dot{e}$ (which follows from 11.18) in both cases.

Robertson[48] has developed a slightly more elaborate version of the Eyring viscosity theory. For simplicity it is considered that there are only two rotational conformations, the *trans* low-energy state and the *cis* high-energy state, which Robertson terms the 'flexed state'. Applying a shear stress τ causes the energy difference between the two stable conformational states of each bond to change from ΔU to $(\Delta U - \tau v \cos \theta)$. $\tau v \cos \theta$ represents the work done by the shear stress in the transition between the two states and θ is the angle defining the orientation of a particular element of the structure with respect to the shear stress.

Prior to application of stress the fraction of elements in the high-energy state is

$$\chi_i = \frac{\exp \left\{ -\Delta U / k\theta_g \right\}}{1 + \exp \left\{ -\Delta U / k\theta_g \right\}} \qquad (11.19)$$

where $\theta_g = T_g$ if the test temperature $T < T_g$ and

$$\theta_g = T \text{ if } T > T_g$$

i.e. below T_g the configurational state 'freezes' at that which exists at T_g. For application of a shear stress τ at a temperature T, the fraction of elements in the upper state with orientation θ is given by

$$\chi_f(\theta) = \frac{\exp - \{(\Delta U - \tau v \cos \theta)/kT\}}{1 + \exp \{-(\Delta U - \tau v \cos \theta)/kT\}} \tag{11.20}$$

Clearly the fraction of flexed elements increases for orientations such that

$$\frac{\Delta U - \tau v \cos \theta}{kT} \leq \frac{\Delta U}{k\theta_g}$$

For one part of the distribution of structural elements, applying the stress tends to make for an equilibrium situation where there are more flexed bonds and this can be regarded as corresponding to a rise in temperature. For the other part of the distribution, the effect of stress can be regarded as tending to lower the temperature. Robertson now argues that the *rate* at which conformational changes occur is very dependent on temperature (cf WLF equation). Hence the rate of approach to equilibrium is much faster for these elements which flex under the applied stress, so that changes in the others can be ignored in calculating the maximum flexed-bond fraction which can occur under a given applied stress. This maximum corresponds to a rise in temperature to a temperature θ_1. The strain rate $\dot{e}$ at θ_1 is calculated from the WLF equation

$$\dot{e} = \frac{\tau}{\eta_g} \exp - \left\{ 2 \cdot 303 \left[\left(\frac{C_1{}^g C_2{}^g}{\theta_1 - T_g + C_2{}^g} \right) \frac{\theta_1}{T} - C_1{}^g \right] \right\} \tag{11.21}$$

where $C_1{}^g$, $C_2{}^g$ are the universal WLF parameters (See Section 7.4.1) and η_g is the 'universal' viscosity of a glass at T_g.

Duckett, Rabinowitz and Ward[49] have modified the Robertson model to include the effect of the hydrostatic component of stress p. It was proposed that p also does work during the activation event and that the energy difference between the two states should therefore be

$$\Delta U - \tau v \cos \theta + p\Omega$$

where Ω is a constant with the dimensions of volume. Figure 11.36 shows that in this modified form the Robertson model can bring consistency to yield data in tension and compression for polymethylmethacrylate, together with the measured effect of hydrostatic pressure. Table 11.1 compares the optimized values of the coefficients used in obtaining this fit with those estimated by Robertson from his fit to tension data only, with no hydrostatic pressure term.

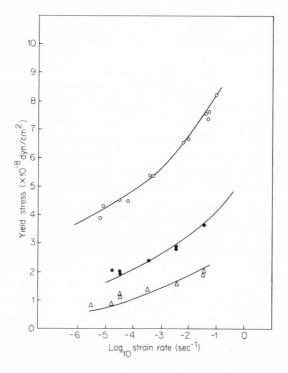

Figure 11.36. Yield stress of polymethylmethacrylate as a function of strain rate. ○ compression at 23°C, △ tension at 90°C, ● tension at 60°C. Curves represent the best theoretical fit (see text).

It can probably only be concluded at this stage that such theories do contain some of the elements which are required for a successful understanding of yield behaviour. More sophisticated molecular theories of yield are undoubtedly required.

Table 11.1. Table of coefficients for PMMA[49].

	Optimized values including hydrostatic pressure term	Values suggested by Robertson
C_1^g	11·1°C	17·44°C
C_2^g	55·9°C	51·6°C
ΔU	0·88 kcal/mole	1·44 kcal/mole
T_g	105°C	105°C
η_g	10^{12} poise	$10^{13} - 10^{14·6}$ poise
v	109 Å³	140 Å³
Ω/v	0·175	0

Free-volume Theories

Bryant[50] and Litt and Koch[51] have proposed that yielding is due to the increase in free volume under stress, the free volume rising until it reaches its value at the glass-transition temperature. At this point flow can occur and the polymer deforms. In this connection Whitney and Andrews[52] have reported that a relative volume dilation does occur in compression yielding.

It is important to note that we cannot conclude that the deformation is similar to that in the rubbery state. This is because the molecular reorientation processes during cold-drawing are known to be very different from those occurring in a rubber (see Section 10.6).

11.8.2   Cold-Drawing

General Considerations

We have seen that strain hardening is a necessary prerequisite for cold-drawing (Section 11.1.3). There are two possible sources of strain-hardening in polymers.

(1) Strain-induced crystallization may occur at the high degree of extension occurring in cold-drawing. This is identical to the crystallization observed in rubbers at high degrees of stretching. It is believed to correspond morphologically to extended-chain crystallization[53]. The occurrence of crystallization on cold-drawing is probably dependent on raising the local temperature in the specimen sufficiently to permit the necessary molecular mobility for the required structural reorganization.

(2) There is a change in the directional properties as the molecular orientation changes during drawing such that the stiffness increases

along the draw direction. This is a general phenomenon, true for both crystalline and amorphous polymers. (Note that the theories of mechanical anisotropy developed in Sections 10.6 and 10.7 apply to the final drawn material and do not relate directly to the strain-hardening effect.)

Cold-drawing occurs in both crystalline and amorphous polymers. Examples of the former are nylon and polyethylene[54]; examples of the latter are polymethylmethacrylate and poly(ethylenemethyl terephthalate)[55-57]. It is also clear that although the general effect of drawing is to produce some degree of molecular alignment parallel to the draw direction, the morphological changes are complex. Whereas some degree of crystallization occurs when amorphous polyethylene terephthalate filaments are cold-drawn, an exactly contrary effect occurs in sodium thymonucleate where the crystalline filaments cold-draw to an amorphous drawn state[58].

Amorphous Polymers and the Natural Draw Ratio

Cold-drawing occurs at temperatures below the glass transition, sometimes as much as $\sim 150°c$ below. It has been suggested by Andrews and others that the yield process and subsequent cold-drawing do not involve long-range molecular flow but are associated with molecular rearrangements between points of entanglement and/or cross-linkage. This view is consistent with the observation of yield, necking and cold-drawing, in highly cross-linked rubbers at temperatures below their glass transition. It is evident that cross-linking does not prevent the required molecular rearrangements.

The natural draw ratio for amorphous polymers is very sensitive to the degree of preorientation, i.e. the molecular orientation in the polymer before cold-drawing. This was reported for polyethylene terephthalate by Marshall and Thompson[13] and for PMMA and polystyrene by Whitney and Andrews[52].

It has been proposed[59] that the sensitivity of natural draw-ratio to preorientation arises as follows. The extension of an amorphous polymer to its natural draw ratio is regarded as equivalent to the extension of a network to a limiting extensibility. This limiting extensibility is then a function of the original geometry of the network and the nature of the links of which it is comprised.

During fibre-spinning, the network forms immediately below the point of extrusion from the small holes, and the fibre is subsequently stretched in the rubber-like state before cooling further and being collected as a frozen stretched rubber. Quantitative stress-optical measurements have confirmed this part of the hypothesis[60]. Cold-drawing then extends the network to its limiting extensibility. The ratio of the extended to unextended lengths of the network is a constant independent of the division of the extension between the spinning, hot-drawing and cold-drawing processes, providing that the junction points holding the network together are not ruptured nor the links in the chain broken.

The dimensions of the unstrained network can be measured by shrinking the preoriented fibres back to the state of zero strain, i.e. isotropy[60]. These results can then be combined with measurements of the natural draw ratio to give the maximum extensibility for the network.

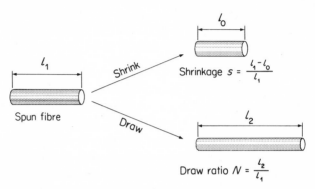

Figure 11.37. A representation of the shrinkage and drawing processes.

Consider the cold-drawing of a sample of length l_1 (Figure 11.37). If the fibre were allowed to shrink back to its isotropic state, length l_0, the shrinkage s would be defined by

$$s = \frac{l_1 - l_0}{l_1} \qquad (11.22)$$

Cold-drawing to a length l_2 gives a natural draw ratio

$$N = \frac{l_2}{l_1} \qquad (11.23)$$

Combining Equations (11.22) and (11.23) we have

$$\frac{l_2}{l_0} = \frac{N}{1-s} \qquad (11.24)$$

Table 11.2 shows collected results for a series of PET filaments. It can

Table 11.2. Value of $N/(1-s)$ for samples of differing amounts of pre-orientation. (Polymer: polyethylene terephthalate, see Reference 59)

Initial birefringence $\times 10^3$	Natural draw ratio N	Shrinkage s	$(1-s)$	$N/(1-s)$
0·65	4·25	0·042	0·958	4·44
1·6	3·70	0·094	0·906	4·08
2·85	3·32	0·160	0·840	3·96
4·2	3·05	0·202	0·798	3·83
7·2	2·72	0·320	0·680	4·01
9·2	2·58	0·378	0·622	4·14

be seen that N varied from 4·25 to 2·58 and s from 0·042 to 0·378, but the ratio l_2/l_0 remained constant at a value of about 4·0.

It is, of course, possible that the natural draw ratio is determined directly by the strain-hardening requirements. This does not invalidate the hypothesis that cold-drawing involves the extension of a molecular network, but suggests that strain-hardening increases very rapidly as the network reaches its limiting extensibility.

Crystalline Polymers

The plastic deformation of crystalline polymers, in particular polyethylene, has been studied intensively from the viewpoint of changes in morphology. Notable contributions to this area have been made by Keller and his coworkers and by Peterlin, Geil and others[61-63]. It is now evident that very drastic reorganization occurs at the morphological level, with the structure changing from a spherulitic to a fibrillar type as the degree of plastic deformation increases. The molecular reorientation processes are very far from being affine or pseudo affine (p. 258) and can also involve mechanical twinning in the crystallites. It is surprising that some of the continuum ideas for mechanical anisotropy are nevertheless still relevant (see p. 260) although they must be appropriately modified.

The natural draw ratio for a highly crystalline polymer such as high-density polyethylene appears to increase with increasing temperature and can become extremely large, say ~ 20. It would seem that this arises from the pulling out of folded molecules, and that the simpler ideas of molecular networks suggested for amorphous polymers are not relevant.

11.9   DEFORMATION BANDS IN ORIENTED POLYMERS

As mentioned above, the molecular reorientation which occurs in a
deformation band is important because it can lead to a molecular understanding of the deformation process.

With this objective in mind, Brown and Ward[64] made an exact analysis
of the geometry of the deformation bands by measuring the deformation
of grids scratched lightly on the surface of oriented sheets of polyethylene
terephthalate. The deformation was shown to be a combination of a pure
shear normal to the sheet and a simple shear parallel to the band.

The molecular reorientation within the deformation band was determined from optical measurements of the rotation of the extinction
direction between the undeformed matrix and the material in the band[65]
and by measurements of refractive indices[66]. The extinction direction
in the band could be predicted exactly in terms of the strain in the deformation band, if it was assumed that the oriented polymer deforms as an

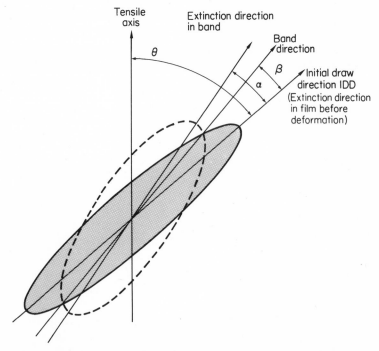

Figure 11.38. The strain ellipsoid in the undeformed and deformed
material (full and dotted lines respectively) for a shear strain of 3 with
no normal strain (after Brown, Duckett and Ward).

oriented continuum. The assumption was made that the refractive index ellipsoid always has principal axes parallel to those of the strain ellipsoid. Figure 11.38 illustrates the situation. The orientation of the strain ellipsoid in the deformation band was calculated from the *total strain*, i.e. it is necessary to add in an appropriate manner the strain introduced by the initial drawing process and that determined from the analysis of the scratched grid. The agreement between measured and predicted extinction directions for a typical case is shown in Table 11.3.

Table 11.3. Comparison of measured and predicted extinction directions for drawn polyethylene terephthalate sheet[65] (for explanation of angles see Figure 11.38)

θ (deg)	β (deg)	Shear strain in band	Extensional strain in band	α (deg) Predicted	α (deg) Measured
21	1·2	1·67	1·00	10·5	9·7
33	3·0	3·43	1·00	7·9	8·5
34	5·2	3·67	1·00	10·5	9·7
60	0	1·34	1·40	6·0	4·2
80	53	0·18	1·10	79·7	79·5

It was later shown[66] that an extension of the aggregate theory (Section 10.6) can be used to predict the refractive indices in the deformation band to a reasonable approximation. The total strain, as described above, is used to determine the principal axes of the strain ellipsoid. The pseudo-affine model (p. 258) can then be used to predict the refractive index ellipsoid.

This simple explanation of the molecular reorientation shows that polyethylene terephthalate undergoes deformation as a continuum. It is strong evidence for a fringed-micelle type model for the structure of the polymer, and suggests that the crystalline regions merely rotate within the deforming material and do not undergo specific deformation processes as, for example, occurs in polyethylene[67].

REFERENCES

1. A. Nadai, *Theory of Flow and Fracture of Solids*, McGraw–Hill, New York, 1950.
2. E. Orowan, *Rept. Prog. Phys.*, **12**, 185 (1949).
3. P. I. Vincent, *Polymer*, **1**, 7 (1960).

4. P. W. Bridgeman, *Studies in Large Plastic Flow and Fracture*, McGraw–Hill, New York, 1952.
5. Reference 1, p. 20.
6. H. Tresca, *C. R. Acad. Sci. (Paris)*, **59**, 754 (1864) and **64**, 809 (1867).
7. R. von Mises, *Göttinger Nachrichten, Math.-Phys. Klasse*, 582 (1913).
8. C. A. Coulomb, *Mém. Math et Phys.*, **7**, 343 (1773).
9. A. H. Cottrell, *Dislocation and Plastic Flow in Crystals*, Clarendon Press, Oxford, 1953.
10. R. Hill, *The Mathematical Theory of Plasticity*, Clarendon Press, Oxford, 1950, p. 50, 318.
11. M. Lévy, *C. R. Acad. Sci. (Paris)*, **70**, 1323 (1870).
12. D. C. Drucker, *Proceedings of the First U.S. National Congress on Applied Mechanics*, *ASME*, New York, 1952, p. 487.
13. I. Marshall and A. B. Thompson, *Proc. Roy. Soc.* **A221**, 541 (1954).
14. F. H. Müller, *Kolloidzeitschrift*, **114**, 59 (1949); **115**, 118 (1949) and **126**, 65 (1952).
15. D. C. Hookway, *J. Text. Inst.*, **49**, 292 (1958).
16. P. Brauer, and F. H. Müller, *Kolloidzeitschrift*, **135**, 65 (1954).
17. Y. S. Lazurkin, *J. Polymer Sci.*, **30**, 595 (1958).
18. S. W. Allison and I. M. Ward, *Brit. J. Appl. Phys.*, **18**, 1151 (1967).
19. W. Whitney and R. D. Andrews, *J. Polymer Sci.*, **C-16**, 2981 (1967).
20. N. Brown and I. M. Ward, *J. Polymer Sci.*, **A-2, 6**, 607 (1968).
21. J. Miklowitz, *J. Colloid Sci.*, **2**, 193 (1947).
22. R. L. Thorkildsen, *Engineering Design for Plastics*, Reinhold, New York, 1964, p. 322.
23. P. B. Bowden and J. A. Jukes, *J. Mater. Sci.*, **3**, 183 (1968).
24. J. G. Williams and H. Ford, *J. Mech. Eng. Sci.*, **6**, 405 (1964).
25. F. C. Frank (private communication).
26. D. A. Zaukelies, *J. Appl. Phys.*, **33**, 2797 (1961).
27. M. Kurakawa and T. Ban, *J. Appl. Polymer Sci.*, **8**, 971 (1964).
28. A. Keller and J. G. Rider, *J. Mater. Sci.*, **1**, 389 (1966).
29. F. C. Frank, A. Keller and A. O'Connor, *Phil. Mag.*, **3**, 64 (1958).
30. N. Brown and I. M. Ward, *Phil. Mag.*, **17**, 961 (1968).
31. N. Brown, R. A. Duckett and I. M. Ward, *Phil Mag.*, **18**, 483 (1968).
32. C. Bridle, A. Buckley and J. Scanlan, *J. Mater. Sci.*, **3**, 622 (1968).
33. R. E. Robertson and C. W. Joynson, *J. Appl. Phys.*, **37**, 3969 (1966).
34. S. B. Ainbinder, M. G. Laka and I. Y. Maiors, *Mekhanika & Polimerov*, **1**, 65 (1965).
35. K. D. Pae, D. R. Mears and J. A. Sauer, *J. Polymer Sci.*, **B6**, 773 (1968).
36. G. Biglióne, E. Baer and S. V. Radcliffe, Paper presented at Brighton Conference on Fracture, 1969.
37. S. Rabinowitz, I. M. Ward and J. S. C. Parry, *J. Mater. Sci.*, **5**, 29 (1970).
38. G. Langford, W. Whitney and R. D. Andrews, *Mat. Res. Lab. Res. Rept. No. R63–49*, M.I.T. School of Engineering, Cambridge, 1963.
39. Y. S. Lazurkin and R. A. Fogelson, *Zh. Tech. Fiz.*, **21**, 267 (1951).
40. R. E. Robertson, *J. Appl. Polymer Sci.*, **7**, 443, (1963).
41. C. Crowet and G. A. Hòmes, *Appl. Mat. Res.* **3**, 1 (1964).
42. C. Bauwens-Crowet, J. A. Bauwens and G. Hòmes, *J. Polymer Sci.*, **A-2, 7**, 735 (1969).

43. J. C. Bauwens, C. Bauwens-Crowet and G. Hòmes, *J. Polymer Sci.*, **A-2**, 7, 1745 (1969).
44. J. A. Roetling, *Polymer*, **6**, 311 (1965).
45. R. N. Haward and G. Thackray, *Proc. Roy. Soc.*, **A302**, 453 (1968).
46. D. L. Holt, *J. Appl. Polymer Sci.*, **12**, 1653 (1968).
47. R. A. Horsley and H. A. Nancarrow, *Brit. J. Appl. Phys.*, **2**, 345 (1951).
48. R. E. Robertson, *J. Chem. Phys.*, **44**, 3950 (1966).
49. R. A. Duckett, S. Rabinowitz and I. M. Ward, *J. Mater. Sci.*, **5**, 909 (1970).
50. G. M. Bryant, *Text. Res. J.*, **31**, 399 (1961).
51. M. H. Litt and P. Koch, *J. Polymer Sci.*, **B5**, 251 (1967).
52. W. Whitney and R. D. Andrews, *J. Polymer Sci.*, **C-16**, 2981 (1967).
53. E. H. Andrews, *Proc. Roy. Soc.*, **A277**, 562 (1964).
54. C. W. Bunn and J. C. Alcock, *Trans. Faraday Soc.*, **41**, 317 (1945).
55. I. M. Ward, *Text. Res. J.*, **31**, 650 (1961).
56. E. A. W. Hoff, *J. Appl. Chem.*, **2**, 441 (1952).
57. R. J. Curran and R. D. Andrews, *Mat. Res. Lab. Res. Rept. No. R63–55*, M.I.T. School of Engineering, Cambridge, 1963.
58. M. F. H. Wilkins, R. G. Gosling and W. E. Seeds, *Nature*, **167**, 759 (1951).
59. S. W. Allison, P. R. Pinnock and I. M. Ward, *Polymer*, **7**, 66 (1966).
60. P. R. Pinnock and I. M. Ward, *Trans. Faraday Soc.*, **62**, 1308 (1966).
61. I. L. Hay and A. Keller, *Kolloidzeitschrift*, **204**, 43 (1965).
62. A. Peterlin, *J. Polymer Sci.*, **69**, 61 (1965).
63. P. H. Geil, *J. Polymer Sci.*, **A2**, 3835 (1964).
64. N. Brown and I. M. Ward, *Phil. Mag.*, **17**, 961 (1968).
65. N. Brown, R. A. Duckett and I. M. Ward, *J. Phys. D: Appl. Phys.*, **1**, 1369 (1968).
66. I. D. Richardson, R. A. Duckett and I. M. Ward, *J. Phys. D: Appl. Phys.*, **3**, 649 (1970).
67. F. C. Frank, A. Keller and A. O'Connor, *Phil. Mag.*, **3**, 64 (1958).

12

Breaking Phenomena

12.1 DEFINITION OF TOUGH AND BRITTLE BEHAVIOUR IN POLYMERS

As we have seen, the mechanical properties of polymers are very greatly affected by temperature and strain rate. In general terms the load–elongation curve at a constant strain rate will change with increasing temperature as shown in Figure 12.1. At low temperatures the load rises approximately linearly with increasing elongation up to the breaking point, when the polymer fractures in a brittle manner. At higher temperatures a yield point is observed, and the load falls before failure, sometimes with the appearance of a neck; this is ductile failure, but still at quite low strains (typically 10–20%). At still higher temperatures, providing that certain conditions are fulfilled, strain-hardening occurs, the neck

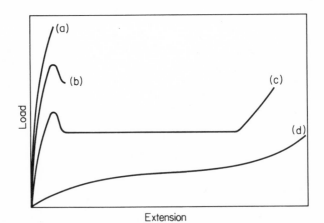

Figure 12.1. Load–extension curves for a typical polymer tested at four temperatures showing different regions of mechanical behaviour. (a) Brittle fracture, (b) ductile failure, (c) necking and cold-drawing and (d) rubber-like behaviour.

stabilizes, and cold-drawing ensues. The extensions in this case are generally very large, i.e. up to 1000 %. Finally, above the glass-transition temperature the load–deformation behaviour of a rubber is observed.

The idea of a brittle–ductile transition is a familiar one in the discussion of the mechanical properties of metals. For polymers the situation is clearly more complicated in that there are in general four regions of behaviour and not two. However, it will still be of considerable value to discuss the factors which influence the brittle–ductile transition in polymers, and then to consider further factors which are involved in the observation of necking and cold-drawing.

The best definition of tough and brittle behaviour comes from the stress–strain curve. Brittle behaviour is designated when the specimen fails at its maximum load (as in Figure 12.1a). It is necessary to exclude rubbers (Figure 12.1d) by adding a corollary that the failure should occur at comparatively low strains (say $\sim 20\%$).

The distinction between brittle and ductile failure is also manifested in two other ways: (1) the energy dissipated in fracture, and (2) the nature of the fracture surface.

The energy dissipated is an important consideration for practical applications, and forms the basis of the Charpy and Izod impact tests (to be discussed further below). At the testing speeds under which the practical impact tests are conducted it is difficult to determine the stress–strain curve. Thus impact strengths are customarily quoted in terms of the fracture energy for a standard specimen.

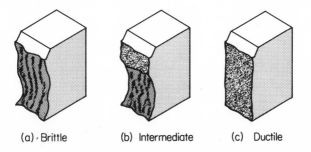

(a) · Brittle   (b) Intermediate   (c) Ductile

Figure 12.2. Types of failure of a fractured specimen.

The appearance of the fracture surface is also an indication of the distinction between brittle and ductile failure. This is illustrated in Figure 12.2. The present state of knowledge concerning the crack propagation is not sufficiently extensive to make this distinction more than an empirical one.

12.1.1   Factors Influencing Brittle–Ductile Behaviour

Many aspects of the brittle–ductile transition in metals, including the effect of notching, which we will discuss separately, have been explained on the basis that brittle fracture occurs when the yield stress exceeds a critical value[1]. This is the Ludwik–Davidenkov–Orowan hypothesis, illustrated in Figure 12.3a. It is assumed that brittle fracture and plastic

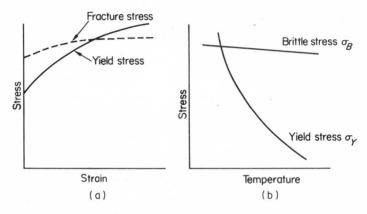

Figure 12.3. (a) and (b) Diagrams illustrating the Ludwig–Davidenkov–Orowan theories of brittle–ductile transitions.

flow are independent processes, giving separate characteristic curves for the brittle fracture stress σ_B and the yield stress σ_Y as a function of temperature at constant strain rate (as shown in Figure 12.3b). Changing strain rate will produce a shift in these curves. It is then argued that whichever process can occur at the lower stress will be the operative one. This can be either fracture or yield. Thus the intersection of the σ_B/σ_Y curves defines the brittle–ductile transition and the material is ductile at all temperatures above this point.

The influence of chemical and physical structure on the brittle–ductile transition can be analysed from this simple starting point, by considering how these factors affect the brittle stress curve and the yield–stress curve respectively.

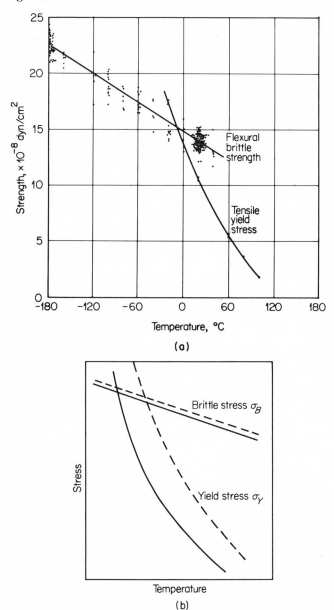

Figure 12.4. (a) Effect of temperature on brittle strength and tensile yield stress of polymethylmethacrylate (after Vincent 1961); (b) Diagram illustrating the effect of strain rate on the brittle–ductile transition: —— low strain rate, – – – high strain rate.

The brittle stress is not much affected by strain rate and temperature (e.g. by a factor of 2 in the temperature range $-180°c$ to $+20°c$). There is not a great deal of definite evidence for this, but Vincent[2] has gathered some data to support it. The yield stress, on the other hand, is greatly affected by strain rate and temperature, increasing with increasing strain rate and decreasing with increasing temperature. (A typical figure would be a factor of 10 over the temperature range $-180°c$ to $+20°c$.) These ideas are clearly illustrated by results for polymethylmethacrylate shown in Figure 12.4a. The brittle–ductile transition will therefore be expected to move to higher temperatures with increasing strain rate (Figure 12.4b). This is a well-known effect with polymers and can be illustrated by varying the strain rate in a tensile test on a sample of nylon at room temperature. At low strain rates the sample is ductile and cold-draws, whereas at high strain rates it fractures in a brittle manner.

There is a further complication in varying strain rate. At low speeds we have seen that within a certain temperature range cold-drawing occurs. It is possible that at high speeds the heat is not conducted away rapidly enough. Strain-hardening is therefore prevented and the specimen fails in a ductile manner. This is an isothermal–adiabatic transition; it does not affect the yield stress and therefore does not affect the brittle–ductile transition; but it does cause a considerable reduction in the energy to break and can be the situation occurring in impact tests, even if brittle fracture does not intervene. It has therefore been proposed that there are two critical velocities at which the fracture energy drops sharply as the strain rate is increased. First there is the isothermal–adiabatic transition, and secondly at higher strain rates, the brittle–ductile transition. As we would expect, changes in ambient temperature have very little effect on the position of the isothermal–adiabatic transition, but a large effect on the brittle–ductile transition.

It was at first thought that the brittle–ductile transition was related to a mechanical relaxation and in particular to the glass transition. This is true for natural rubber, polyisobutylene and polystyrene, but is not true for most thermoplastics. It was then proposed[3] that where there is more than one mechanical relaxation, the brittle–ductile transition may be associated with a lower temperature relaxation. Although again it appeared that there might be cases where this is correct, it was soon shown that this hypothesis has no general validity. Because the brittle–ductile transition occurs at fairly high strains, whereas the dynamic mechanical behaviour is measured in the linear low-strain region, it is unreasonable to expect that the two can be directly linked. It is certain that fracture, for example, depends on several other factors such as the presence of flaws which will not affect the low-strain dynamic mechanical behaviour.

Effect of Basic Material Variables on the Brittle–Ductile Transition[2]

Molecular Weight

Molecular weight does not appear to have a direct effect on the yield strength, but it is known to reduce the brittle strength. As long ago as 1945, Flory[4] proposed that the tensile strength of a polymer was related to the number average molecular weight $\overline{M}_n$ by the relationship

$$\text{tensile strength} = A - \frac{B}{\overline{M}_n}$$

and Vincent[2] has given evidence to suggest that this relationship holds to a rough approximation for the brittle strengths of a wide range of polyethylene samples (Figure 12.5). It is remarkable that the brittle strengths of both linear and branched polyethylenes appear to be the same function of molecular weight.

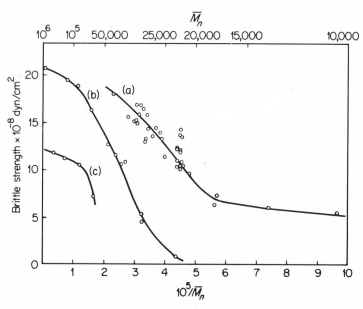

Figure 12.5. Effect of number average molecular weight, $\overline{M}_n$, on brittle strength for (a) polythene; (b) polymethylmethacrylate; (c) polystyrene (after Vincent).

The yield stresses of the different polyethylenes could differ appreciably with the degree of branching (which affects the crystallinity) so that the temperature of the brittle–ductile transition would be a complex function of at least molecular weight and branch content.

Side-Groups

Vincent[2] quotes evidence to suggest that rigid side groups increase both the yield strength and the brittle strength, whereas flexible side-groups reduce the yield strength and the brittle strength. There can thus be no general rule regarding the effect of side-groups on the brittle–ductile transition.

Cross-linking

Cross-linking increases the yield strength but generally does not increase the brittle strength very much. The brittle–ductile transition is therefore usually raised in temperature.

Plasticizers

Plasticizers can decrease the chance of brittle failure because they usually reduce the yield stress more than they reduce the brittle strength.

Molecular Orientation

Molecular orientation is a very different basic variable from the others which we have considered. It introduces anisotropy of mechanical properties. Assuming the Orowan hypothesis of a distinction between brittle strength and yield stress, both will now depend on the direction of the applied stress. It is generally considered that on this viewpoint the brittle strength is more anisotropic than the yield stress. Thus a uniaxially oriented polymer is more likely to fracture when the stress is applied perpendicularly to the symmetry axis, than an unoriented polymer at the same temperature and strain rate. 'Fibrillation', as this is termed, is the basis of commercial processes for manufacturing synthetic fibres from polymer films.

Notch Sensitivity

It is well known that the presence of a sharp notch can change the fracture of a metal from ductile to brittle, and similar considerations apply to the behaviour of polymers. For this reason a standard impact test for a polymer is the Charpy or Izod test, where a notched bar of polymer is struck by a pendulum and the energy dissipated in fracture calculated.

A very simple explanation of the effect of notching has been given by Orowan[1]. For a deep, symmetrical tensile notch, the slip-line field is identical with that for a flat frictional punch indenting a plate under conditions of plane-strain[5] (Figure 12.6).

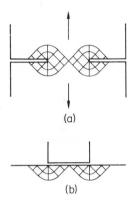

(a)

(b)

Figure 12.6. The slip-line field for a deep symmetrical notch (a) is identical with that for the frictionless punch indenting a plate under conditions of plane-strain (b) (after Cottrell).

The compressive stress on the punch required to produce plastic deformation can be shown to be $(2 + \pi)\mathbf{K}$, where $\mathbf{K}$ is the shear yield stress.

This is $2.57\sigma_Y$ or $2.82\sigma_Y$ (where σ_Y is the tensile yield stress) on either the Tresca or von Mises yield criterion respectively. This shows that for an ideally deep and sharp notch in an infinite solid, the plastic constraint raises the yield stress to a value of approximately $3\sigma_Y$. This gives us the following classification for brittle–ductile behaviour, first proposed by Orowan[1].

(1) $\sigma_B < \sigma_Y$ the material is brittle.

(2) If $\sigma_Y < \sigma_B < 3\sigma_Y$, the material is ductile in an unnotched tensile test, but brittle when a sharp notch is introduced.

(3) If $\sigma_B > 3\sigma_Y$, the material is fully ductile, i.e. ductile in all tests, including those in notched specimens.

Vincent and others have recognized that this distinction can be applied to polymers. Their arguments have, however, been based on the more qualitative ideas of Parker[6]. Parker argued that the effect of the notch is to produce a triaxial stress system. The constraints in the contraction of a notched bar produce transverse tensions σ_2 and σ_3 in both the width and thickness directions (Figure 12.7).

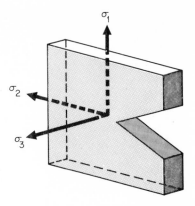

Figure 12.7. The stress system near the tip of a notch (after Parker).

On the basis of the Tresca criterion, plastic flow occurs when

$$\tau_{max} = \frac{\sigma_1 - \sigma_3}{2}$$

For an unnotched bar $\sigma_3 = 0$ and

$$\tau_{max} = \frac{\sigma_1}{2}$$

i.e. The yield stress $\sigma_Y = 2\tau_{max}$. For a notched bar σ_3 is finite. Parker supposes that it may be $\frac{2}{3}\sigma_Y$, i.e. $\frac{4}{3}\tau_{max}$. Yield occurs when

$$\tau_{max} = \frac{\sigma_1 - \sigma_3}{2}$$

$$= \frac{\sigma_1 - \frac{4}{3}\tau_{max}}{2}$$

or $\sigma_1 = \frac{10}{3}\tau_{max} \sim 1\cdot7\sigma_Y$.

Vincent then follows Parker in distinguishing two types of failure: triaxial tensile failure which is brittle, and shear failure which is tough or ductile. A sharp notch increases the triaxial tension relative to the shear stress, thus accentuating the possibility of b.ittle fracture. This explanation is similar to that given by Orowan but does not explicitly state that the brittle fracture remains unaltered by the notch and that only the yield behaviour is affected.

Vincent's σ_B–σ_Y Diagram

We may ask how relevant these ideas are to the known behaviour of polymers. Vincent[7] has constructed a σ_B–σ_Y diagram which is very instructive in this respect (Figure 12.8).

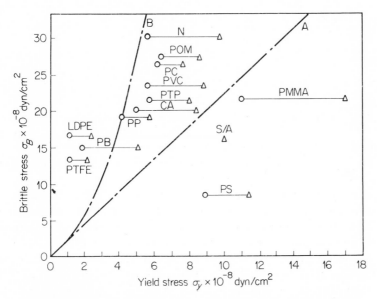

Figure 12.8. Plot of brittle stress at about $-180°$C against a line joining yield-stress values at $-20°$C ($\triangle$) and $20°$C ($\bigcirc$) respectively for various polymers. Line A divides polymers which are brittle unnotched from those which are ductile unnotched but brittle notched, and line B divides polymers which are brittle notched but ductile unnotched, from those which are ductile even when notched (after Vincent).

PMMA	Polymethylmethacrylate	PVC	Polyvinylchloride
PS	Polystyrene	PTP	Polyethyleneterephthalate
S/A	Copolymer of styrene and acrylonitrile	CA	Cellulose acetate
		PP	Polypropylene
N	Nylon 66	LDPE	Low-density polyethylene
POM	Polyoxymethylene	PB	Polybutene-1
PC	Polycarbonate	PTFE	Polytetrafluoroethylene

The term σ_Y was taken as the yield stress in a tensile test at a strain rate of about 50% per minute; for polymers which were brittle in tension, σ_Y was the yield stress in uniaxial compression, and σ_B was the fracture strength measured in flexure at a strain rate of $18\,\text{min}^{-1}$ at $-180°$C.

The yield stresses were measured at $+20°c$ and $-20°c$, the idea being that the $-20°c$ values would give a rough indication of the behaviour in impact at $+20°c$, i.e. lowering the temperature by $40°c$ is assumed to be equivalent to increasing the strain rate by a factor of about 10^5.

In the diagram the circles represent σ_B and σ_Y at $+20°c$; the triangles σ_B and σ_Y at $-20°c$. As we have already discussed, both σ_Y and σ_B are affected by subsidiary factors such as molecular weight and the degree of crystallinity so that each point can only be regarded as of first-order significance.

From the known behaviour of the 13 polymers shown in this diagram, two characteristic lines can be drawn. Line A divides the brittle materials on the right from the ductile materials on the left. Line B divides the materials on the right which are brittle when notched from those on the left which are ductile even when notched. Both these lines are approximations, but they do summarize the existing knowledge.

Line A is not the line $\sigma_B/\sigma_Y = 1$, but is $\sigma_B/\sigma_Y \sim 2$. This difference can be accounted for by the measurement of σ_B at very low temperatures and possibly by the measurement of σ_B in flexure rather than in tension. (The latter may reduce the possibility of fracture at serious flaws in the surface.) It is encouraging that even an approximate relationship holds along the lines of the Ludwig–Davidenkov–Orowan hypothesis. Even more encouraging is the fact that the line B has a slope $\sigma_B/\sigma_Y \sim 6$.

Thus we find that introducing the notch changes the ratio of σ_B/σ_Y for the brittle–ductile transition by a factor of 3 as expected on the basis of the plastic constraint theory.

The principal value of the $\sigma_B–\sigma_Y$ diagram is that it may guide the development of modified polymers or new polymers. Together with the ideas of the previous section on the influence of material variables on the brittle strength and yield stress, it can lead to a systematic search for improvements in toughness.

12.2   BRITTLE FRACTURE OF POLYMERS

12.2.1   The Continuum Approach: Application of the Griffith Theory

Benbow and Roesler[8] initiated a most fruitful approach to the brittle fracture of polymers, when they reported their studies on the slow cleavage of polystyrene and polymethylmethacrylate. They interpreted these results using the Griffith theory of rupture[9] which has been applied extensively to the brittle fracture of glass and metals.

Griffith based his theory of rupture on two ideas. First, he considered that rupture produces a new surface area and postulated that for rupture to occur the increase in energy required to produce the new surface must be balanced by a decrease in elastically stored energy. Secondly, to explain the large discrepancy between the measured strength of materials and those based on theoretical considerations, Griffith proposed that the elastically stored energy is not distributed uniformly throughout the specimen but is concentrated in the neighbourhood of small cracks.

Fracture thus occurs due to the spreading of cracks which originate in pre-existing flaws. The condition for the growth of a crack by an amount dc is

$$-\frac{dU}{dc} = \gamma \frac{dA}{dc} \tag{12.1}$$

where γ is the surface free-energy per unit area of surface, dA is the associated increment of surface area and $-dU$ is the change in elastically stored energy.

Griffith calculated the change in elastically stored energy, using a solution obtained by Inglis[10] for the problem of a plate, pierced by a small elliptical crack, which is stressed at right angles to the major axis of the crack. Equation (12.1) then allows the tensile strength σ_B of the material to be defined in terms of the crack length $2c$ by the relationship

$$\sigma_B = \left(\frac{2\gamma E}{\pi c}\right)^{1/2} \tag{12.2}$$

where E is the Young's modulus

Later workers have shown that a relationship of similar form to (12.2) results when the elastic problem is generalized and extended to three dimensions[11,12].

Much recent work on fracture uses an alternative formulation of the problem due to Irwin[13]. This approach considers the stress field near an idealized crack of length $2c$. In two-dimensional polar coordinates for $r \ll c$, with the x-axis as the crack axis,

$$\sigma_{xx} = \frac{K}{(2\pi r)^{1/2}}(\cos \theta/2)(1 - \sin \theta/2 \sin 3\theta/2)$$

$$\sigma_{yy} = \frac{K}{(2\pi r)^{1/2}}(\cos \theta/2)(1 + \sin \theta/2 \sin 3\theta/2)$$

$$\sigma_{zz} = \nu(\sigma_{xx} + \sigma_{yy}) \text{ for plane strain}$$

$$\sigma_{zz} = 0 \text{ for plane stress}$$

$$\sigma_{xy} = \frac{K}{(2\pi r)^{1/2}} \cos \theta/2 \sin \theta/2 \cos 3\theta/2$$

$$\sigma_{yz} = \sigma_{zx} = 0$$

In these equations θ is the angle between the axis of the crack and the radius vector. K is called the 'stress–intensity factor' and is determined by the details of the model.

The decrease in strain energy for unit increase in crack length can be shown to be K^2/E for plane stress. This quantity is called G, the 'strain-energy release rate', and it is assumed that fracture occurs when G reaches a critical value G_c which is equal to 2γ, in the Griffith formulation.

For an infinite sheet with a central crack of length $2c$ subjected to a uniform stress σ

$$K = \sigma(\pi c)^{1/2}$$

The tensile strength σ_B is then given by $\sigma_B = K_c/(\pi c)^{1/2}$, where K_c is the critical stress-intensity factor for failure and this is identical to

$$\sigma_B = \left(\frac{G_c E}{\pi c}\right)^{1/2} = \left(\frac{2\gamma E}{\pi c}\right)^{1/2}$$

The Griffith and Irwin formulations of the fracture problem can therefore be seen to be equivalent. The usefulness of the critical stress-intensity factor lies in the engineering applications of fracture, whereas for physical understanding of fracture the Griffith formulation is probably the more instructive.

There is extensive evidence to show that for the brittle fracture of glass and metals the form of the Griffith relationship holds to a good approximation. Except in the case of glass the values for the surface energy are greatly in excess of those calculated on the basis of separating planes of atoms to form the fracture surface. This led Orowan and others to propose that the surface free energy may also include a term which arises from the plastic work done in deforming the metal near the fracture surface as the crack propagates.

Experimental studies of the brittle fracture of polymers have been undertaken by cleavage methods, which do not necessarily involve an analysis of the state of stress and examine the relevance of equation (12.1), and by tensile tests, which are directed at an examination of equation (12.2)[14].

In its simplest form, the Griffith theory ignores any contribution to the energy balance arising from the kinetic energy associated with movement of the crack. It has therefore been considered by many workers that a basic study of the brittle fracture of glassy polymers would be likely to be most rewarding if care were taken to ensure that fracture takes place so slowly that a negligible amount of energy is dissipated in this way. With this in mind Benbow and Roesler[8] devised a method of fracture in which flat strips of polymer were cleaved lengthwise by gradually propagating a crack down the middle.

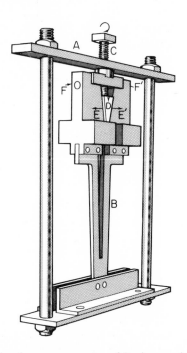

Figure 12.9. The cleavage apparatus of Benbow and Roesler. Flexure in the bar A compresses the sample B and stabilizes the crack direction. Turning the screw C moves the wedge D forward, forcing the clamps E, E' apart. Rotation of the clamps, in the plane of the specimen, is prevented by the sliding bearings F, F'.

Figure 12.9 shows a drawing of the Benbow and Roesler apparatus. The specimens were cut from $\frac{1}{4}$-in thick sheets of commercial plastic; their widths varied from 1 to 4 in and their lengths from 6 to 12 in. Figure 12.10 shows a diagram of the system. A major difficulty was to ensure

that the crack propagated straight along the strip. It was found empirically that the crack could be kept straight by applying a preset lengthwise
compression (the force Q in Figure 12.10). This experimental finding of
Benbow and Roesler has recently been explained theoretically by
Cottrell[15].

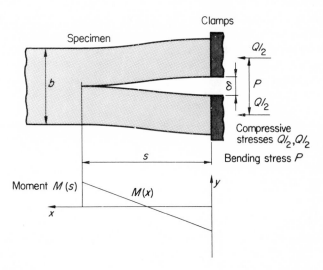

Figure 12.10. The system of the two beams and the moment distribution.

The essence of Benbow and Roesler's analysis of cleavage is to equate
the increase in Griffith surface energy to the decrease in elastically stored
energy as the crack propagates to a larger crack length. The form of the
energy balance can be found from consideration of similarity, any fixed
value of δ/b defining a manifold of geometrically similar systems.

It can then be shown that

$$\frac{\gamma}{E} = \frac{\delta^2}{s}\phi\left(\frac{b}{s}\right)$$

(symbols as in Figure 12.10 and ϕ is
a numerical function)

This equation was tested by plotting experimental values of $f(\delta^2/s)$
against $g(b/s)$ where f and g are convenient functions for a variety of
different samples. For small values of b/s, i.e. for large crack length,
the elastic energy can be calculated from classical elasticity formulae for
the bending of beams and shown to be $\gamma/E = 3\delta^2 b^3/64s^4$. This gives an

asymptote for plots of $(\delta^2/s)^{1/3}$ against s/b which passes through the origin. Knowing a value for the Young's modulus E, the surface energy γ can then be found.

In a subsequent paper[16], Berry adopted a slightly different approach and assumed that the force P required to bend the beam is given by an empirical formula $P = as^{-n}\delta$ where a is a constant.

The energy-balance equation can then be written in terms of P, δ, γ, s and n, with n being determined by a subsidiary experiment. The surface energy can thus be obtained without a value for Young's modulus E, since this is effectively eliminated by measuring the force P. Berry's experimental procedure was simpler than that of Benbow and Roesler

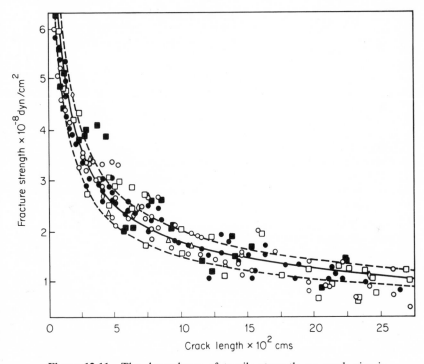

Figure 12.11. The dependence of tensile strength on crack size in polymethylmethacrylate samples for various sample cross-sections (○) 0·42 × 0·19 in., extension rate 0·5 %/min.; (●) 0·42 × 0·19 in., extension rate 5·0 %/min.; (□) 0·42 × 0·063 in., extension rate 0·5 %/min.; (■) 0·42 × 0·063 in., extension rate 5·0 %/min.; (△) 0·98 × 0·19 in., extension rate 0·5 %/min.; (◇) 0·98 × 0·063 in., extension rate 0·5 %/min. (after Berry).

in that the crack direction was determined by routing an initial groove in the sample.

Both Berry and Benbow–Roesler showed that the surface energy was independent of the sample dimensions, suggesting that it is a basic material property. Berry also examined the validity of the equation for the fracture of polymethylmethacrylate and polystyrene by measuring the tensile strength of the samples containing deliberately introduced cracks of known magnitude (see Figure 12.11). The relationship appeared to hold within the fairly large experimental errors experienced in fracture investigations, and gave values for the surface energy comparable with those obtained by cleavage tests. Berry[16] has summarized his own results and those of other workers; these are shown in Table 12.1 below.

Table 12.1. Fracture Surface Energies ($10^5 \gamma$ erg cm^{-2}).

Method	Polymer	
	Polymethylmethacrylate	Polystyrene
Cleavage (Benbow[17])	$4.9 + 0.5$	25.5 ± 3
Cleavage (Svensson[18])	4.5	9.0
Cleavage (Berry[16])	1.4 ± 0.07	7.13 ± 0.36
Tensile (Berry[14])	2.1 ± 0.5	17 ± 6

At this point it is interesting to attempt a theoretical estimate of the surface energy. If it is assumed that the energy required to form a new surface originates in the simultaneous breaking of chemical bonds only, this gives an upper theoretical limit for the surface energy. Let us assume that the bond-dissociation energy is 100 kcals/mole and that the concentration of molecular chains is 1 chain/20 Å^2, giving 5×10^{14} molecular chains/cm^2. It can then be shown that to form 1 cm^2 of new surface requires about 1.5×10^3 erg, which is two orders of magnitude less than that obtained from the cleavage and tensile measurements.

This large discrepancy between experimental and theoretical values for the surface energy is comparable with that found for metals where, as has been mentioned, it has been proposed that although the form of the Griffith theory is correct, the surface-energy terms should be modified to include the work of plastic deformation, which is, in fact, much larger than the surface energy. Andrews[20] has suggested that the quantity measured in the brittle fracture of polymers should be described by Υ, the 'surface-work parameter', to distinguish it from a true surface energy.

Berry came to a similar conclusion, remarking that the largest contribution to the surface energy appears to come from a viscous flow process. He later suggested that in polymethylmethacrylate this was related to the interference bands observed on the fracture surfaces, as seen in Figure 12.14a (see Plate VII). Berry proposed that the large surface-energy term arises from work expended in the alignment of polymer chains ahead of the crack, the subsequent crack growth leaving a thin, highly oriented layer of polymeric material on the fracture surface. This aspect will shortly be discussed in greater detail.

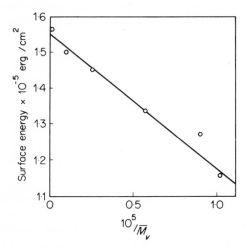

Figure 12.12. Dependence of fracture surface energy on reciprocal molecular weight ($\overline{M}_v$ is viscosity average molecular weight) (after Berry).

In a later paper[19] Berry showed the influence of molecular weight on the surface-energy parameter of polymethylmethacrylate. His results (Figure 12.12) fitted an approximately linear dependence of the fracture-surface energy on reciprocal molecular weight. We have already noted that Flory[4] proposed that the brittle strength is related to the number–average molecular weight.

Berry found that a good fit to his results was obtained if he assumed that $\gamma = A' - B'/\overline{M}_v$ where $\overline{M}_v$ is the viscosity-average molecular weight.

It was suggested that this dependence on molecular weight arises as follows. The measured surface energy parameter includes plastic deformation of the material in the region at the tip of the crack (Figure 12.13). In polymethylmethacrylate and polystyrene this is the material which forms the crazes. (For a detailed discussion of crazes see Section 12.3 below.) These crazes are of thickness ~ 5–6×10^2 Å and Berry calculated that the molecular weight of a fully extended polymethylmethacrylate molecule of this length would be $\sim 2 \times 10^5$. He argued that the smallest molecule which can contribute fully to the surface energy will have its

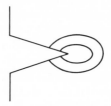

Figure 12.13. The plastic zone at the tip of a notch (after Sternstein and Cessna 1965).

ends on the boundaries of the yielded region, on opposite sides of the fracture plane, and be fully extended between these points.

The results showed that the surface-energy parameter becomes markedly dependent on molecular weight when this is less than 4×10^5. The details of Berry's explanation may be open to doubt, but it seems very likely that his basic ideas are correct, viz. that there is appreciable orientation of the material adjacent to the fracture surface as the crack propagates and that the amount of work done in this process relates to the molecular weight. It may be that a molecular weight of about 4×10^5 is required to produce a molecular network by entanglements and thus give rise to strain-hardening.

12.2.2 The Molecular Approach

It has long been recognized that oriented polymers (i.e. fibres) are much less strong than would be predicted on the basis of elementary considerations by assuming that fracture involves simultaneously breaking the bonds in the molecular chains across the section perpendicular to the applied stress. Calculations of this nature were undertaken by Mark[21] on cellulose and more recently by Vincent[22] on polyethylene. It was found that in both cases the measured tensile strength was at least an order of magnitude less than that calculated.

We have seen one possible explanation of this discrepancy—the Griffith flaw theory of fracture. It has also been considered that there may be a general analogy between this difference between measured and calculated strengths and the difference between measured and calculated stiffnesses (Section 10.8). A general argument for both discrepancies could be that only a small fraction of the molecular chains are supporting the applied load. In a crystalline polymer these are considered to be the interlamellar ties (molecules which are not chain-folded but thread from one crystal lamella to a neighbouring lamella as in a modified fringed-micelle structure). Peterlin[23] has attempted to quantify modulus data for polyethylene on this basis.

A second approach to the molecular understanding of fracture in polymers notes the time and temperature dependence of the fracture processes. Zhurkov and his collaborators[24] have measured the lifetime of polymers as a function of tensile stress at various temperatures. It was proposed that the relationship between the lifetime, the tensile stress σ_B and the absolute temperature T could be represented by an Eyring-type equation

$$\tau = \tau_0 \exp \frac{U_0 - \beta \sigma_B}{kT}$$

where τ_0, U_0 and β are constants determining the strength characteristics of a polymer.

The parameter U_0 has the dimensions of energy and it is suggested that it corresponds to the height of the activation barrier which has to be surmounted for fracture to occur. Zhurkov and his coworkers showed that for a wide range of polymers U_0 was approximately equal to the activation energies obtained for thermal breakdown. They then showed by the technique of electron-spin resonance that free radicals are produced in the fracture process in polymers and moreover that correlations can be established between the radical-formation rate and the time to break. Similar studies have also been undertaken by Peterlin[25].

Recent work has added the techniques of infrared spectroscopy[26] and mass spectroscopy[27] to studies of the chemical nature of the fracture processes. In addition Zhurkov and his collaborators[28] have used low-angle X-ray diffraction to measure the formation and growth of cracks.

12.3 CRAZE FORMATION

Craze formation in polymers manifests itself in two ways. First, there are the brilliant interference effects observed on the fracture surfaces of, for example, polymethylmethacrylate, as in Figure 12.14a (see Plate VII). Secondly, when certain polymers, notably polymethylmethacrylate and polystyrene, are subjected to a tensile test in the brittle state above a certain tensile stress, opaque striations appear in planes whose normals are the direction of tensile stress, as in Figure 12.14b (see Plate VIIIA).

The surface effects in polymethylmethacrylate were first observed by Higuchi[29] and Berry[30]. Berry, and later Kambour[31] have done much to establish the nature of these surface crazes and their variation with specimen and conditions of fracture. Kambour confirmed that PMMA fracture-surface layers were qualitatively similar to the internal crazes of this polymer, by showing that the refractive indices were the same. Both surface layers and bulk crazes appear to be oriented-polymer structures of low density, produced by orienting the polymer under conditions of abnormal constraint where it is not allowed to contract in the lateral direction, while being extended locally to strains of the order of unity, i.e. the polymer has undergone inhomogeneous cold-drawing.

It is particularly interesting to note that polymers containing internal crazes can still withstand very high loads ($\sim 2 \times 10^5$ dyn cm^{-2} in polystyrene), even when the crazes extend across the whole cross-section.

It has been suggested[32] that the fracture process in a glassy polymer is two-fold:

(1) A thin layer of material is expanded to form a craze.
(2) The crazed layer is then extended to its breaking point.

An estimate by Kambour[31] of the plastic work involved in forming the craze suggests that this is $\sim 2 \times 10^4$ ergs/cm^2 which is only about 15% of the measured surface-energy parameter. He remarks that craze *fracture* may require large expenditure of plastic work.

12.3.1   Stress Criterion for Craze Formation

There is considerable interest in attempting to obtain a stress criterion for craze formation analagous to that for yield behaviour described in Chapter 11. To this end Sternstein and his collaborators[33] have used the ingenious technique of examining the formation of crazes in the vicinity of a small circular hole ($\frac{1}{16}$-in diameter) punched in the centre of polymethylmethacrylate strips ($0.5 \times 2 \times \frac{1}{32}$ in) when the latter are pulled in tension. A typical pattern is shown in Figure 12.15a (see Plate VIIIB). The solutions for the *elastic* stress field in the vicinity of the hole were compared with the craze pattern. It was found that the crazes grew parallel to the minor principal stress vector. As the contours of the minor principal stress vector are orthogonal to those of the major principal stress vector this shows that the major principal stress acts along the craze plane normal and therefore parallel to the molecular orientation axis of the crazed material.

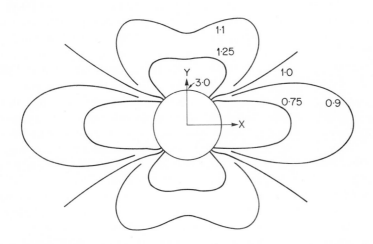

Figure 12.15(b). Major principal stress contours (σ_1) for an elastic solid containing a hole. The specimen is loaded in tension in the X direction. Contour numbers are per unit applied tensile stress (after Sternstein *et al.*).

The boundary of the crazed region coincided to a good approximation with contour plots showing lines of constant major principal stress σ_1. These are shown in Figure 12.15b where the contour numbers are per unit of applied stress. It should be noted that at low applied stresses it is not possible to discriminate between the contours of constant σ_1 and contours showing constant values of the first stress invariant $I_1 = \sigma_1 + \sigma_2$. However, the consensus of the results is in accord with a craze-stress criterion based on the former rather than on the latter, and as we have seen the *direction* of the crazes is consistent with the former.

Sternstein, Paterno and Ongchin[34] carried this investigation one stage further by examining the formation of crazes under biaxial stress conditions. It was found that the stress conditions for crazing involved both the principal stresses σ_1 and σ_2. The results were represented in the following manner. The general biaxial stress was represented by two quantities, the first stress invariant $I_1 = \sigma_1 + \sigma_2$ and a stress bias $\sigma_b = |\sigma_1 - \sigma_2|$. The criterion for craze formation was then proposed to be

$$\sigma_b = |\sigma_1 - \sigma_2| = A + \frac{B}{I_1}$$

where A and B are constants which depend on temperature.

While the exact form of the stress criterion for craze formation may perhaps be questioned, the experimental data do indicate that the hydro-static component of stress is important in craze formation as it is important in yield behaviour (Section 11.6).

12.4 IMPACT STRENGTH OF POLYMERS

In the definition of tough and brittle behaviour in polymers it was noted that the energy dissipated in fracture, which can indicate the distinction between brittle and tough behaviour, is a property of considerable technological importance. It should be emphasized here that high-impact strength does not necessarily imply tough failure, nor does brittle fracture necessarily imply low impact strength. If we are prepared to extend the discussion to composite materials, a glass-reinforced polyester can have an extremely high impact strength, even if it fractures in a brittle manner.

The measurement of impact strength is most commonly made by the Izod or Charpy Impact Test (Figure 12.16). The principle of both methods is to strike a small bar of polymer with a heavy pendulum swing. In the Izod tests the bar is held vertically by gripping one end in a vice and the other free end is struck by the pendulum. In the Charpy test the bar is supported at its ends in a horizontal plane and struck in the middle by the pendulum.

Because the technological assessment of impact strength demands that the severest criteria should be adopted, the impact bars are normally notched. For compression-moulded specimens the notch can be intro-duced either by machining after moulding or by having a notch in the mould. For injection-moulded specimens it is essential to machine the notch, as injection-moulding introduces small amounts of molecular orientation which would entirely alter the stress distribution near a notch and hence invalidate the test. We have discussed the likely effect on brittle–ductile failure for the use of an infinitely deep notch. In practice the impact strength depends on the sharpness of the notch in a progressive manner (Figure 12.17). Notching also reduces the scatter of results, because it reduces the effect of random flaws.

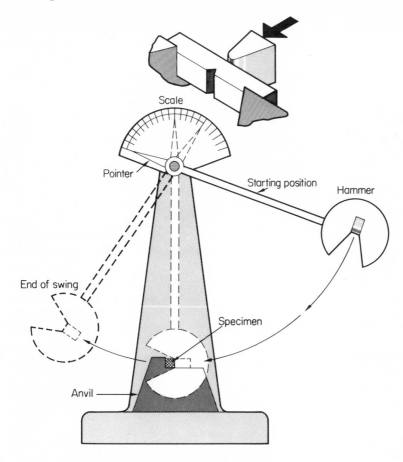

Figure 12.16. Schematic drawing of Charpy impact tester.

An alternative impact test which is sometimes used is the falling weight-impact test. In this test a dumb-bell test piece is mounted vertically at the end of a large vertical rod. Annular weights fall down the rod and the minimum energy required to produce fracture is estimated by making measurements on about 20 test pieces.

In all these tests there are obvious difficulties, e.g. in correcting for the kinetic energy of the broken parts of the polymer after impact. The reader is referred to the literature[35] for accounts of the attempts to make corrections of this type.

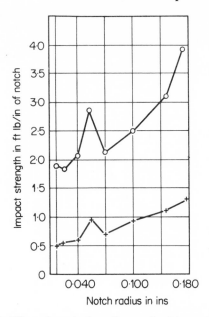

Figure 12.17. Effect of notch tip radius on Izod impact strength (after Adams and Jackson). ○ Styrene/Rubber; + Polystyrene.

We have seen that as we change temperature and strain rate in a polymer the nature of the stress–strain curve can change remarkably. It is therefore natural to seek a correlation between the area beneath the stress–strain curve and the impact strength. The obvious difficulty in making such a comparison is that impact-strength tests are carried out at very high strain rates. In particular, at the tip of the sharp notch very high strain rates can be reached which are difficult to estimate quantitatively.

Using specially constructed machines, tensile tests have been performed on polymers at strain rates in the impact-test range[36]. Bucknell has also described a technique of mounting a pressure transducer behind an Izod specimen to determine the stress–strain curve to fracture in an impact test.

Using these high-speed tensile machines, Evans, Nara and Bobalek[38] showed that there was a poor correlation between the falling-weight impact-test energy (unnotched specimen) and the area under the stress–strain curve at low strain rates, but a good correlation at medium strain rates (Figure 12.18). However, the correlation became poorer at very high strain rates, suggesting that such comparisons should be regarded

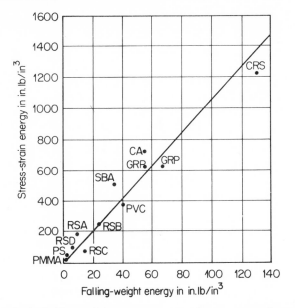

Figure 12.18. Falling-weight energy compared to stress–strain energy determined at 0·333 in./sec (after Evans *et al.*). *Key*: RSA, RSB, RSC, RSD rubber/polystyrene blends; PVC polyvinyl chloride, PS polystyrene, CA cellulose acetate, SBA terpolymer, GRP glass fibre composite, CRS styrene rubber blend.

as unsatisfactory in the absence of an understanding of the failure mechanism involved.

Bucknell makes an interesting comparison between fracture-surface energy and impact-strength figures for polystyrene. He shows that the Izod impact-strength value of $0\cdot3$ ft lb/in^2 notch is equivalent to $1\cdot02 \times 10^6$ ergs for a standard $\frac{1}{4}$-in specimen, i.e. $8\cdot5 \times 10^5$ erg/cm^2. This can be compared with the Benbow and Roesler–Berry values of $3\cdot0$–25×10^5 erg/cm^2. The closeness of these values suggests that the majority of the energy in an impact strength test is dissipated in a fracture process very similar to that measured in cleavage experiments under more closely defined conditions.

12.4.1 High-Impact Polyblends

Recent work on the production of high-impact polymers has included the rubber-modified thermoplastics, e.g. high-impact polystyrene in which a rubber is dispersed throughout the polystyrene in the form of small aggregates or balls. Nielsen[39] lists three conditions which are required for an effective polyblend:

(1) The glass temperature of the rubber must be well below the test temperature.

(2) The rubber must form a second phase and not be soluble in the rigid polymer.

(3) The two polymers should be similar enough in solubility behaviour for good adhesion between the phases.

There are several theories which have been proposed for the action of rubbers in improving the impact strength. One is that the rubber becomes stretched during the fracture process and absorbs a great deal of energy. Another theory proposes that the rubber particles act to introduce a multiplicity of stress-concentration points. Thus there are many cracks which propagate during the fracture process rather than a single crack. It is proposed that this requires greater energy due to the production of more new surfaces (relating the fracture energy to the new surface area by Griffith's theory of fracture).

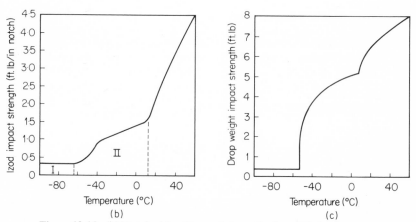

Figure 12.19. (b) Notched Izod impact strength of modified polystyrene as a function of temperature, showing the limits of the three types of fracture behaviour; (c) Drop weight impact strength of ·08 in. high-impact polystyrene sheet as a function of temperature (after Bucknell).

Recent studies by Bucknell[32,37] have linked high-impact strength in modified polystyrene polymer to craze formation. Bucknell and Bucknell–Smith[40] compared the force–time curves for impact specimens over a range of temperatures, with both the notched Izod impact strength and the falling-weight impact strength and the nature of the fracture surface.

The force–time curves, such as in Figure 12.19a (see Plate IX) showed three regions similar to those observed for a homopolymer as discussed in the introduction above. Both impact-strength tests also showed three regions (Figures 12.19(b) and (c)). The fracture surfaces at the lowest temperature were quite clean, whereas at high temperatures stress-whitening or craze formation occurred. These three regions were explained as follows:

(1) Low temperatures. The rubber is unable to relax at any stage of fracture. There is no craze-formation and brittle fracture occurs.

(2) Intermediate temperatures. The rubber is able to relax during the relatively slow build-up of stress at the base of the notch, but not during the fast-crack propagation stage. Stress-whitening occurs only in the first (precrack) stage of fracture, and is therefore confined to the region near the notch.

(3) High temperatures. The rubber is able to relax even in the rapidly forming stress field ahead of the travelling crack. Stress-whitening occurs over the whole of the fracture surface. Bucknell and Smith[40] report similar results for other rubber-modified impact polymers.

12.4.2 Crazing and Stress-whitening

Bucknell and Smith[40] have remarked on the connection between crazing and stress-whitening. High-impact polystyrenes are manufactured by the incorporation of rubber particles into the polystyrene. It was observed that the fracture of this material is usually preceded by opaque whitening of the stress area. Figure 12.14(b) (Plate VIIIA; please see p. 176) shows a stress-whitened bar of high-impact polystyrene which failed at an elongation of 35%.

A combination of different types of optical measurements (polarized light to measure molecular orientation and phase-contrast microscopy to determine refractive index) showed that these stress-whitened regions are similar to the crazes formed in unmodified polystyrene. They are birefringent, of low refractive index, capable of bearing load and are healed by annealing treatments.

Bucknell and Smith concluded that the difference between stress-whitening and crazing exists merely in the size and concentration of the craze bands, these being of much smaller size and greater quantity in stress-whitening. Thus the higher conversion of the polymer into crazes accounts for the high breaking elongation of toughened polystyrene. It is suggested that the effect of the rubber particles is to lower the craze-initiation stress relative to the fracture stress, thereby prolonging the crazing stage of deformation. The crazing stage appears to require the relaxation of the rubber. The function of the rubber particles is not, however, merely to provide points of stress concentration. It is known that there must be a good bond between the rubber and polystyrene, and this is achieved by chemical grafting. The rubber must bear part of the load at the stage when the polymer has crazed but not fractured. Bucknell and Smith suggest that the rubber particles may be constrained by the surrounding polystyrene matrix so that their stiffness remains high. These ideas lead directly to an explanation of the three regimes for impact testing, as discussed above. At low temperatures there is no stress-whitening because the rubber does not relax during the fracture process and we have low impact strengths. At intermediate temperatures, stress-whitening occurs near the notch, where the crack initiates and is travelling sufficiently slowly compared with the relaxation of the rubber. Here the impact strength increases. Finally at high temperatures, stress-whitening is observed along the whole of the crack, and the impact strength is high. It seems likely that these ideas have a greater generality, and will apply to the fracture of other polymers, including, for example, impact-modified polyvinyl chloride.

12.5   THE NATURE OF THE FRACTURE SURFACE IN POLYMERS

The work of Irwin and his coworkers[41] demonstrated that the fracture surfaces in polymers bear many similarities to those in metals. In particular there is often a clearly distinguishable mirror area (Figure 12.20), where

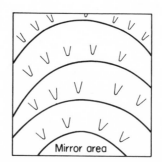

Figure 12.20. The fracture surface (after Zandman).

the fracture commences, which is surrounded by parabolic markings due to the interference of the main crack with new cracks nucleating ahead of it (Reference 5, p. 350).

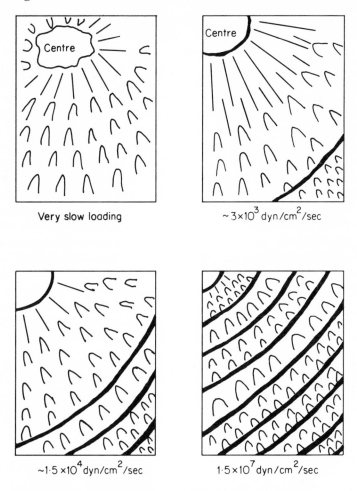

Figure 12.21. Effect of rate of loading on fracture appearance of polymethylmethacrylate (after Wolock, Kies and Newman).

A review article[42] shows that changes in both the rate of loading and molecular weight can produce identical changes in the fracture surface of polymethylmethacrylate. These results are summarized in Figures 12.21 and 12.22 (see Plate X) taken from this article[42]. It can be seen that the mirror-like area where the crack initiates becomes more extensive either as the rate of loading is *decreased* or as the molecular weight is *increased*.

This corresponds to an increase in the work done during fracture according to the idea of Bucknell and Smith for high impact polystyrene and to the experimental results of Berry for polymethylmethacrylate. It is also clear that changing the test temperature produces similar changes in the fracture surface. These changes again occur in the way which we would expect from the high-impact polystyrene results, viz. that the mirror area increases in extent with increasing temperature.

12.6   CRACK PROPAGATION

For brittle fracture of materials, Mott[43] has shown that if the kinetic energy of the moving crack is included in the Griffith energy-balance equation, the crack propagates with a limiting velocity proportional to the velocity of sound in the material. This has been verified for metals and polycrystals by Schardin[44] and others. It was extended to polymers by Bueche and White[45] and Schardin[44] but the results illustrated in Figure 12.23 are not so precise as for metals. Bueche and White[45] commented that although their results are consistent with the Griffith theory of preexisting cracks, an alternative theory due to Poncelet[46] assuming that the cracks are produced by stress (the 'flaw genesis' theory), gives an almost identical numerical value for the limiting velocity.

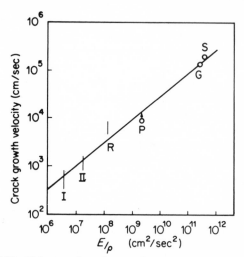

Figure 12.23. Velocity of crack growth, versus modulus E/density ρ. I and II are silicone rubbers. R, P, G and S are irradiated polyethylene, polymethylmethacrylate, glass and fused silica, respectively (after Bueche and White).

It should be noted, however, that the mechanisms of crack propagation may be different from that of crack initiation, so these experiments do not necessarily give information regarding the latter.

12.7 BRITTLE FRACTURE BY STRESS PULSES

Considerable support for the Griffith theory of fracture came from the observation that, in glass, cracks originate at the surface, and are absent in freshly prepared material, giving rise to very high observed strengths in the latter case.

The investigations of Bueche and White[45] on crack propagation in polymers were based on rapid photographic examination of the fracture. A typical set of results for a silicone rubber is shown in Figures 12.24a and b (see Plates XIA and XIB). It is seen that the fracture can commence either at the surface or in the interior of the polymer. No obvious imperfection could be found from microscopic examination.

The question of the origin of cracks has been studied in more detail by using stress pulses to produce internal fractures. By this technique Kolsky[47] has compared the 'internal strengths' of polystyrene and polymethylmethacrylate with that of soda glass. Internal fractures were produced in cylindrical rods of polymer by detonating a small explosive charge at the centre of one of the end faces, as in Figure 12.25 (see Plate XII). Identical experiments on glass cylinders could not produce internal fracture, the glass fracturing completely across or not at all. This showed that the internal strength of glass is much greater than the surface strength.

Similar conclusions were obtained by Bowden and Field[48] in a comparison of the brittle fracture of glass and polymethylmethacrylate. In glass many changes in the fracture behaviour could be obtained by etching procedures, whereas in polymethylmethacrylate surface modifications did not affect fracture, confirming that in this case the cracks originate at flaws throughout the material.

12.8 THE TENSILE STRENGTH AND TEARING OF POLYMERS IN THE RUBBERY STATE

12.8.1 The tearing of rubbers: Application of Griffith theory

The tearing of rubbers has been extensively studied by Rivlin and Thomas[49] and by Thomas[50] and his collaborators. The Griffith theory implies that the quasi-static propagation of a crack is a reversible process. Rivlin and Thomas recognized, however, that this may be unnecessarily restrictive, and that the reduction in elastically stored energy due to the crack propagation may be balanced by changes in energy other than that due to an increase in surface energy. Their approach was to define a quantity termed the 'tearing energy', which is the energy expended per unit thickness per unit increase in crack length. The tearing energy includes surface energy, energy dissipated in plastic flow processes and energy dissipated irreversibly in viscoelastic processes. Providing that all these changes in energy are proportional to the increase in crack length and are primarily determined by the state of deformation in the neighbourhood of the tip of the crack, then the total energy will still be independent of the shape of the test piece and the manner in which the deforming forces are applied.

In formal mathematical terms if the crack increases in length by an amount dc, an amount of work Tt dc must be done, where T is the tearing energy per unit area and t is the thickness of the sheet. Equating the work done to the change in elastically stored energy we have

$$-\left[\frac{\partial U}{\partial c}\right]_l = Tt \qquad (12.3)$$

The suffix l indicates that differentiation is carried out under conditions of constant displacement of the parts of the boundary which are not force-free. Equation (12.3) is similar to (12.1) above but T is now not to be interpreted as a surface free-energy.

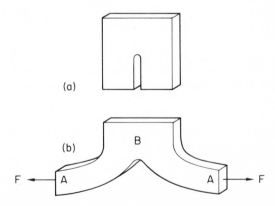

Figure 12.26. The standard 'trousers'-tear experiment showing the sample before stretching (a) and when under test (b) (after Bueche).

It is possible to choose particularly simple cases where the equation can be immediately evaluated. For example, consider the so-called 'trousers'-tear experiment shown in Figure 12.26. After making a uniform cut in a rubber sheet (a) the sample is subjected to tear under the applied forces F. The stress distribution at the tip of the tear is complex, but providing that the legs are long, is independent of the depth of the tear.

If the sample tears a distance Δc under the force F, the work done is given by $\Delta W = 2F\Delta c$. This ignores any changes in extension of the material between the tip of the tear and the legs.

Since the tearing energy

$$T = \frac{\Delta W}{t\Delta c}, \qquad T = \frac{2F}{t}$$

and can be measured easily.

Rivlin and Thomas[49] found that two characteristic tearing energies could be defined, one for very slow rates of tearing ($T = 3.7 \times 10^6$ ergs/cm^2) and one for catastrophic growth ($T = 1.3 \times 10^7$ ergs/cm^2) and that both these quantities were independent of the shape of the test piece.

It is important to note that the tearing energy of a rubber does not relate directly to tensile strength. The tearing energy is the energy required to extend the rubber to its maximum elongation. It depends on the shape of the stress–strain curve together with the viscoelastic nature of the rubber. For example, we may contrast two different rubbers, the first possessing a high tensile strength but a very low elongation to fracture and very low viscoelastic losses, and the second possessing a low tensile strength but a high elongation to fracture and high viscoelastic losses. In spite of its comparatively low tensile strength the second rubber may still possess a high tearing energy.

12.8.2 The Tensile Strength of Rubbers

Bueche and Berry[51] have considered the relevance to rubbers of the Griffith equation

$$\sigma_B = \left(\frac{2\gamma E}{\pi c}\right)^{1/2}$$

relating tensile strength to surface energy, modulus and crack length. This equation was modified to take into account the large strain conditions occurring at the breaking extensions of rubbers, and used to derive values for the surface energies of a series of silicone elastomers assuming a constant critical crack size of 10^{-3} cm. As in the calculations above for glassy polymers the measured surface energies were two orders of magnitude greater than those calculated. Looked at in another way, the calculated surface energies would imply crack sizes of 10^{-5} cm, which were much smaller than those indicated by direct examination of the sample. It was therefore concluded that most of the energy involved in the breaking process is dissipated in viscoelastic and flow processes.

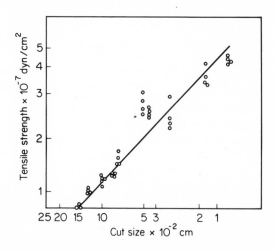

Figure 12.27. The relationship between cut size and tensile strength of a filled vulcanized silicone elastomer (after Beuche and Berry).

Bueche and Berry[51] also made a direct test of the Griffith criterion by examining the relationships between tensile strength, Young's modulus and the size of the deliberately introduced cracks. Their results are summarized in Figures 12.27 and 12.28. They found a linear, rather than a square root, dependence of tensile strength on both crack length and modulus. This led them to reconsider their data in terms of the critical stresses developed at the breaking points for a series of irradiated silicone elastomers. It was found that the values for the critical stress were very similar, in spite of the very different ultimate extensions of these polymers. This, taken in conjunction with the surface-energy calculations, led to the conclusion that, for elastomers, a critical-stress criterion for rupture is preferable to the Griffith criterion.

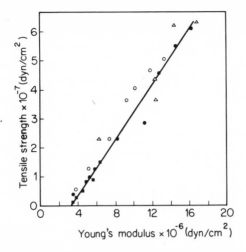

Figure 12.28. The relationship between tensile strength and Young's modulus for a series of silicone elastomers containing various fillers (after Bueche and Berry). ○ santocel; ● aerosil; △ treated santocel; ■ carrara.

12.8.3 Molecular Theories of the Tensile Strength of Rubbers

Most molecular theories of the strength of rubber treat rupture as a critical-stress phenomenon. It is accepted that the strength of the rubber is reduced from its theoretical strength in a perfect sample by the presence of flaws. Moreover, it is assumed that the strength is reduced from that of a flawless sample by approximately the same factor for different rubbers of the same basic chemical composition. It is then possible to consider the influence on strength of such factors as the degree of cross-linking and the primary molecular weight.

Bueche[52] has considered the tensile strength of a model network consisting of a three-dimensional net of cross-linked chains (see Figure 12.29).

Consider the cube of the material with edges of length 1 cm parallel to the three chain directions in the idealized network. Assume that there are v chains in this unit cube and that the number of chains in each strand of the network is n. There are then n^2 strands passing through each face of the cube. To relate the number n to the number of chains per unit volume of the network (which forms the link with rubber-elasticity theory) we note that the product of the number of strands passing through

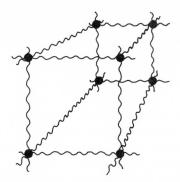

Figure 12.29. Model network of cross-linked chains.

each cube face and the number of chains in each strand will be $\frac{1}{3}v$, since there are three strand directions. Thus

$$n^3 = \tfrac{1}{3}v$$

$$n = (v/3)^{1/3} \tag{12.4}$$

Now consider a stress σ applied to the specimen parallel to one of the three strand directions. Further consider that the specimen fractures such that the strands break simultaneously at an individual fracture stress σ_c. Then

$$\sigma_B = n^2 \sigma_c$$

which from Equation (12.4) can be written as

$$\sigma_B = \left(\frac{v}{3}\right)^{2/3} \sigma_c$$

For a real network v is the number of effective chains per unit volume. It is given in terms of the actual number of chains per unit volume v_a by the Flory relationship

$$v = v_a \left[1 - \frac{2\overline{M}_c}{\overline{M}_n} \right]$$

where $\overline{M}_c$ and $\overline{M}_n$ are the average molecular weight between cross-links and the number-average molecular weight of the polymer, respectively. (Note that for a network there must be at least two cross-links per chain, i.e. $\overline{M}_n > 3\overline{M}_c$.)

This gives

$$\sigma_B \propto \left[1 - \frac{2\overline{M}_c}{\overline{M}_n}\right]^{2/3}$$

Bueche remarked that the variation of tensile strength with the polymer molecular weight $\overline{M}_n$, found by Flory[53] for butyl rubber, follows the predicted $[1 - 2\overline{M}_c/\overline{M}_n]^{2/3}$ relationship. The variation of tensile strength with degree of cross-linking was also studied by Flory et al[54] for natural rubber. Although there was the expected increase in tensile strength with increasing degree of cross-linking, it was also found that the strength decreased again at very high degrees of cross-linking. Flory attributed this decrease to the influence of cross-links in the crystallization of the rubber. However a similar effect was observed for the non-crystallizing SBR rubber by Taylor and Darin[55], which led Bueche[56] to suggest an alternative explanation. He proposed that the simple model described above fails because of the assumption that each chain holds the load at

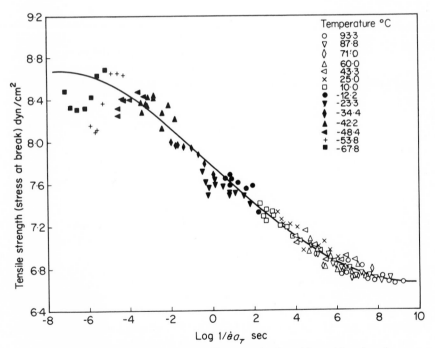

Figure 12.30. (a) Variation of tensile strength of a rubber with reduced strain rate $\dot{e}a_T$. Values measured at various temperatures and rates and reduced to a temperature of 263°K (after Smith).

fracture. Although this may be a good approximation at low degrees of cross-linking, it can be shown to be less probable at high degrees of cross-linking.

It is of considerable technological importance that the tensile strength of rubbers can be much increased by the inclusion of reinforcing fillers such as carbon black and silicone. These fillers increase the tensile strength by allowing the applied load to be shared amongst a group of chains, thus decreasing the chance of a break to propagate[57].

12.9 EFFECT OF STRAIN RATE AND TEMPERATURE

Another area in the fracture of polymers which has been studied extensively concerns the influence of strain rate and temperature on the tensile properties of elastomers and amorphous polymers. The principal experimental contribution to this area has been made by Smith and his co-workers[58-60], who measured the variation of tensile strength and ultimate strain as a function of strain rate for a number of elastomers. It was found

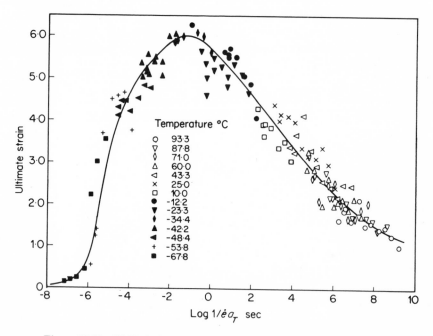

Figure 12.30. (b) Variation of ultimate strain of a rubber with reduced strain rate $\dot{e}a_T$. Values measured at various temperatures and strain rates and reduced to a temperature of 263°K (after Smith).

that the results for different temperatures could be superimposed, by shifts along the strain-rate axis, to give master curves for tensile strength and ultimate strain as a function of strain-rate. Results of this nature are shown in Figures 12.30a and b which summarize Smith's data for an unfilled SBR rubber. Remarkably, the shift factors obtained from super-position of both tensile strength and ultimate strain took the form predicted by the WLF equation for the superposition of low-strain linear viscoelastic behaviour of amorphous polymers (Figure 12.31). The actual value for T_g agreed well with that obtained from dilatometric measurements.

This result suggests that, except at very low strain rates and high temperatures, where the molecular chains have complete mobility, the fracture process is dominated by viscoelastic effects. Bueche[61] has treated this problem theoretically and obtained the observed form of the dependence of tensile strength on strain rate and temperature. Later theories have attempted to obtain the time dependence of both tensile strength and ultimate strain, or the time to break at a constant strain rate[62,63].

A final point is that Smith used these and other similar data to predict what he termed the 'failure envelope' for elastomers. The failure envelope is obtained by plotting $\log \sigma_B/T$ against $\log e$ and was found to be a unique curve for all strain rates and test temperatures. It was also found[60]

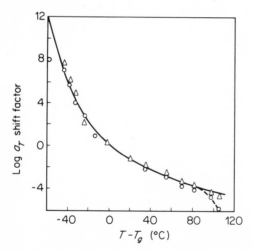

Figure 12.31. Experimental values of $\log a_T$ shift factor obtained from measurement of ultimate properties compared with those predicted using the WLF equation (after Smith). $\triangle$ from tensile strength; $\bigcirc$ from ultimate strain; — WLF equation with $T_g = 263°$K.

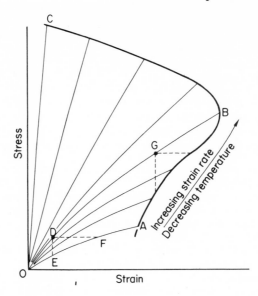

Figure 12.32. Schematic representation of the variation of stress–strain curves with strain rate and temperature. Envelope connects rupture points and the dotted lines illustrate stress relaxation and creep under different conditions.

that the failure envelope can represent failure under more complex conditions such as creep and stress relaxation. In Figure 12.32 such failure can take place by starting from the initial stage G and progressing parallel to the abscissa (constant stress, i.e. creep) or parallel to the ordinate (constant strain, i.e. stress relaxation) until a point is reached on the failure envelope ABC, as indicated by the progress along the dotted lines.

REFERENCES

1. E. Orowan, *Rept. Prog. Phys.*, **12**, 185, 1949.
2. P. I. Vincent, *Polymer*, **1**, 425 (1960).
3. E. A. W. Hoff and S. Turner, *Bull. Am. Soc. Test. Mat.*, **225**, TP208 (1957).
4. P. J. Flory, *J. Amer. Chem. Soc.*, **67**, 2048 (1945).
5. A. H. Cottrell, *The Mechanical Properties of Matter*, Wiley, New York, 1964, p. 327.
6. E. R. Parker, *Brittle Behaviour of Engineering Structures*, Wiley, New York, 1957.
7. P. I. Vincent, *Plastics*, **29**, 79 (1964).
8. J. J. Benbow and F. C. Roesler, *Proc. Phys. Soc.* **B70**, 201 (1957).

9. A. A. Griffith, *Phil. Trans. Roy. Soc.*, **221**, 163 (1921).
10. G. E. Inglis, *Trans. Inst. Naval Arch.*, **55**, 219 (1913).
11. R. A. Sack, *Proc. Phys. Soc.*, **58**, 729 (1946).
12. H. A. Elliott, *Proc. Phys. Soc.*, **59**, 208 (1947).
13. G. R. Irwin, *J. Appl. Mech.*, **24**, 361 (1957).
14. J. P. Berry, *J. Polymer Sci.*, **50**, 313 (1961).
15. B. Cottrell, *Int. J. Fract. Mech.*, **2**, 526 (1966).
16. J. P. Berry, *J. Appl. Phys.*, **34**, 62 (1963).
17. J. J. Benbow, *Proc. Phys. Soc.*, **78**, 970 (1961).
18. N. L. Svensson, *Proc. Phys. Soc.*, **77**, 876 (1961).
19. J. P. Berry, *J. Polymer Sci.*, **A2**, 4069 (1964).
20. E. H. Andrews, *Proceedings of the Conference on the Physical Basis of Yield and Fracture*, Oxford, 1966, p. 127.
21. H. Mark, *Cellulose and its Derivatives*, Interscience publishers, New York (1943).
22. P. I. Vincent, *Proc. Roy. Soc.*, **A282**, 113 (1964).
23. A. Peterlin, *Poly. Eng. Sci.*, **9**, 172 (1969).
24. S. N. Zhurkov and E. E. Tomashevsky, *Proceedings of the Conference on the Physical Basis of Yield and Fracture*, Oxford, 1968, p. 200.
25. D. Campbell and A. Peterlin, *J. Polymer Sci.*, **B6**, 481 (1968).
26. S. N. Zhurkov, I. I. Novak, A. I. Slutsker, V. I. Vellegren, V. S. Kuksenko, S. I. Velier, M. A. Gezalov and M. P. Vershina, *Proceedings of the Conference on the Yield, Deformation and Fracture of Polymers*, Cambridge, 1970.
27. V. R. Regel, T. M. Muinov, and O. F. Pozdnyakov, *Proceedings of the Conference on the Physical Basis of Yield and Fracture*, Oxford, 1966, p. 194.
28. S. N. Zhurkov, V. S. Kuksenko, and A. I. Slutsker, *Proceedings of the Second International Conference on Fracture*, Brighton, 1969, p. 531.
29. M. Higuchi, *Rept. Res. Inst. Appl. Mech.* (*Japan*), **6**, 173 (1959).
30. J. P. Berry, *Fracture*, Wiley, New York, 1959, p. 263.
31. R. P. Kambour, *Gen. Electric Res. Est.*, Rept. No. 64-RL-3776C (1964).
32. G. B. Bucknell, *British Plastics*, **40**, 84 (1967).
33. S. S. Sternstein, L. Ongchin and A. Silverman, *Appl. Polymer Symp.*, **7**, 175 (1968).
34. S. S. Sternstein and L. Ongchin, American Chemical Society Polymer Preprints, **10**, **(2)**, 1969, p. 1117.
35. P. I. Vincent, *Physics of Plastics*, Iliffe Books, Ltd., London, 1965, p. 174.
36. S. Strella and L. Gilman, *Mod. Plastics*, **34**, (8), 158 (1957).
37. G. B. Bucknell, *British Plastics*, **40**, 118 (1967).
38. R. M. Evans, H. R. Nara and R. G. Bobalek, *Soc. Plastics Engrs. J.*, **16**, 76 (1960).
39. L. E. Nielsen, *Mechanical Properties of Polymers*, Reinhold, New York, 1962.
40. G. B. Bucknell and R. R. Smith, *Polymer*, **6**, 437 (1965).
41. J. A. Kies, A. M. Sullivan and G. R. Irwin, *J. Appl. Phys.*, **21**, 716 (1950).
42. I. Wolock, J. A. Kies and E. B. Newman, *Fracture*, Wiley, New York, 1959, p. 250.
43. N. F. Mott, *Engineering*, **165**, 16 (1948).
44. H. Schardin, *Fracture*, Wiley, New York, 1959, p. 297.
45. A. M. Bueche and A. V. White, *J. Appl. Phys.*, **27**, 980 (1956).
46. E. F. Poncelet, *Metals Technology*, **11**, 1684 (1944).

47. H. Kolsky, *Fracture*, Wiley, New York, 1959, p. 281.
48. F. P. Bowden and J. E. Field, *Proc. Roy. Soc.*, **A282**, 331 (1964).
49. R. S. Rivlin and A. G. Thomas, *J. Polymer Sci.*, **10**, 291 (1953).
50. A. G. Thomas, *J. Polymer Sci.*, **18**, 177 (1955).
51. A. M. Bueche and J. P. Berry, *Fracture*, Wiley, New York, 1959, p. 265.
52. F. Bueche, *Physical Properties of Polymers*, Interscience publishers, New York, 1962, p. 237.
53. P. J. Flory, *Industr. Eng. Chem.*, **38**, 417 (1946).
54. P. J. Flory, N. Rabjohn and M. C. Shaffer, *J. Polymer Sci.*, **4**, 435 (1949).
55. G. R. Taylor and S. Darin, *J. Polymer Sci.*, **17**, 511 (1955).
56. F. Bueche, *J. Polymer Sci.*, **24**, 189 (1957).
57. F. Bueche, *J. Polymer Sci.*, **33**, 259 (1958).
58. T. L. Smith, *J. Polymer Sci.*, **32**, 99 (1958).
59. T. L. Smith, *Soc. Plastics Engrs. J.*, **16**, 1211 (1960).
60. T. L. Smith and P. J. Stedry, *J. Appl. Phys.*, **31**, 1892 (1960).
61. F. Bueche, *J. Appl. Phys.*, **26**, 1133 (1955).
62. F. Bueche and J. C. Halpin, *J. Appl. Phys.*, **35**, 36 (1964).
63. J. C. Halpin, *J. Appl. Phys.*, **35**, 3133 (1964).

Index

Aggregate model for mechanical anisotropy 253
Alfrey approximation 102
Anisotropic mechanical behaviour,
   generalized Hooke's law 24

Bauschinger effect 312
Blend, definition of 7
   glass transition in 173
Boltzmann Superposition Principle 84
Brittle–ductile transition 331

Charpy impact test 352
Compliance, definition of complex com-
   pliance 97
Condensation polymers, chemical struc-
   ture 1
Considère construction 274
Copolymer, definition of 7
   glass transition in 173
Crankshaft mechanism 176
Crazing 349
Creep, conditioning procedures 111
   general features in a linear viscoelastic
      solid 79
   measurement of extensional creep
      111
   measurement of torsional creep 115
Cross-linking, effect on failure behaviour
   336
   effect on glass transition 172
Crystallinity 9

Fractional recovery, definition of 200
Free volume, interpretation of WLF
   equation in terms of 152
   relationship to yield behaviour 322
   theory of Cohen and Turnbull 156

Glass transition, general features 16
Glass transitions, in amorphous poly-
   mers 168
Graft, glass transition in 173
Graft polymer, definition of 7

Impact behaviour 352
Isochronous stress–strain curve, in poly-
   propylene 197
Izod impact test 352

Kelvin model 90

Linear viscoelasticity, normal mode
   theories of 158

Maxwell model 89
Modulus, definition of complex modulus
   95
   measurement of dynamic extensional
      modulus 125
Molecular weight, definition of number
   average 5
   definition of weight average 5
   effect on brittle strength 335
   effect on glass transition 171
   effect on surface energy of fracture
      347

Natural draw ratio, definition of 272
   effect of preorientation on 323
Nylon, mechanical anisotropy in 243
Nylon 66, cold drawing in 292
   deformation bands 302
Nylon 610, deformation bands 302

Orientation 9

Plasticizers, effect on failure behaviour
   336

effect on glass transition 175
Plastic potential 289
Polychlorotrifluorethylene, linear visco-
    elastic behaviour 139
Polyethylene, chain branching 2
    deformation bands in high-density
        polyethylene sheets 302
    effect of pressure on yield behaviour
        314
    mechanical anisotropy in low density
        and high density 243
    mechanical anisotropy in low-density
        polyethylene 240
    relaxation transitions in 181
Polyethylene terephthalate, cold drawing
    in 290
    deformation bands in drawn films
        304
    effect of crystallinity on relaxation
        transitions 180
    effect of pressure on yield behaviour
        314
    linear viscoelastic behaviour 139
    mechanical anisotropy in 243
    relaxation transitions 168
    yield drops in 296
Polyisobutylene, linear viscoelastic
    behaviour 133
Polymethylmethacrylate, effect of pres-
    sure on yield behaviour 314
    relaxation transitions 166
    stress criterion for craze formation in
        350
    stress-relaxation behaviour 136
    surface crazes in 350
    variation in yield stress with tempera-
        ture and strain rate 316
    yield behaviour 299
    yield behaviour in tension and com-
        pression 321
Poly n-octyl methacrylate, storage com-
    pliance as a function of temperature
    and frequency 148
Polypropylene, mechanical anisotropy
    in 243
    non-linear creep and recovery in 215
    viscoelastic relaxations in 138
Polystyrene, effect of pressure on yield
    behaviour 314

yield behaviour 297
Polytetrafluoroethylene, relaxation
    transitions in 179
Polyvinyl chloride, creep and recovery
    in plasticized 210
Polyvinylfluoride, linear viscoelastic
    behaviour 139

Relaxation strength, definition of 107
Rotational isomerism 7
Rubber elasticity, internal energy con-
    tribution 72
    statistical theory of 63
    thermodynamic considerations 61
    thermoelastic inversion effect 62
Rubbers, effect of stress rate and temper-
    ature on failure 368
    tearing behaviour of 361
    tensile strength of 363

Side groups, effect on failure behaviour
    336
Site-Model Theory 145
    application to relaxation transitions
        in polyethylene 188
Spectrum of relaxation times 94
Spectrum of retardation times 95
Standard linear solid 92
Stereoregularity 5
Strain, engineering components of small
    strain 21
    generalized definition of 29
    invariants of 45
    tensor components of small strain 23
Strain-energy function, for finite strains
    46
    for small strains 44
    thermodynamic considerations of 41
Strain hardening 275
Strain softening 275
Stress, components of stress in finite
    elasticity 35
    components of stress in small-strain
        theory 19
Stress-intensity factor 342
Stress-relaxation, general features in a
    linear viscoelastic solid 82
    measurement of 115

Stress–strain relations, for finite strain elasticity 48
Surface energy of fracture 341

Takayanagi model 265
Time–temperature equivalence, general principles 140
in amorphous polymers 148

Vinyl polymers, chemical structure 1
Voigt model 90

WLF equation 151
statistical thermodynamic theory of 156

Yield criterion, definition of Coulomb Yield Criterion 283
definition of Tresca Yield Criterion 281
definition of Von Mises Yield Criterion 281
for anisotropic materials, definition of 287
geometrical representations of 283